POLITICS AND ECONOMICS
OF TROPICAL HIGH FOREST MANAGEMENT

FORESTRY SCIENCES

Volume 68

The titles published in this series are listed at the end of this volume.

Politics and Economics of Tropical High Forest Management

A case study of Ghana

by

Thorsten Treue

The Royal Veterinary and Agricultural University,
Department of Economics and Natural Resources,
Copenhagen, Denmark

SPRINGER-SCIENCE+BUSINESS MEDIA, B.V.

A C.I.P. Catalogue record for this book is available from the Library of Congress.

ISBN 978-0-7923-6931-8   ISBN 978-94-010-0706-1 (eBook)
DOI 10.1007/978-94-010-0706-1

Originally published by Kluwer Academic Publishers in 2001

Contents

Foreword

For the last two decades the loss of, in particular, tropical rainforest has alarmed the public in the developed parts of the world. The debate has been characterised by a lack of understanding of the causes and effects of the process, leading to the prevailing reaction being unqualified condemnation. Such attitude has even been observed among scientists, claiming supremacy to biodiversity conservation. Many scientific analyses are available, but the basis for sober debates and appropriate actions is still highly insufficient. Two recent international initiatives[1] will hopefully lead to improved knowledge of deforestation and forest degradation as they recognise the need for studies to critically investigate those issues. This book will provide useful input to the initiatives.

In my opinion, the scientific analyses have not sufficiently promoted the understanding that the fate of tropical forests is first and foremost a concern of the governments of the countries in which the forests are situated. Tropical forests may be important to the global environment and their rich biodiversity may be a human heritage. But their main importance is their potential contribution to improving livelihood in the countries in question.

Most scientific analyses do not immediately contribute to such understanding because their scope is narrow. They often deal with specific technical questions, e.g. how to carry out reduced impact logging, or investigate ecological or silvicultural issues, e.g. dynamics of a certain forest type or the sustainable felling intensity of certain tree species. More recently, social questions have entered the scene, e.g. the participation of a local community in the management of its forest resources. I do not try to devalue such studies. They are necessary, but seldom thrown in a national context relief.

The issue of deforestation or forest degradation in the tropics is extremely complex. Many different forces are simultaneously at work, e.g. of cultural, economic and political nature. Some recent scientific analyses acknowledge this and try to elucidate the intricate pattern of causes and effects[2]. However, most of these analyses are general in the sense that they deal with the issue in an abstract or overall way, even though some restrict themselves to a certain region. There is a conspicuous lack of in-depth and comprehensive case studies, i.e. studies that deal with a specific country. To the best of my knowledge, only two such studies have recently been published: Wunder (2000)[3] and the present study.

This study provides an in-depth analysis of the national-level forest policy and economic forces behind the development of high forest management in Ghana in the period 1991-1999, tracing the development back to 1986 or even further. Ghana is one of the most advanced countries in terms of natural high forest utilisation, but has not reached a state of sustainable forest and timber management.

The study is based on literature, semi-structured interviews and direct field observations. It is impressive how much unpublished literature the author was provided access to – key persons have shown him an uncommon confidence that he does not abuse even if making many frank statements. Similar confidence is demonstrated in interviews on sensitive subjects. It goes

[1] Inter-governmental Forum on Forests 1999: *Valuation of Forest Goods and Services: Report of the Secretary General.* United Nations Commission on Sustainable Development, New York, Report E/CN.17/IFF/1999.12. World Commission on Forests and Sustainable Development (WCFSD) 1999: *Our Forests Our Future.* Cambridge University Press, Cambridge.

[2] For example, Brown, K. and D. Pearce (eds.) 1994: The Causes of Tropical Deforestation. The Economic and Statistical Analysis of Factors Giving Rise to the Loss of Tropical Forests. University of London Press, London.

[3] Wunder, S. 2000: *The Economics of Deforestation. The Example of Ecuador*, Macmillan Press Ltd., London.

without saying that the study benefits significantly from that – in this respect the study has similarities to that carried out by Judge Barnett in Papua New Guinea[4], however on a voluntary basis. Also impressive is the comprehensive data material, much of which has been far from easily available. Further, the author demonstrates great acquaintance with not only Ghanaian forestry but also with society and culture in general.

The subject of the study is complex. Technical aspects, e.g. forest inventory, must be analysed for interpreting economic data – which again must be investigated from a policy point of view. Competent analyses are made of the national forest inventories, demonstrating design errors and unclear objectives that for a long time produced data about the standing volume and growth of timber species which was either over-optimistic or of limited value in a forest policy context. Among the economic analyses made, two should be emphasised: (i) Ghana appears to be foregoing substantial export revenue earnings by promoting the export of value added products, and (ii) Economic incentives in the forest sector have kept domestic log prices artificially low and opened the opportunity for log processing companies to earn windfall profits, cover high informal costs, and operate profitably at very low recovery rates.

Among the main findings of the study is that in Ghana trees outside forest reserves are the key to sustainable forest and timber resource management in the future. However, despite recent legislative reforms emphasis remains on control of the off-reserve timber exploitation rather than on creating incentives for increasing the off-reserve timber production. It is questioned whether this approach, in combination with an attempt to control the domestic timber market, will actually work when in the foreseeable future the supply of raw logs cannot meet the combined export and domestic demand. In the author's opinion, "the obvious alternative, over-exploitation of the remaining resource, will definitely be a risk."

Many of Ghana's past and present forestry related problems reflect rather general issues of managing tropical high forests, and by investigating these problems the study may prove useful in national forest management planning in other tropical countries. And being a thorough and well-documented analysis should make the study interesting to politicians and forest sector stakeholders in and outside Ghana, to the international development assistance community, and to students and scholars in the field of tropical forestry.

FINN HELLES
Professor of Forest Economics and Policy
Royal Veterinary and Agricultural University, Denmark

---

[4] Marshall, G. (1990). The political economy of logging: The Barnett inquiry into corruption in the Papua New Guinea Timber Industry. *The Ecologist* 20(5): 174-181.

Preface

My sincere thanks for providing funding for this study go to the Danish Research Academy and Danida. I am pleased to return these institutions' belief in what was then a PhD research proposal with this publication of the results. In this respect I am also indebted to the two anonymous peer reviewers of Kluwer Academic Publishers.

During my collection of data in Ghana, I was at all levels met by friendly open-minded people who from the outset gave me access to a wealth of information and took time (and pains) to answer my endless line of questions. For this remarkable support, without which the study would probably not have been possible, I would like to warmly thank:

Mr K. Smith, Technical Director, Ministry of Lands and Forestry
Mr E. Siisi Wilson, Deputy Technical Director, Ministry of Lands and Forestry
Mr J. Francois, Former Chief Conservator of Forests
Mr E. O. Nesnkyire, Chief Conservator of Forests
Mr G. A. Symes, Chief Technical Advisor/Sector Planner, FAO
Mr Timothy Nolan, Team-leader, Forest Inventory and Management Project
Ms Julia Falconer, NTFP Specialist, Forest Inventory and Management Project
Ms Jane Cronow, Collaborative Forest Management Specialist, Forest Inventory and Management Project
Ms Jenny Wong, Forest Inventory Specialist, Forest Inventory and Management Project

Last, but not least, I would like to thank Professor Finn Helles and Chief Secretary Lizzie Rohde, Unit of Forestry, KVL for the tedious work of re-arranging appendices and for giving the final touch to the manuscript.

THORSTEN TREUE

Executive Summary

The present study investigates developments of high forest management in Ghana during the period 1991-99. Focus is on national level forest policy and economics. Overall, the study findings may be summarised by three main points:

- The re-vitalisation of an export oriented timber industry during the early 1980s, has led to an undermining of its own resource base. However, design errors and unclear (political) objectives of national forest inventories, which were carried out during the period 1985-97, for a long time produced data about the standing volume and growth of timber species, which was either over-optimistic or of limited value in a forest policy context. This, in combination with strong financial interests to maximise timber export revenues, gave the timber industry the benefit of the doubt when sustainability of the recorded timber harvest was questioned.
- The package of (i) restrictions on log exports, (ii) forest taxes and fees and (iii) the modalities of granting rights to standing timber has, until 1999, produced rather unfortunate incentives. It has (i) created an over-capacity in the timber industry, (ii) resulted in massive official revenues foregone and (iii) promoted a notorious over-harvest of the most valuable timber species. Poor felling control has failed to counteract the latter effect.
- Trees outside forest reserves are the key to sustainable forest and timber resources management in the future. Yet, recent legislative reforms of forest taxation and modalities of granting rights to standing trees indicate that decision-makers and the official administration maintain an anachronistic (colonial) attitude to the management of renewable natural resources. An unfortunate official and constitutional preoccupation with extraction of revenue from commercial harvest of naturally occurring trees off-reserves is undermining financial incentives for off-reserve timber production by farmers and land owning communities who are the real managers of such trees.

The Timber Resource and Harvest

Commercial utilisation of timber from the high forest zone was strongly promoted by the World Bank and the IMF during the Economic Recovery and Export Rehabilitation Programmes in the mid 1980s. In reality, the timber export was rehabilitated with little technical knowledge about the resource base although the then Chief Conservator of Forests, J. Francois often tried to raise the issue of matching the capacity of the fast growing timber industry with the production capacity of the country's forest resource.

A first attempt to estimate the annual allowable cut (AAC) from forest reserves was made by Silviconsult (1985). Based on rather meagre growth and inventory data the sustainable on-reserve harvest was estimated to be around 1.1 million m^3 per year. Later this proved too optimistic. In fairness, it should be mentioned that Silviconsult (1985; v) at the time did emphasise that knowledge about the resource was insufficient and incomplete for efficient planning and utilisation.

In 1985 the British Government decided to fund a forest inventory in Ghana's high forest zone, the Forest Inventory Project (FIP), which ran from 1985 to 1989. The mechanisms of identifying this project are unclear, but it appears likely that the possible negative effects of rehabilitating the timber industry were increasingly becoming a matter of concern in ODA (now DFID). Unfortunately, FIP suffered from a lack of clearly and consistently defined objectives which ended up confusing the required level of sampling intensity. It was clear that estimation of a national AAC from forest reserves should be produced, but whether inventory results were supposed to be used at management unit level remained unclear.

As a result, the static inventory design, which involved measurement of 1,332 one ha temporary sample plots, ended up yielding overly accurate estimates of the standing volume at the national level, but rather inaccurate estimates at forest management unit level. Acceptable national level estimates of standing volumes above 110 cm diameter at breast hight (1.3 m above ground level) for the 55 species that had been exported during the period 1973-88 (referred to as FIP class 1 species) could have been produced by measuring only around 120 plots (Flint and Hardcastle, 1993; 19).

Although the national level static inventory results of FIP appeared statistically very accurate, the volume available for harvest was actually over-estimated because the area of forest suitable for timber production was not sufficiently investigated. Accordingly, the area of production forest was over-estimated by at least 260,000 ha (see below). Indeed, it is surprising and criticisable that the logical sequence, of first identifying areas suitable for logging and then undertaking a timber inventory for these areas only, was not followed by the FIP.

The static inventory under FIP also failed to take another politically rather obvious issue into account. It was never investigated how much af the recorded timber harvest originated from areas outside forest reserves. Later IIED (1993; 15) could reveal that more than 50% of the annual harvest actually came from off-reserve areas. Had this been realised at an earlier stage, it is likely that an off-reserve timber inventory had formed part of the FIP design.

An AAC of 1.5 million m^3 from forest reserves was estimated and published by FIP. However, this estimate, like the one produced by Silviconsult (1985), also suffered from inadequate primary growth data. In reality, the 1.5 million m^3 per year was mainly the result of a theory based computer simulation. In addition, this simulation assumed the area of production forest to be 1.3 million ha, instead of the 1.16 million ha used to generate the static inventory results. Given the available growth data, it was clearly over-ambitious by FIP to assume it would be possible to produce a scientifically sound estimate of the AAC.

Surprisingly, the computer simulated AAC of 1.5 million m^3 was, under FIP, never related to the actual principles of selecting trees for harvest within forest reserves. During the FIP, the "Interim Yield Allocation Formula" was developed. This formula says $Z = 0.2X + 0.5Y$, where Z is the number of trees above felling limit allocated for felling, Y is the number of trees above felling limit and X is the number of trees in the diameter class (20 cm) immediately below felling limit. The formula was applied on each tree species within a 120 ha compartment, with the intension to ensure that, after a felling cycle of 40 years, the number of trees above felling limit would be re-established. Later, $Z = 0.2X + 0.5Y$ was named the "Normal Formula" because the "Reduced Formula", $Z = 0.2X + 0.25Y$ was eventually applied for over-harvested species in all vegetation zones and for all species in fire prone areas.

Had the Normal Formula been applied on the national level static inventory data for FIP class 1 species, an estimated *allowed* annual cut of 0.94 million m^3, or only 62% of the computer simulated AAC, would have been the result. This was, however, never done. Instead, FIP's rather ill informed policy analysis concluded that the then recorded annual harvest of 1.14 million m^3 stayed within the AAC.

In 1988, the FIP was, in practice, continued under a new name, the Forest Inventory and Management Project (FIMP). This project "completed" the forest inventory such that, during the period 1988-94, a total of 3,101 one ha plots had been sampled from 128 forest reserves. In addition, the FIMP finally began to make use of collected information about site conditions within the temporary sample plots, such that the total on-reserve area suitable for protection and for timber production, respectively could be estimated. On this basis, the extent of possible hill sanctuaries, swamp sanctuaries, special biological protection sites, shelter belts, intact forest sanctuaries, provenance protection sites, fire protection zones, convalescence forest and conversion forest was estimated. It would have been better (and more logical) to define and make direct area measurements of such areas on the basis of maps and field visits. Yet, as a second best solution, the definition of forest categories and area estimates based on sampling data, must be considered sufficient for policy level considerations.

Depending on how many of the above categories are set aside for protection and how strictly logging, for environmental reasons, is controlled, this realistically leaves a production forest area between 0.65 million ha (fine-grained forest protection) and 0.76 million ha (large-grained forest protection). Quite a reduction in comparison with Silviconsult's (1985) and FIP's production forest estimates of 1.18 and 1.16 million ha, respectively.

According to rarity and harvest intensity, FIMP divided commercial timber species into three groups. Scarlet, Red and Pink Star species with Scarlet Stars being those under highest harvest pressure. Furthermore, FIMP finally abolished the rather fruitless attempts to estimate the on-reserve AAC through analyses of the meagre primary data on forest growth. Instead, the Normal and the Reduced Yield Allocation Formulae were applied on the FIMP static inventory data. Table E.1 summarises the FIMP estimated AACs from forest reserves.

Table E.1 **Estimated AACs in the forest reserves of Southern Ghana**

Species Group	**AAC under large-grained protection only** (m^3)	**AAC under large- and fine-grained protection** (m^3)
Scarlet Stars	132,437	120,817
Red Stars	224,195	201,353
Pink Stars	534,877	479,518
TOTAL	**891,509**	**801,688**
Total production area (ha)	762,400	645,080
Annual coupe (ha)	19,060	16,127
Average harvest (m^3/ha/yr)	1.17	1.24
Average harvest on coupe (m^3/ha)	46.77	49.71

Source: Based on FIMP (1995b).

It is seen from table E.1 that the previously estimated on-reserve AACs of 1.1 and 1.5 million m^3 were much too optimistic. Had the FIP been designed more carefully, the figures in table E.1 could have been made available five to six years earlier. During the FIMP it became possible to compare the recorded on-reserve timber harvest with the above estimated AACs. Table E.2 summarises this exercise.

Table E.2 **Estimated on-reserve AACs compared with the 1996 recorded on-reserve extraction by species groups**

SPECIES GROUP	AAC total volume[1] (m³)	1996 on-reserve extraction (m³)	1996 extraction as % of AAC	AAC total volume[2] (m³)	1996 on-reserve extraction (m³)	1996 extraction as % of AAC
Scarlet Stars	132,437	183,869	139	120,817	183,869	152
Red Stars	224,195	66,470	30	201,353	66,470	33
Pink Stars	534,877	33,420	6	479,518	33,420	7
TOTAL	**891,509**	**284,932**	**32**	**801,688**	**284,932**	**36**

Notes: [1] Under large-grained protection only (762,400 ha of production forest).
[2] Under fine- and large-grained protection (645,080 ha of production forest).
Sources: FIMP (1995b) and FPIB LMC Totals (1996).

The general message from table E.2 is clear. Scarlet Star species are being over-harvested, whereas the harvest of Red and Pink Stars could, in general, be increased substantially. In other words, the problem of sustainability is that market forces are creating an excess harvest of certain (the commercially most valuable) species. Luckily, other species are, however, "under-utilised". This represents an option for maintaining or even increasing the total on-reserve harvest, provided the present species composition of the harvest could be changed. Yet, when one looks at the off-reserve harvest and inventory data, the picture becomes more gloomy. Table E.3 summarises an off-reserve resource lifetime scenario under a continuation of the 1996 recorded timber extraction off-reserves.

Table E.3 **Off-reserve standing timber resource above 70 cm DBH and estimated lifetime of resource if the 1996 recorded extraction is continued**

Species Group	Standing vol. > 70 cm RME of good trees[1] (m³)	1996 recorded extraction[2] (m³)	Lifetime of resource[3] (Years)
Scarlet Stars	19,863,000	515,349	39
Red Stars	38,489,000	240,633	160
50% of Pink Stars	28,344,000	65,068	436
Scarlet + Red Stars	58,352,000	755,982	77
Scarlet + Red + 50% of Pink Stars	86,696,000	821,050	106

Notes: [1] Reliable minimum estimates (lower value of 95% confidence interval) for trees of FIP quality codes 1 and 2 (good trees).[2] Includes recorded *extraction* of all Pink Star species (not only 50%).
[3] Assumes that the resource is effectively mined, ie. growth of standing timber is cancelled out by felling of trees for other purposes than timber (firewood, clearing of land for agriculture, etc.).Sources: FIMP (1997b) and FPIB LMC Totals (1996).

It is seen that Scarlet Star species make up the lion's share (63%) of the off-reserve harvest, while this group represents the smallest resource base. Furthermore, Scarlet Star species are, within the coming 40 years, likely to become economically extinct off-reserves, unless the harvesting pattern is radically changed.

Tables E.2 and E.3 taken together show that the total recorded 1996 timber extraction was approximately 1.1 million m^3 of which Scarlet Star species accounted for approximately 0.7 million m^3 or 63%. It is, in other words, clear that the timber industry is heavily dependent on Scarlet Star species. The FIMP data has finally made it equally clear that the resulting harvesting pattern is distinctly unsustainable for this group of species.

An analysis of the wood export reveals that the annual harvest for export during the period 1986-96 was approximately 0.8 million m^3. Of this, approximately 0.6 million m^3 (75%) was Scarlet Star species with Red and Pink Star species making up approximately 0.14 and 0.06 million m^3, respectively. The documented over-exploitation of Scarlet Star species is, therefore, to a large extent export driven, and a rather unfortunate (but foreseeable) result of the above mentioned Export Rehabilitation Programme. Accordingly, the much praised increase in annual timber export revenue from approximately DM 110 million in 1986 to DM 220 million in 1996 has had a serious "cost-side" to it. Depletion of the stock of Scarlet Star species.

A comparison between the estimated harvest for export and the recorded extraction of timber from on- and off-reserve areas indicates that the annual domestic timber consumption (excluding charcoal and firewood) is at least 0.7 million m^3 roundwood equivalents. In the near future this figure might increase to approximately one million m^3. Together with and an export-driven harvest demand of about 0.8 million m^3 roundwood equivalents, this gives an annual timber demand of some 1.8 million m^3 roundwood equivalents. Since the on-reserve AAC is only between 0.8 and 0.9 million m^3, this comprises a serious problem. Naturally, the domestic timber demand is likely to be satisfied either legally or illegally, and the timber industry will obviously try very hard to maintain its export earnings. Meanwhile, off-reserve timber resources are being steadily depleted.

Accordingly, the national timber inventory results, the AAC estimates and analyses of the timber harvest, point to two core policy priorities:

- Diversifying the timber harvest on-reserves
- creating a sustained production of timber off-reserves.

The Economics of Timber Exploitation

Past regulatory mechanisms and financial incentives for harvesting different timber species obviously play an important role in understanding why the timber harvest has become unsustainable. Since 1962, the Government has, both on- and off-reserves, regulated access to standing timber through a system of concessions or concession-like arrangements (allocation of long- and short-term rights to standing timber). In this process rights to standing timber have never been allocated according to a system of competitive bidding. Largely, the official costs of acquiring rights to standing timber have been and still are confined to the costs of processing applications. Yet, the Government is officially influencing domestic timber prices through:

- Timber royalties
- stumpage fees (substituted royalties in December, 1998)

- export levies
- forest fees and other taxes
- log export bans.

Other factors, which, in addition to the official charges, influence the domestic log prices are:

- FOB export prices on wood products
- costs of logging and further processing
- informal payments.

The resource conserving effect of forest fees and taxes has been almost negligible. In general, they have been too low and differed too little among species to produce any significant financial incentive, for holders of timber rights, to harvest Pink or Red rather than Scarlet Star species. In theory, this could have been checked through strict felling control, but a glance at the official figures for timber extraction from forest reserves proves that this has not been the case. Direct inquirations into the practicalities of timber harvest in forest reserves also show that, especially for Scarlet Star species, under-sized trees and higher volumes than the yield selection formulae prescribe have been extracted. Furthermore, there is evidence that this has happened (and may still be going on) as a result of collusion between forestry officials and timber companies.

In combination with a progressive restriction on log exports, which began in 1979 by the banning of log exports for 14 species and became complete in 1995, the low forest fees and taxes have kept domestic log prices at 20-50% of FOB log prices. In this way, domestic wood processing has been heavily subsidised, because export levies on processed wood have always been and still are quite low. Furthermore, it has created a significant over-capacity in the wood industry. Current estimates suggest that the processing capacity of the industry is around 3 million m^3 annually, which compares rather unfavourably with the on-reserve AAC.

An investigation of the actual willingness to pay for standing timber has show that, until 1999, this was significantly higher than forest taxes and fees. As a result, DM 20 million or more in official forest revenue was annually foregone during the period 1993-96. Accordingly, huge sums have been available for informal payments, windfall profits and keeping inefficient industries financially afloat. Considering the magnitude of funds being available for informal payments, it is no wonder that the procedures for allocating timber concessions completely messed up in the early 1990s.

From the beginning of 1999, the system of forest fees and taxes was finally reformed. The main achievement of this reform has been that timber royalties were substituted by stumpage fees, which almost equalled the industry's willingness to pay for standing timber. This has significantly narrowed the financial room for windfall profits and informal payments. However, in combination with the log export ban, the new forest fees and taxes will still keep domestic log prices far below the world market level.

Development of the domestic timber industry may, due to the Government controlled "prices" of standing timber, mainly be pursued through a gradual increase of forest taxes and fees. The industry cannot be considered fully "developed" before the difference between domestic and world market log prices is close to zero. Future adjustments (increases) of stumpage fees and other forest taxes should, therefore, be assessed against their effect on domestic log prices. In addition, the difference in forest fees and taxes between Scarlet, Red and Pink Star species should be maintained or increased in order to conserve over-utilised and increase the harvest of "under-utilised" species.

If the current harvest for export should be maintained and the expected domestic demand for timber should be satisfied on a sustainable basis, there is a rather urgent need to promote sustained yield off-reserve timber production. Yet, the subsidy provided to the timber industry through log export bans and low forest fees and taxes is a serious obstacle in this respect. An unfortunate, but rather important, effect of keeping domestic log prices artificially low, is that this completely undermines financial incentive for establishing plantations on a commercial basis. Attempts to attract foreign investments for establishing large-scale plantations have stranded specifically on this condition.

Presumably, the low official forest fees and taxes have, during the period 1986-93, generated revenues which were below the actual costs of forest management. From 1994 onwards, forest fees and taxes have generated a surplus, but at an unsustainable harvest level. This is so, because most of the revenue came from harvest of timber outside forest reserves.

Institutional studies indicate that the costs of sustainable forest management through a semi-private Forest Service (to substitute the Forestry Department) may be estimated to DM 7.5 million per year. In this respect it is an interesting fact that, if the on-reserve harvest is reduced to the AAC for over-utilised species and kept at the 1996-level for other species, forest fees and taxes could, if they equalled the industry's willingness to pay, generate an annual revenue of DM 10-12 million. The revenue form a sustainable on-reserve harvest would, with the 1999 forest fees and taxes, be around DM 8-9 million annually. It should, in other words, be financially possible to finance a sustainable forest management through a sustainable harvest from forest reserves alone.

Rights to Benefits from Timber Exploitation

It is, however, not only the official forestry authorities that are entitled to a share of the forest revenue. When forest reserves were constituted, the ownership to land did not change. Only the management authority was shifted from the land owning local communities (stools) to the State (the Forestry Department). Off-reserves the situation is and has, since 1962, essentially been the same, because the 1962 concession legislation transferred the right to all naturally occurring timber trees from the land owning communities to the State. In short, local communities have lost the right to manage timber on their land, but they have maintained a legal right to a share of revenues from timber exploitation.

Prior to 1962, land owning communities were legally entitled to no less than two thirds of the gross revenue from forest reserves. This rule was cancelled by the 1962 concession legislation, according to which all revenues from forest reserves were paid into the Forest Improvement Fund. This fund has, since then, been used to finance the Forestry Department. After covering the Forestry Department's running costs, any excess money in the fund should each year be transferred back to the land owning communities and local authorities from whose land timber had been extracted. In reality, very small amounts, if any, have been paid from the Forest Improvement Fund to the land owners and local authorities.

Off-reserve timber revenue was collected by the Lands Department, and presumably local communities actually received some share of this. How big this share has been is, however, impossible to say, because the Lands Department has never been able (willing) to present any accounts for the collection and disbursement of off-reserve timber revenues.

Payment of on-reserve timber revenue into the Forest Improvement Fund probably became illegal with the 1992 Constitution, which stipulates that:

> "Ten per cent of the revenue accruing from stool lands shall be paid into the office of the Administrator of Stool Lands to cover administrative expenses; and the remaining revenue shall be disbursed in the following proportions:
>
> a) twenty-five per cent to the stool through the traditional authority for the maintenance of the stool in keeping with its status;
> b) twenty per cent to the traditional authority[1];
> c) fifty-five per cent to the District Assembly, within the area of authority of which the stool lands are situated."

The 1992 Constitution does not include any provision for covering administrative expenses of the Forestry Department through (a share of) timber royalties. Accordingly, the Forestry Department has, de jure, been illegally financed from 1992 to December 1998. The Timber Resources Management Regulations (1998) provides for financing the Forestry Department out of revenue arising from timber. This is legally possible because the Constitution does not specify the term "revenue" as *gross* revenue. The Timber Resources Management Regulations (1998) do, however, not specify whether timber revenue used to finance the Forestry Department should stem from forest reserves alone, or whether it could accrue from stool land in general. Nor is there a specified maximum share of timber revenues, that may be used to finance the Forestry Department.

Considering that the annual revenue from a sustainable on-reserve timber harvest is hardly going to be higher than DM 8-9 million (at least in the short-term) and that the running costs of sustainable forest management is estimated to DM 7.5 million, it is rather convenient for the Forestry Department that the legislative provision for covering its running costs are held in unspecific terms. Conversely, the legislation provides little guarantee that local communities will enjoy significant financial benefits from exploitation of naturally occurring trees on their land.

This is hardly in the spirit of the 1992 Constitution, and it is weakening the incentives for land owning communities to protect and tend naturally occurring timber trees off-reserves, despite the fact that such trees would attain high commercial values within say 10-30 years. Consequently, off-reserve timber trees might end up as firewood rather than high value export commodities.

Unfortunately, the problem of creating clear financial incentives for people to grow timber trees off-reserves is compounded by the fact that farmers do, in general, not own the land they cultivate. Practically all land in Southern Ghana is stool land. Accordingly, any revenue from such land must be shared between the stool, the traditional authority and the District Assembly (see above). The individual, who cultivates the land from which timber is extracted, has no legal rights to a share of the generated revenue.

Summing up, the legislative and financial structures are still favouring timber exploitation rather than timber production. Extraction of government revenue from timber exploitation and creation of favourable financial conditions for the timber industry is dominating the legislative background

---

[1] House of Chiefs or any council or body established or recognised as such under customary law (Office of the Administrator of Stool Lands Act, 1994; 18).

for forest and timber resources management. This is rather unfortunate considering that Ghana could easily face a timber deficit and declining timber export revenues in a not so distant future. Furthermore, farmers and local communities are deprived an otherwise obvious opportunity to generate income from timber production.

Exchange Rates Used

Exchange Rates	
Year	Cedis per DM
1971	0.30
1972	0.40
1973	0.40
1974	0.46
1975	0.49
1976	0.45
1977	0.48
1978	0.75
1979	1.48
1980	1.49
1981	1.30
1982	1.16
1983	8.24
1984	13.61
1985	16.69
1986	39.75
1987	93.91
1988	127.37
1989	149.05
1990	197.54
1991	244.62
1992	279.50
1993	369.05
1994	615.00
1995	780.00
1996	1,065.00
1997	1,131.00
1998	1,436.00
1999	1,550.00

Price Indexes Used

Consumer Price Index in Germany 1986-96

Year	Change in per cent from previous year	Index (1996=100)
1986	-0.1	129.14
1987	0.2	128.88
1988	1.3	127.22
1989	2.8	123.76
1990	2.7	120.51
1991	3.6	116.32
1992	5.1	110.67
1993	4.5	105.91
1994	2.7	103.12
1995	1.7	101.40
1996	1.4	100.00

Notes: For the period 1986-91 the table covers the former West Germany only.
For the period 1992-96 the table covers the re-united Germany.

Source: Statistisches Bundesamt Deutschland (1999)

Acronyms

AAC: Annual allowable cut.
CCF: Chief Conservator of Forests (Head of the Forestry Department).
CITES: Convention on International Trade in Endangered Species.
Danida: Danish International Development Agency.
DFO: District Forest Officer.
DM: German Marks (Deutsche Mark).
DS: Dry semi-deciduous zone (sometimes called the dry semi-deciduous forest zone (DSFZ)).
DSFZ: Dry semi-deciduous forest zone (sometimes called the dry semi-deciduous zone (DS)).
DSIZ: Dry Semi-deciduous Inner Zone.
DBH/dbh: Diameter at breast height (diameter 1.3 metres above ground level).
EPC: Export Promotion Council.
FAO: United Nations Food and Agricultural Organisation.
FD: Forestry Department in Ghana.
FIP: Forest Inventory Project.
FIMP: Forest Inventory and Management Project.
FMU: Forest Management Unit.
FOB: Free On Board (price of export commodities when loaded and ready for shipment).
FORIG: Forestry Research Institute of Ghana.
FPIB: Forest Products Inspection Bureau.
FPRI: Forest Products Research Institute.
FRMP: Forest Resource Management Project.
FROGGIE: Forest Reserves Graphical Information Exhibitor.
GATT: General Agreements on Tariffs and Trade.
GCC: Ghana Chamber of Commerce.
GSC: Ghana Shippers Council.
IDA: International Development Agency (the World Bank's soft loan agency).
IMF: International Monetary Fund.
IUCN: International Union for Conservation of Nature (The World Conservation Union).
LC: Lands Commission.
LMC: Log Measurement Certificate (issued at road check points by FPIB. In principle, the LMC system covers all logs extracted both on- and off-reserves).
LDC: Less Developed Country (developing country).
ME: Moist Evergreen zone.
MLF: Ministry of Lands and Forestry.
MLG: Ministry of Local Government.
MLNR: Ministry of Lands and Natural Resources.
MSNW Moist Semi-deciduous North West type.
MSSE: Moist Semi-deciduous South East type.
NA: Not Available/Applicable.

NIC:	National Investigation Committee.
NTFP:	Non-timber forest product.
ODA:	Overseas Development Administration, UK (now called DFID, the Department for International Development).
PNDC:	Provisional National Defence Council.
PSP:	Permanent Sample Plot.
RFO:	Regional Forest Officer.
RME:	Reliable Minimum Estimate (lower value of 95% confidence interval).
RWE:	Roundwood Equivalent.
SM:	Southern Marginal.
SO:	Southern Outlier.
TEDB:	Timber Export Development Board.
TSP:	Temporary Sample Plot.
UE:	Upland Evergreen.
UK:	United Kingdom.
UNDP:	United Nations Development Programme.
WE:	Wet Evergreen zone
WTO:	World Trade Organisation.

1 Introduction

The present study is the result of studying the technical, political and economic development of high forest management in Ghana during the period 1991-99. In addition to the collection of data specifically for the thesis we have undertaken or participated in four consultancies in Ghana, which were all related to political and economic issues of high forest management (IIED, 1993; Bach, Gram, Treue, 1995; Treue and Armstrong-Mensah; 1995 and Treue and Odoom, 1996). Eight visits totalling 13 months have been made to Ghana in connection with the studies.

Originally our interest in high forest management in Ghana was associated with the early 1990s' growing popular concern over tropical deforestation. At that time environmental (green) organisations in Europe and North America were quite successful in placing the links between tropical deforestation and international trade in tropical timber on the political agenda, demanding bans on import of tropical timber[1]. Despite the rather weak links between international trade in tropical timber and tropical deforestation in general, timber trading companies were, for political reasons, forced to look at the forest management behind the products they traded. In this connection the Danish timber trading company, Dalhoff Larsen and Horneman Ltd. requested an assessment of forest management in Ghana as the company was (and still is) exporting substantial volumes of timber from Ghana to overseas markets (Enevoldsen, Kronslev and Treue, 1991).

1.1 Objective of the Study

The first visit to Ghana clarified that management of the country's high forest resource was very much linked to commercial timber harvest. There were concerns over the green organisations' call for banning tropical timber on especially North American and some European markets, but other markets in the Middle and Far East were not affected by this trend. It was also doubtful whether any tropical timber importing country would unilaterally implement bans on tropical timber imports since this was likely to be against the then GATT (now WTO).

External forces springing directly out of green organisations' campaign against the use of tropical timber were, in other words, not likely to radically influence forest management in Ghana. It could disturb but hardly undermine the timber export. In any case, the management of timber and forest resources in Ghana would primarily be an internal matter.

In fact, a process on improving the management of Ghana's timber and high forest resources was already ongoing when the international debate on banning and/or labelling tropical timber took off. Essentially, the problem in Ghana was that, following a World Bank and IMF supported (designed) Economic Recovery Programme, the timber harvest and export began to pick up tremendously in the mid 1980s, while knowledge about the timber resource was very limited (see section 3.2.1). Clearly, this held little guarantee for a sustainable development.

The primary forest policy problem in Ghana was (and still is) to manage the development of an export oriented wood industry such that it does not undermine the timber and forest resources. Initially, the response has been a major piece-meal implemented forest inventory. This was accompanied by numerous analyses of economic and political issues influencing forest management and timber harvest. During this process it transpired that the fundamental forest policy issue in Ghana could be summarised as follows:

---

[1] Today this demand has generally been modified to a promotion of labelling products of tropical timber originating from forests under sustainable management such that end-users, in addition to price and quality, may choose in accordance with environmental concerns.

How to create a decision-making environment that is genuinely in favour of sustainable forest and timber resources management and willing to adjust rules and regulations even if this goes against the (short term) interests of the timber industry?

Accordingly, this study is concerned with national level forest management and policy. The overall objective is to document and analyse recent developments in the strive towards sustainable management of the country's natural high forest and timber resource. Among tropical countries Ghana is probably one of the most advanced in terms of natural high forest management although one can hardly say it has reached a state of sustainable forest and timber utilisation.

It is hoped that the analyses presented may be useful in national forest management planning for other tropical countries, since many of Ghana's past and present forestry related problems are likely to reflect general issues of managing tropical high forests. In addition, chapters 12 and 13 should provide relevant analyses for Ghana on the necessity to improve local communities' incentives for tending naturally occurring trees on farmland.

More specifically, the study attempts to:

- Document the links between the World Bank and IMF supported Economic Recovery Programme and over-exploitation of the high forest timber resource;
  document the political background for the various timber inventories;
  analyse and assess the technical results and political impacts of the timber inventories;
- document the notorious over-exploitation of certain (the most valuable) tree species;
- document that the over-exploitation of certain species is largely export driven;
- document that the annual allowable cut from forest reserves cannot satisfy the current harvest for export as well as the apparent domestic consumption;
- document that in the near future forest reserves alone cannot even satisfy the expected domestic demand for timber;
- assess the apparent successful development in wood exports and prove that the development in export revenue, when counted per roundwood equivalent, is not so impressing after all;
- estimate the "real" financial value of standing timber as opposed to the official forest taxes and fees, and prove that annually millions of DM in forest revenue have been foregone by the Government;
- analyse and show that policy failures have until 1999 created economic incentives which have worked against sustainable forest and timber resource management;
- analyse the rights to timber and benefits from timber exploitation on- and off-reserves, and make probable that sustainable forest and timber resource management in the high forest zone may only be achieved if local communities are given extensive rights to off-reserve timber resources.

1.2 Methodology

The study methodology has mainly been based on two sources of information:Literature;

- semi-structured interviews.

In addition, direct field observations on- and off-reserves have been used to build an understanding about the realities on the ground. In particular, discussions held in the forest with personnel from the Forestry Department and logging companies have, due to the informal setting and the obvious observations one can make, provided valuable background information about practical realities and how these sometimes differ from theory and official versions.

1.2.1 Literature

The literature studies have mainly been based on official sources like statistics and legislation combined with so-called "grey" literature, ie. reports of various kinds that have never been published. Almost all literature used has, in practice, only been available in Ghana. As a result, getting access to files and reports has been crucial for the study. The Ministry of Lands and Forestry, the Forestry Department and the Forest Inventory and Management Project have all been very generous in this respect.

1.2.2 Semi-structured Interviews

During the early stages of the study it became clear that things in the forestry sector were moving quite fast. In 1992 it was easy to point out many weaknesses in the system, but it was equally obvious that data and analyses, which were likely to trigger important changes, would soon become available.

During the 1990s an impressive amount of data and reports about the forestry sector in Ghana have been produced. Understanding the background for and observing the impact of the data and reports produced, therefore, became more relevant for a national level tropical high forest management study than pinpointing issues which were likely to change in a few years anyway. In this respect it has, over the years, been useful to interview (have discussions with) centrally placed people in the forestry sector. Again, employees of the Ministry of Lands and Forestry, the Forestry Department, and the Forest Inventory and Management Project have generously made time for such interviews/discussions. In addition, many owners and managers of logging companies and sawmills have provided useful and important information.

2 Brief Introduction to Ghana's High Forest Zone

Ghana is situated on the West African Continent. To the North, West, East and South the country is bordered by Burkina Faso, Côte d'Ivoire, Togo and the Gulf of Guinea, respectively.

2.1 Forest Type and Vegetation Zones

The high forest zone[2] in Ghana covers approximately 82,000 km^2, or about one third of the land area. It is situated in the South Western corner of the country and forms part of the Guineo-Congolean phytogeographical region. The flora and fauna have strong affinities with the forests of Côte d'Ivoire, Liberia and Sierra Leone and a lesser affinity with the Nigerian forests from which they are separated by the arid "Dahomey gap".

In general, the Ghanaian high forests are wettest in the South West, turning increasingly drier as one moves to the North and East. According to Hawthorne (1993b), the high forest zone may be divided into nine vegetation zones, each with distinct associations of plant species and corresponding rainfall and soil conditions. These are:

- WE Wet Evergreen zone
- ME Moist Evergreen zone
- MSSE Moist Semi-deciduous South East type
- MSNW Moist Semi-deciduous North West type
- DSIZ Dry Semi-deciduous Inner Zone
- DSFZ Dry Semi-deciduous Fire Zone
- UE Upland Evergreen
- SM Southern Marginal
- SO Southern Outlier

Figure 2.1 shows the division of the high forest zone into vegetation zones as well as the administrative regions of southern Ghana.

---

[2] Where the natural climax vegetation is forest with high stems (above 20 metres).

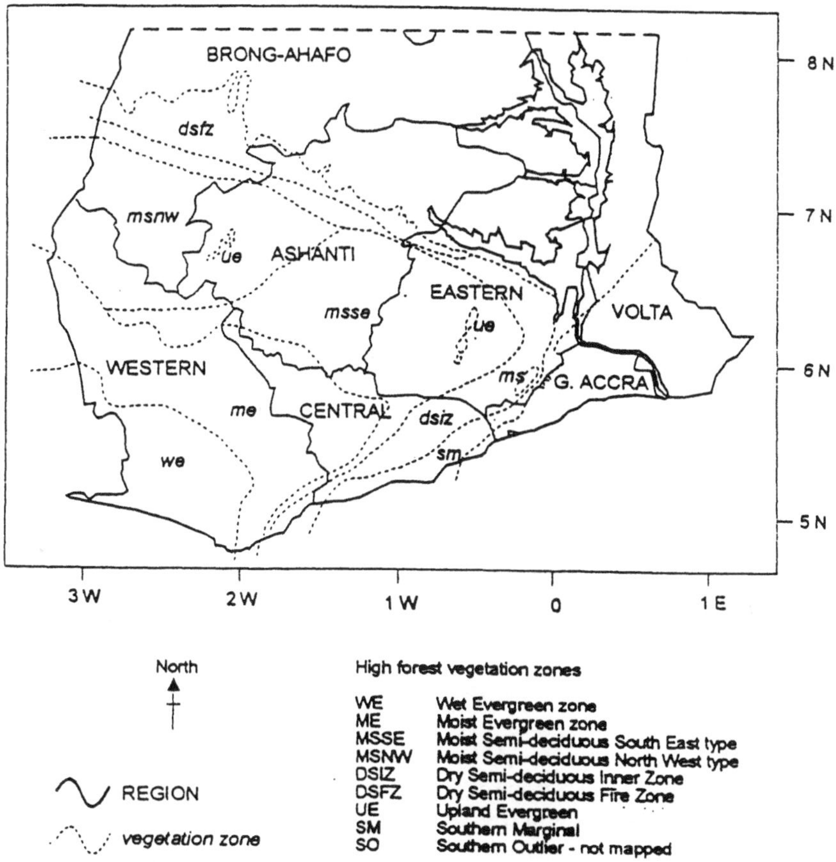

Figure 2.1 **Forest vegetation zones and administrative regions of southern Ghana**

Source: Hall and Swaine (1981), modified by Hawthorne (1993b)

2.2 Biodiversity

More than 2,100 plant species, of which 23 are endemic, have been found in the high forest (Hall and Swaine, 1981). In total, 730 tree species, of which 680 attain a diameter of 5 cm or more at breast hight (1.3 metre above the ground) have been found in the high forest (Hawthorne, 1989). The wet evergreen and moist semi-deciduous forest zones are the most floristically rich, while the drier forest types are relatively species poor.

The fauna of the high forest zone includes over 200 species of mammal, many of which are rare or endangered. These include the bongo (*Tragelaphsus euryceros*), Ogilby's duiker (*Cephalophus ogilbyi*), golden cat (*Felis aurata*), chimpanzee (*Pan troglodytes*), forest elephant (*Loxodonta africana cyclotis*) and the pygmy hippopotamus *(Choeropsis liberiensis*) (Mensah-Ntiamoa, 1989). The high forest zone also supports 74 species of bats, 37 rodents, a variety of reptiles and over 200 bird species (IUCN, 1992).

3 General Issues on Trees and Forests in the High Forest Zone

Today the high forest zone is dominated by farmland (including fallow and some plantations), with forest and wildlife reserves covering approximately 20% of the zone (Hawthorne and Musha, 1993; 4). However, in terms of timber the standing volume is not confined to the forest reserves only. Farm and fallow land in the high forest zone carry a substantial timber resource in the form of scattered trees on agricultural fields, providing shade for the crops, sacred groves (pockets of closed-canopy forest serving religious and spirituel purposes) and a few larger areas of closed-canopy forest outside forest reserves.

Despite the substantial timber resource on farmland, forest reserves clearly stand out on satellite images as irregular shaped dark polygons from the surrounding farm and fallow land (Hawthorne, 1989). This tells:

- Since the establishment of forest reserves a substantial off-reserve deforestation has taken place;
- the straight-line boundaries of the forest reserves have largely been enforced by the Forestry Department and respected by the local populations.

The latter point does, however, cover some conflict areas where reserve boundaries and reserved forests have been heavily encroached. Furthermore, the apparent success of protecting forest areas through reserves may now begin to face its ultimate test, where off-reserve forests and timber resources are becoming depleted (Hawthorne and Musha, 1993; 9).

Figure 3.1 shows the distribution of forest reserves in the high forest zone. The figure indicates which reserves were inventoried by FAO and the Forestry Department 1980-83 and the Forest Inventory Project 1985-88 (see chapters 4 and 5). Eventually, all reserves were inventoried by the Forest Inventory and Management Project (see chapter 6).

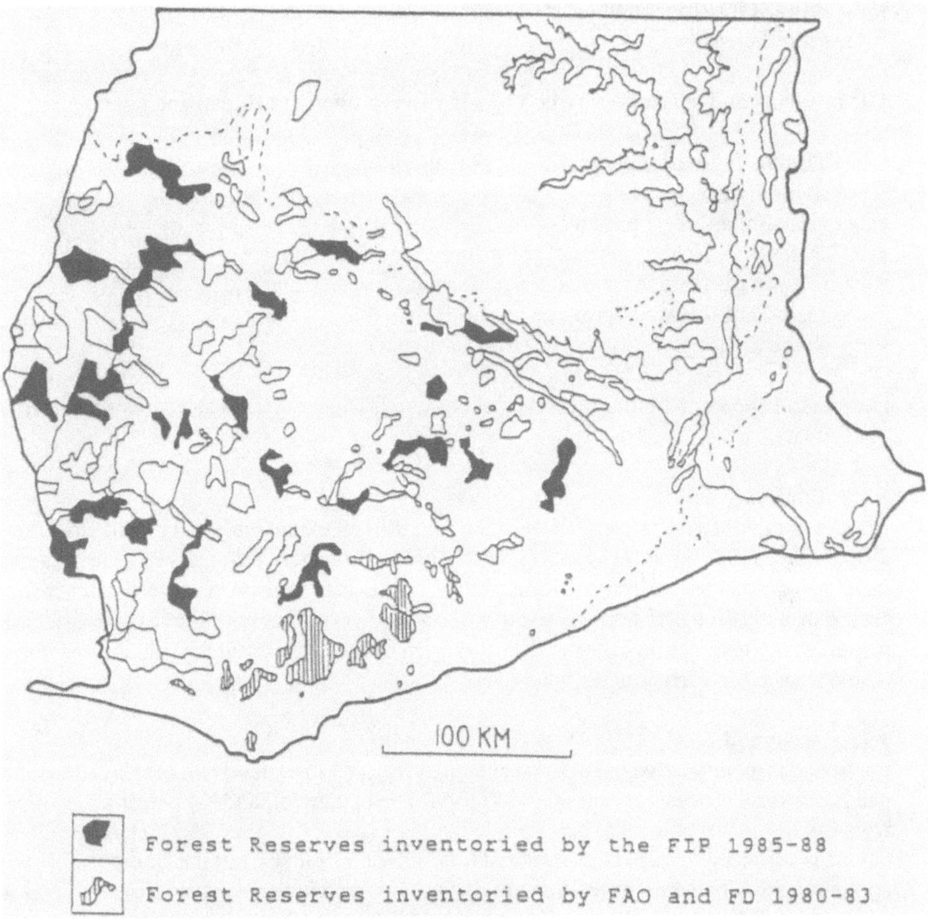

Figure 3.1 **Position of forest reserves in the high forest zone**

3.1 Functions of Trees and Forests the High Forest Zone

Forests and trees on farmland in the high forest zone serve a multitude of environmental and social functions. In comparison to the vast forests of, eg. The Democratic Republic of Congo and Brazil, the fragmented structure of Ghana's high forest together with a fairly well developed road infrastructure makes human access relatively easy. On the one hand, this gives the forests of Ghana a high social value, since many people derive many goods and services from them. On the other hand, easy access for many interested groups (not least logging companies) generally means high risk of over-exploitation.

The products and services provided by forest reserves include (FRMP, 1988; 5-8):

- Timber for export and domestic consumption;
- fuelwood;
- non-timber forest products;
- watershed protection;
- wind shelter (some reserves have "shelter belt" as part of their names);

- refuges for flora and fauna;
- sacred areas.

The products and services provided by off-reserve trees and forests include:

- Timber for export and domestic consumption;
- fuelwood;
- non-timber forest products;
- watershed protection (patches of forests on hill tops);
- refuges for flora and fauna (sacred groves in particular);
- sacred areas (sacred groves);
- shade on crops (scattered trees on farmland).

Despite the apparent similar functions of on- and off-reserve trees and forests there are important differences.

3.1.1 Timber

Timber for industrial purposes has been and still is extensively harvested off-reserves, even though the stocking is obviously much lower off- than on-reserves (see chapter 7). The development in the pattern of timber harvest indicates that on-reserve timber resources have served as a buffer to off-reserve resources and vise versa, but with the distinct effect that timber resources off-reserves have been steadily depleted since the establishment of forest reserves. This is investigated in further detail in chapter 7.

3.1.2 Fuelwood

Fuelwood is mainly collected off-reserves on farm and fallow land to meet rural household energy needs, for rural processing enterprises (food for sale, palm oil and "akpeteshie", a local gin made from distillation of palm sap) and for sale to urban areas (Falconer, 1992). It is widely argued that Ghana is suffering from a fuelwood crisis, eg. FRMP (1988; 6), but it is questionable whether this applies to the high forest zone in general. Through an extensive study of six villages in the high forest zone Falconer (1993; 81-82) found no signs of fuelwood shortages, and many villagers even found questions about fuelwood scarcities difficult to understand. However, the same study found that some rural enterprises (food, palm oil and akpeteshie producers) were having difficulties with their fuelwood supply, which raises the, yet unexamined, question of possible negative long-term effects urban fuelwood consumption may have on rural income generation (Falconer, 1993; 82). At the national level, fuelwood and charcoal is of major economic importance. It is estimated that the annual household expenditure on charcoal in Ghana is US$ 66.49 millions (1992 prices), and the figure for fuelwood is US$ 21.98 millions (Townson, 1995; 20).

3.1.3 Non-timber Forest Products

Non-timber forest products include a multitude of commodities collected/hunted for household consumption and for income generation. Roughly, NTFPs may be divided into:

- Forest foods;
- forest medicines;
- items for house construction;
- items for household goods and agricultural equipment;
- fodder for livestock.

Forest foods (uncultivated plants and animals) include seeds, fruit, leaves, roots and tubers, mushrooms and bushmeat (forest snails, game and freshwater fish and crabs). They are collected both inside and outside reserves and contribute significantly to the rural and urban diet in Ghana, both as a buffer when agricultural production fails and as a normal part of the daily diet. To children and lactating women, bush foods are probably highly important sources of vitamins and minerals during the season where food stocks dwindle and new crops are being planted (Falconer, 1993). However, bushmeat is regularly consumed by 75% of the population and it is presumably, in general, a vital source of protein (FRMP, 1988; 7).

There is a strong demand for bushmeat in urban markets, where it seems to become a luxury item, and this is a significant source of income generation. In a survey stratifying the high forest zone into five vegetation zones, Townson (1995) covers approximately 258,000 people who obtain income from NTFPs. He estimates that, in the area surveyed, approximately 20% of the economically active people derive an income from NTFPs, of which a little less than 20% do so from forest snails and bushmeat (Townson, 1995; 6-7). In 1992, the mean annual expenditure per household on bushmeat and forest snails was, according to the third Ghana Living Standards Survey, US$ 6.77. Extrapolated to all households in Ghana, this indicates an annual value, measured by 1992 consumer prices, of approximately US$ 22.56 millions. This may even be on the low side since purchases by businesses, government institutions, etc. are not included (Townson, 1995; 20).

Akpeteshie is among all NTFPs the top scorer in terms of consumer value and numbers of people deriving income from its production and trade. Measured by 1992 consumer prices, the annual value of akpeteshie production in Ghana is estimated to be more than US$ 63 million (Townson, 1995; 20). In Townson's survey it is estimated that almost five per cent of the economically active population derive an income from akpeteshie (Townson, 1995; 6-7).

Forest medicines are undoubtedly the most commonly used medicines in rural areas. It is noticeable that the medicines are not confined to specialist healers. The most common use is, in fact, as self-administered first aid (Falconer, 1992; 7). Western medicines are used in rural areas, but generally only if treatment with forest medicines has not been successful. In urban areas the choice between forest and western medicines tends to be the other way round, although it seems that both rural and urban people believe that certain illnesses (piles, boils, mental and spiritual problems) are best treated by traditional plant cures (Falconer, 1992; 8). The market for forest medicines in Ghana is not described in detail and much trade is likely to go unrecorded (Townson, 1995; 7).

Items for house construction include poles from tree saplings, raphia palm leaves, bamboo and cane. The nature of these items highlights the importance of forests although they are also collected off-reserves. A very important characteristic of these NTFPs is that they are extensively used in every Ghanaian village. A village house is typically built from standing poles and raphia (leaf petioles) or bamboo for the lattice structure, tied to the poles with cane or other climbers, and mud to fill out the spaces between the poles and the bamboo or raphia. Metal sheets are often used for roofing, but bamboo and raphia leaves are used by the poorest households. People generally collect their own building materials which means the items are rarely subject to market transactions, although the demand is high. Many chiefs and elders feel strongly that the collection of building materials in forest reserves should be an inalienable right (Falconer, 1992; 10).

Items for household goods and agricultural equipment include furniture, sleeping mats, carved stools, mattresses and pillows, bath houses, pit latrines, ladders, sponges, brooms, tool handles, baskets, fish traps, dugout canoes and, most valued by the villagers, pestles for pounding the national dish, fufu[3]. Many NTFPs for making such goods are collected off-reserves, but forest reserves appear to become increasingly important in this respect as some raw materials (eg. *Celtis* spp. and *Blinghia sapida* for pestles) are becoming scarce on farm and fallow land (Falconer, 1993; 80). The items are generally made for own use and not subject to much trade. However, the multitude of products and their extensive use in rural households and agricultural production obviously indicate that their social value is substantial.

Fodder for livestock is not a major issue, since livestock production is not a major activity in the high forest zone. However, those who do keep livestock collect fodder (various types of leaves) every day on farm and fallow land (rarely in forests) even though the animals are often free ranging during part of the day (Falconer, 1993; 80).

3.1.4 Watershed Protection and Shelter Belts

Forest reserves serve an important function as shelter belts and in terms of protecting large watersheds. In fact, the original legal instrument for gazetting forest reserves focused very much on the protective functions of forests as it empowered the Governor, on the advise of the Conservator of Forests, to constitute a forest reserve on lands to (Forest Ordinance (Cap 157), 1927; 4 (4)):

- Safeguard the water supply of a district;
- assist the well-being of the forest and agricultural crops grown on the land [...to become a forest reserve] or in the vicinity thereof;
- secure the supply of forest produce to the inhabitants of villages situated in the lands [...to become a forest reserve] or in the vicinity thereof.

Based on these powers, forest reserves have been defined in all major hills and watersheds. Barrier and shelterbelt forest reserves were also established to hold back bush fires and to maintain local rainfall and humidity respectively (Hawthorne and Musha, 1993; 3).

For obvious reasons, the protective functions of small off-reserve forests are much less important on the national level compared to the forest reserves. However, many such patches of forest in the farm and fallow landscape are found on hill tops near villages, and apart from being sacred groves they clearly protect local agricultural land from water erosion.

3.1.5 Refuge for Flora and Fauna

Forest reserves as well as pockets of forests on farm and fallow land serve as important refuges for wild flora and fauna. Comprehensive inventories and ecological studies of the fauna and its dependency on forest resources in southern Ghana are still to be carried out, which makes it difficult to specifically assess the zoological importance of forests and their management. However, bearing in mind that forests in botanically good condition have come to represent only islands of the original landscape in Southern

---

[3] Fufu is made from starchy stables (cassava, plantain and yams) which are boiled, then pounded with a pestle and mortar to a sticky (but tasty) substance that is served with different soups.

Ghana, it may be assumed that the best strategy for protecting faunistic biodiversity is to protect as large areas as possible from all disturbance, since species which are tolerant to disturbances in the forest cover should not be endangered - at least not through loss of habitat.

In contrast to zoological data, Ghana has a wealth of forest botany studies. All forest reserves in the high forest zone have recently been botanically surveyed with respect to their condition (level of disturbance) and their content of nationally and globally rare tree species, expressed by a so-called genetic heat index (Hawthorne and Musha, 1993). In particular, the forests in the wet evergreen zone (the South Western corner of the country) are in good condition and have a high density of rare tree species. This botanical information about the forest reserves provides a unique basis for taking forest protection adequately into account in management planning from the national down to the individual forest level.

Off-reserve forests have not been subject to extensive botanical studies, but nevertheless there is reason to believe they are refuges for flora and fauna that has become extinct from the surrounding farm and fallow land. In many cases (eg. on the costal plains) off-reserve forests represent the remnants of forest types which are now long gone (Hawthorne and Musha, 1993; 54).

3.1.6 Sacred Areas

Sacred forest areas are found both inside and outside forest reserves, but obviously the off-reserve sacred groves form the most conspicuous evidence of the cultural importance of forests in Southern Ghana. Inside forest reserves the location of sacred areas are sometimes secret, although the District Forest Office often knows of their existence and location (Hawthorne and Musha, 1993; 54).

The sacred groves and forest areas serve a number of different purposes which often vary from community to community. However, the spiritual use of some plants and their use by traditional healers, who are also believed to derive their special powers from forest spirits, is closely linked to the sacredness of the groves (Falconer, 1992; 8). In addition, sacred groves and forest areas are sometimes used as burial grounds.

Access to and collection of products from sacred groves may be strictly limited, but again this differs from community to community. There is a tendency for the younger generations and immigrants to ignore taboos related to sacred groves and their spirits. However, neither Christianity nor the level of education seem to influence people's belief in these traditional institutions. Being a well educated Christian, or Muslim for that matter, certainly does not mean that you brake taboos deliberately and discard traditional beliefs and powers as superstitious nonsense (Falconer, 1992; 8 and Personal observations). Even expatriates are well advised and generally do make an effort not to brake taboos and enter sacred places without permission from spiritual leaders. Anecdotal information about the mysterious and often horrible consequences faced by people who deliberately did so anyway is plentiful.

3.1.7 Trees on Farmland

Farm and fallow in the high forest zone is more often than not a landscape of scattered trees with various degrees of agricultural cultivation underneath. Broadly, trees on farmland may be divided into (i) those that are left standing because the work required to remove them is too heavy, (ii) those that are tolerated because they have few negative impacts on crops and (iii) those that are actively tended or planted by farmers (IIED, 1993; Appendix 4 (1)).

Depending on the agricultural crops trees on farmland serve a number of purposes. "Good" trees are favoured for their high or shallow crown providing a light shade, deep roots which compete less with crops, and the quality of mulching material and soil enrichment provided by their leaves, fruits and seeds. "Bad" trees have the opposite characteristics and may in addition also harbour pests (IIED, 1993; Appendix 4 (3)).

Food crops are not often grown under trees, although Naynkyerene *(Ficus exasperata)* is widely grown in association with plantain. Trees are mostly found in association with cocoa, the prime cash crop in the high forest zone. In a survey by IIED (1993) it was found that farmers considered many valuable timber species like Odum, Mahogany, Sapele, Emire and Ofram[4] "good" trees in association with cocoa. However, Wawa, the by volume most important timber species, was by cocoa farmers considered to be a "bad" tree (IIED, 1993; Appendix 4 (3-11)).

3.2 Deforestation in the High Forest Zone

Given that the remaining forest in the high forest zone is today largely confined to forest and game and wildlife reserves, it is clear that a massive deforestation has taken place in Southern Ghana. Most of this deforestation has been legal, intentional and arguably necessary for development of the country. As mentioned above, the detrimental effects of uncontrolled deforestation was already a concern for the colonial administration, which initiated the national constitution of forest reserves under the 1927 Forest Ordinance. It was an unofficial target to reserve approximately 25% of the high forest (Foggie, 1962 in Hawthorne and Musha, 1993; 4), and most of the forest reserves were gazetted in the first half of the 20th century. Today approximately 20% of the high forest zone is covered by forest and wildlife reserves[5].

Deforestation outside forest reserves to make way for agricultural development was encouraged in the 1948 Forest Policy (Francois, 1989a). The following expansion of farming (including cocoa) together with an increasing demand for pit props in newly mechanised gold mines tolled significantly on the off-reserve forest resources. During the period 1955-72 alone, it is estimated that about one third of the forests in Ghana have disappeared (Hall, 1987, in Hawthorne and Musha, 1993; 3).

3.2.1 The Economic Recovery Programme and Timber Exploitation

Ghana's economy plummeted during the 1970s and early 1980s. Huq (1989; 16) characterises the years from 1972 to 1983 as "complete mismanagement". During this period price controls, deficit financing and maintenance of a highly over-valued Cedi (Ghana's currency) were pursued with vigour (Huq, 1989; 17). This undermined the entire economy and not least the export sector. Payments in hard currencies for export commodities like timber and cocoa became pretty worthless to the exporters after the Bank of Ghana had changed the money into Cedis at the official rate. There was little or no allowance for exporters to retain some payments in hard currencies and changing Cedis to overseas currencies was obviously equally impossible.

---

4 See appendix 1 for botanical names.

5 The Game and Wildlife Department was set up (separated from the Forestry Department) in 1965 and has since then taken over the management of some previous forest reserves (Bia, Nini Suhien, Ankasa, Kakum and parts of Shai Hills Game Production Reserve).

As a consequence, the export of timber along with other commodities fell drastically and with it investments in maintaining the production capacity (logging and saw milling equipment) and the supporting infrastructure (harbour facilities, road and rail network). Ghana's timber harvest and export during the periods 1960-96 and 1972-96 are shown in figures 3.2 and 3.3, respectively.

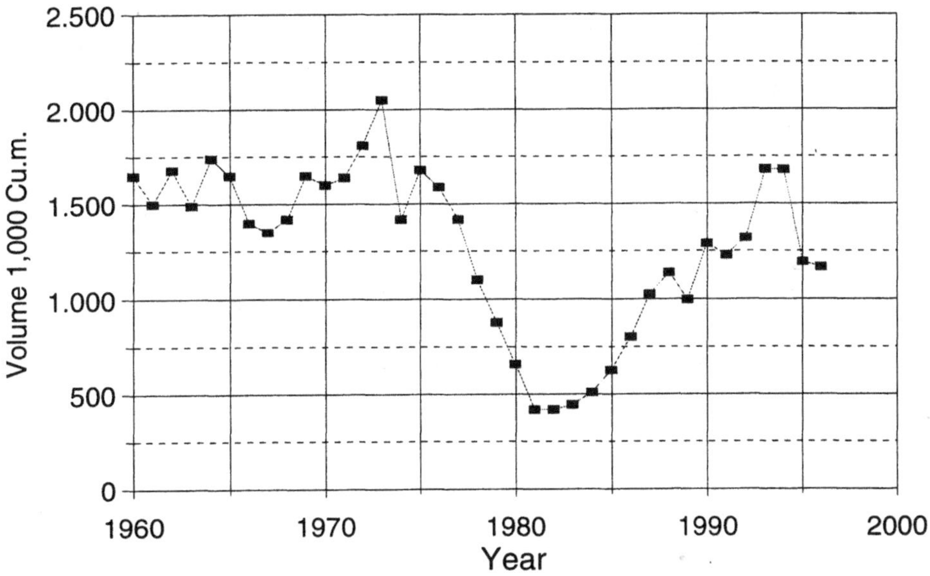

Figure 3.2 Total recorded log harvest by volume 1960-96

Sources:

General Woods (1992; Working Document No. 3, App. 3)

FPIB LMC Totals (1986-96)

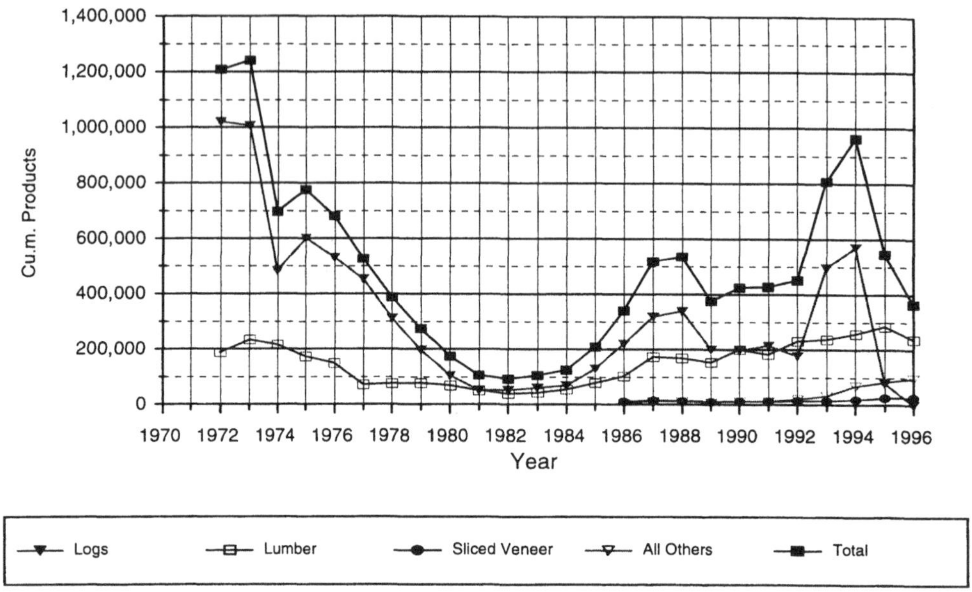

Figure 3.3 **Total wood exports by volume of product groups 1972-96**
Sources:
General Woods (1992; Working Document No.3, App.3)
FPIB Export Permits (1986-96)

Finally in 1984, the Government of Ghana, the IMF and the World Bank agreed on the so-called Economic Recovery Programme (Huq, 1989; 31). This programme encompassed a number of components including, of course, a strategy for enhancing the export of commodities from Ghana (the Export Rehabilitation Programme). Timber was an obvious candidate for this programme.

During the mid 1980s, massive investments went into the Ghanaian timber industry. Precisely how much and under which circumstances money was transferred from overseas donors and suppliers of credits appears to be one of the misty chapters in Ghana's recent forest history. According to the Financial Times (1987) more than US$ 140 millions from the World Bank (IDA), British[6] and Canadian bilateral aid and credit lines went into the Ghanaian timber industry during the period 1984-86. However, of this US$ 90 millions apparently went into transportation machinery, which should be written off the books after three years - an average annual debt repayment of US$ 30 million on transport alone (Financial Times, 1989).

As the curves in figures 3.2 and 3.3 indicate, the timber industry responded very quickly to the incentives provided under the Export Rehabilitation Programme. The annual recorded timber harvest

6 A UK£ 149.5 million credit was established through ODA for the purchase of logging and saw milling equipment (Frimpong-Mensah in FIP, 1989; 76).

and wood export went up from approximately 400,000 m^3 and 100,000 m^3, respectively in 1982 to 1,200,000 m^3 and 530,000 m^3, respectively in 1988.

In the pursuit of economic development, it often becomes an objective to convert available natural capital into human capital. In the case of forests, standing trees are converted into logs and processed wood products for export and domestic consumption. Logging does not require a high level of technical knowledge and equipment and it creates job in rural areas. In addition, the strong forward linkages of the timber industry with other sectors should make natural forests in LDCs attractive assets from which to initiate economic development (Westoby, 1962, in Westoby, 1987; 44).

Although Westoby later on almost totally abandoned this theory as being politically naive and, in practice, leading to severe socio-economic underdevelopment and uncontrolled deforestation in most LDCs (Westoby, 1978, in Westoby, 1987; 246-254), he originally warned against the danger of mobilising the forest resource for industrial development without controlling the process through adequate and enforceable planning in the public interest. On how to mobilise the forest resource, he wrote:

> *The technical and economic conditions for establishing new forest industries in the developing countries are maturing fast. In the course of the coming years many new areas of forests are inevitably going to be brought into use. The choice is between mobilisation in the public interest based on sound planning and with adequate safeguards and with the forest services taking an active part and being built up in the process, and mobilisation taking place in an uncontrolled and haphazard way, while weak forest services stand by helpless. This is the real choice.* (Westoby, 1962, in Westoby, 1987; 69-70)

During the early stages of the Export Rehabilitation Programme in Ghana the then Chief Conservator of Forests, John Francois often tried to raise the issue of matching the capacity of the fast growing timber industry with the production capacity of the country's forest resources (Personal communication J. Francois, CCF, 1993 and Financial Times, 1987). Yet, such basic principles of caution in planning the utilisation of natural forest and timber resources were largely ignored by decision makers. The consequences of this attitude soon began to materialise, and since then these have been dominating policy issues within the forestry sector.

4 The Timber Resource Base, Early Estimates

Under the forestry component of the Export Rehabilitation Project provision was actually made for two special studies to look into (Silviconsult, 1985; 1):

- Strategy Proposals for the Forestry Sector;
- Forestry Department Review and Study of the requirements of the Forest Products Inspection Bureau (FPIB) and the Timber Export Development Board (TEDB).

The studies were undertaken by a private consultancy company, Silviconsult, and a focus area was to determine the annual allowable cut (AAC) of timber. To this end the results of the 1980-82 FAO/UNDP inventory of 16 forest reserves and permanent sample plot data from the Forests Products Research Institute (FPRI) were analysed with a view to estimate the growing stock and annual increment within forest reserves in the high forest zone. Furthermore, as there was no data on off-reserve timber resources, guesstimates were made concerning the off-reserve growing stock and increment (Silviconsult, 1985; 6-16).

4.1 The Growing Stock in 1985

The 16 forest reserves included in the FAO/UNDP inventory cover an area of approximately 200,000 ha, and the sampling intensity was 0.46% (Silviconsult, 1985; 153). However, from a national planning perspective the inventory results are limited by the fact that reserves inventoried are found in the southern part of the Central and Western Region only.

Of the approximately 1.72 million ha forest reserves in the high forest zone, approximately 1.18 million ha were designated as production working circles[7] (production forest). Estimates of the growing stock within production working circles of the entire high forest zone were arrived at by re-grouping FAO/UNDP data according to vegetation zones as classified by Hall and Swaine (1976), and then extrapolating the results to account for all forest reserves in each vegetation zone (Silviconsult, 1985; 151-153). The FAO/UNDP inventory did not include the Dry Semi-deciduous zone which accounts for approximately 8% of the total area of production working circles. As a proxy for this zone the mean volume per ha for the other three zones (the Wet Evergreen, the Moist Evergreen and the Moist Semi-Deciduous) was used (Silviconsult, 1985; 153).

To make inventory results meaningful at the policy level all tree species were classified according to commercial importance. The groups were based on the following criteria (appendix 1 details species groups for different inventories):

Group A: The 40 species which, at the time, had an established export or domestic market.

[7] Production working circles is a term used in the forest working plans of the 1950s. The primary management objective within production working circles was production of timber on a sustained yield basis. These plans were designed for 10 year periods, so formally they expired during the 1960s. However, since then many of the management prescriptions in these plans have, in the absence of new management objectives and prescriptions, been considered valid. The estimate of 1.18 million ha of production working circles was in 1985 changed to approximately 1.16 million ha under the ODA/FD Forest Inventory Project (see chapter 5).

Group B: Twenty potentially commercial species which, at the time, were only harvested to supply specific orders.

Group C: All other species.

4.1.1 National Estimates in 1985 of the Growing Stock

The growing stock within forest reserves of the high forest zone was estimated as bole volume (presumably above bark) for species groups (cf. above) and girth classes. The silvicultural system was (and still is) based on selective felling of trees above a species specific minimum girth. At the time all trees above minimum girth were allocated for felling, which, in principle, meant that the standing stock above felling limit could be considered available for timber harvesting. The inventory estimates are summarised in table 4.1.

Table 4.1 **Estimates in 1985 of bole volume within forest reserves by vegetation zone and species group (1,000 m³)**

Vegetation zone	Area of production W.C. (1,000 ha)	Group A		Group B		Group C		Total	
		Above min. girth	Total volume	Above min. girth	Total volume	Above min. girth	Total volume	Above mini. girth	Total volume
Wet Evergreen	137.41	1,291	3,358	1,113	2,067	3,066	8,788	5,470	14,213
We/Me Transition	80.95	1,065	2,196	1,415	2,402	2,419	5,898	4,899	10,496
Moist Evergreen	327.27	9,384	17,938	6,840	9,847	8,783	17,794	25,007	45,579
Me/Ms Transition	119.15	3,058	5,639	3,493	4,715	2,790	5,424	9,341	15,778
MoistSemi-deciduous	389.62	14,618	29,106	7,461	11,062	7,606	14,790	29,685	54,958
Ms/Ds Transition	33.64	938	1,858	648	960	787	1,681	2,374	4,499
Dry Semi-Deciduous	94.69	2,642	5,230	1,825	2,702	2,215	4,732	6,682	12,665
TOTAL	**1,182.73**	**32,996**	**65,325**	**22,795**	**33,756**	**27,666**	**59,107**	**83,457**	**158,188**
Mean Vol/ha (m³)		27.90	55.23	19.27	28.54	23.39	49.98	70.56	133.75

Source: Silviconsult (1985; 163-167)

4.1.2 National Estimates in 1985 of Annual Growth and AAC

Estimates of the annual growth were established by analysis of permanent sample plot data, However, the quality and in particular the quantity of data apparently left much to be desired. Periodic girth measurements had only been made for some of the species within Group A (groups I and II in a previous classification system used by Forestry Department) and measurements were biased towards trees of small diameters. (Silviconsult, 1985; 155).

The actual number of data on which the growth estimates were based are not stated by Silviconsult (1985), neither is there any information on the location of permanent sample plots in relation to vegetation zones. Furthermore, the crucial question of mortality rates was not accounted for. If the growth of trees in a natural forest is estimated without taking natural mortality into account, the calculated annual increment is bound to overestimate actual growth and hence the number and volume of trees which actually reach timber size (the minimum girth). Obviously, the potential error in this respect is particularly high when measurement is done mainly on trees of small diameter, since their chance of reaching maturity (the minimum girth or more) is smaller than for trees close to the mature state.

Nevertheless, growth estimates on Group A species were calculated and even extrapolated to account for Group B and C species, such that national estimates of the total annual increment could be produced. The annual allowable cut was then recommended to be 70% of the annual increment to allow for wastage in the form of felling damage, road construction etc. The results are summarised in table 4.2

Table 4.2 **National estimates in 1985 of annual increment and AAC within forest reserves (1,000 m³)**

Species group	Total stock	Stock above minimum girth	Annual increment	Estimated AAC
Group A	65,325	32,996	1,024.20	719.10
Group B	33,756	22,795	551.10	385.60
Group C	59,107	27,666	1,197.70	837.40
TOTALS	**158,188**	**83,457**	**2,773.00**	**1,942.10**

Source: Silviconsult (1985; 11)

In addition to the estimated AAC of approximately 1.9 million m^3 some considerations were made with respect to the potentials for off-reserve timber harvest.

Basically, no data on off-reserve timber resources existed and guesstimates from so-called informed sources varied between 15 and 50 million m^3. Concessionaires said that the off-reserve yield per ha was generally one quarter to half of what they usually could harvest on concessions within forest reserves. Based on this it was assumed that the off-reserve forest resource was equivalent to 0.75 million ha of forest reserve. Thus, the off-reserve growing stock should be 100 million m^3 of all species, of which 53 million m^3 is above the minimum girth. Accordingly, for group A species (the commercially most desired species) the stock above minimum girth should be approximately 21 million m^3 (Silviconsult, 1985; 15-16).

However, the off-reserve stock was at the time correctly not considered a sustainable timber resource, since clearings for agriculture were continuously taking place outside forest reserves. Rather, the off-reserves timber resource was considered a going resource of which maximum harvest for industrial purposes should be pursued while it remained (Silviconsult, 1985; 18).

4.2 Recommendations on the Timber Harvest in 1985

With an AAC of approximately 720,000 m^3 for group A species and 380,000 m^3 for group B species within forest reserves alone, the long-term sustainable AAC of commercial and potentially commercial species is, therefore, estimated to 1.1 million m^3. By comparing this estimate and the guesstimated off-reserve resources with the annual timber harvest of 400-600,000 m^3 in the early 1980s (see figure 3.2) it appeared as if there was scope for a significant expansion of the annual harvest.

All in all this message on the timber resource and the AAC must have suited those promoting investments in the timber industry very well. Concerns over the sustainability of increasing the timber harvest could be calmed, although investments in the timber industry were already being undertaken and the investment policy was hardly subject to change according to the findings of Silviconsult. It is certainly not mentioned in Silviconsult's terms of reference for the study that investments in rehabilitating of the timber industry should be considered in relation to the estimated AAC (Silviconsult, 1985; 129-130).

4.2.1 The Stated Warnings and Unstated Risks

However, Silviconsult (1985; 13) emphasised that the data on growth dynamics of Ghana's forests was inadequate to make sufficiently reliable estimates of the annual growth and hence the AAC. Furthermore, it was brought to attention that an estimated 90% of the annual harvest fell within the group A species. A significant but still sustainable increase of the timber harvest should, therefore, first and foremost be concentrated on group B species, which made it essential to develop markets for these species (Silviconsult, 1985; 18).

A glance at harvest rates during the 1970s should have resulted in additional warnings against the risks of boosting the capacity of the timber industry (see figure 3.2). Annual harvest rates between 1.5 and 2.0 million m^3 was clearly above the estimated AAC of the most desired and harvested species (group A), but also above the AAC of group A and B species taken together (se table 4.2).

The declining timber production during the late 1970s and early 1980s was hardly the result of a deliberate policy to reduce the timber harvest. Rather, it was the result of a general economic decline in Ghana. Accordingly, it was likely that the timber industry would return to harvest rates of the 1970s once the capacity to do so was reestablished through the Export Rehabilitation Programme.

Clearly, the Forestry Department was not able to efficiently enforce harvest regulations within its jurisdiction, the forest reserves. Off-reserve the timber harvest was hardly subject to much regulation by the Lands Department. Both departments were found to be under-equipped, under-staffed and under-funded to live up to their responsibilities and duties in relation to the forest and timber resources (Silviconsult, 1985; 32-62).

It may be considered easy to say afterwards, but three major warnings on rehabilitating the timber industry were actually raised by Silviconsult (1985; v):

- The knowledge of the resource was incomplete and insufficient for efficient planning and utilisation;
- the field administration seriously lacked facilities for efficient management and control;
- revenues from forests could, at the time, cover only a small fraction of the required budget for proper resource management, but they could be considerably increased.

These were important findings suitable for policy level considerations, but the scientifically questionable estimate of the AAC probably had more policy effect at the time, although the general message cannot have been overheard. Ironically, more reliable estimates of the AAC have later revealed that the estimates of Silviconsult (1985) were dangerously optimistic (see section 7.5). Yet, it took a long time before this was realised.

5 National Forest Inventory (1985-89)

In 1985, the British Government signed a UK£ 1.146 million aid project with the Government of Ghana to undertake a forest inventory in the high forest zone of Ghana, "the Forest Inventory Project" (FIP). This project was implemented during the period 1985-89 by the Forestry Department, Ghana (FD) with ODA employed expatriate assistance (long and short term advisors and consultants).

5.1 Background of the Forest Inventory Project

As mentioned in chapter 4, one of the main findings of Silviconsult (1985) was that the knowledge about the growing stock and the annual increment in Ghana's forests was insufficient to allow for adequate national level forest management planning. Also, the then Chief Conservator of Forests, Mr John Francois had in various fora insisted that the rehabilitation of the timber industry under the Export Rehabilitation Project had to be linked to the sustainable supply of wood from the forests of Ghana. However, as a result of the general economic decline in Ghana during the 1970s and early 1980s it became impossible for the Forestry Department to continue ongoing forest inventories as well as to implement planned activities in this field, so in 1984 a request was made to ODA for assistance to establish the capacity to conduct a national forest inventory (Francois, 1989b in FIP, 1989a; 6).

On its part ODA is likely to have realised that the support provided to the timber industry in Ghana produced an urgent need to strengthen the management of Ghana's high forest. If such management was not supported by ODA and timber exploitation in Ghana proved uncontrollable, ODA could easily be accused of promoting tropical deforestation for quick financial profits. ODA foresters must to some extent have been involved in designing and implementing the ODA supported rehabilitation of the Ghanaian timber industry during the period 1985-89 (or at least aware of the project). Accordingly, it is likely that the concern this development raised among foresters in Ghana, as expressed by the then CCF, J. Francois, has found its way back to decision makers in ODA, London.

However, the mechanisms of identifying the FIP are unclear. A formal ODA evaluation of the FIP could not document the identification process or the criteria applied. It was found likely that the project was a follow-on to the FAO/UNDP inventory project, for which Ghana initially failed to get funding until ODA expressed interest (Flint and Hardcastle, 1992; 3).

5.2 Objectives of the Forest Inventory Project

The primary objective of the FIP was through a reconnaissance survey to determine the growing stock and the annual increment to provide a macro-scale planning figure of the AAC (Howard, 1989 in FIP, 1989a; 59)[8]. In further detail, the FIP had three main objectives of which in particular the first was considered critical because the rehabilitated timber industry was putting increasing pressure on the forest resource (Blackett, 1989 in FIP, 1989a; 14). The objectives were:

- Static timber Inventory;
- dynamic timber inventory;
- training in all aspects of timber inventory.

---

8 The ODA evaluation documents that while there was never doubt about this overall objective there was serious confusion whether FIP should produce inventory data which was useful at reserve or management unit level. This limited the success of the project (Flint and Hardcastle, 1992).

The FIP developed three species classes reflecting the present and potential commercial importance of tree species in Ghana. The species classes were defined as follows (FIP, 1989a; 32):

FIP class 1: Species at least exported once from Ghana during the period 1973-88.
FIP class 2: Species attaining 70 cm diameter at breast hight[9] (DBH) and occurring at a frequency greater than 1 tree per km^2.
FIP class 3: Species not attaining 70 cm DBH or occurring at a frequency less than 1 tree per km^2 (all other species than FIP class 1 and 2 species).

Appendix 1 details the species in FIP classes 1 and 2 together with the previous classification of species made by Silviconsult. All but one of Silviconsult's class A species and all but six of Silviconsult's class B species are included in FIP class 1.

5.2.1 Design of the FIP Static Timber Inventory

Initially it was the objective to sample approximately 500,000 ha of forest reserves throughout the high forest zone. In practice, this meant that temporary sample plots (TSPs) were established in about half of the forest reserves in the high forest zone. At a later stage, under the Forest Inventory and Management Project (FIMP) which followed the FIP, the remaining reserves were covered such that by 1994 practically all reserves in the high forest zone had been inventoried.

Within the four major ecological zones (Wet evergreen, Moist evergreen, Moist semi-deciduous and Dry semi-deciduous)[10] a random selection of forest reserves was made such that the area surveyed was directly proportional to the area of forest reserves within each zone. It was decided to apply a 0.25% sampling intensity as recommended by FAO (Flint and Hardcastle, 1992; 15). In addition, an analysis of the 0.5% sampling intensity applied under the previous FAO/UNDP inventory had indicated that satisfactory accuracy on volume estimates could be obtained using a lower sampling intensity (Blackett, 1989 in FIP, 1989a; 17). Apparently, a statistical check of the sampling error was conducted only once (in 1986) during the FIP. This proved to be 12% (presumably the 95% confidence interval of the volume estimate for all diameter classes within the FIP class 1 species) which was considered acceptable against a target of 20% (Flint and Hardcastle, 1992; 15).

The TSP layout was done by randomly superimposing a 2 x 2 kilometre grid upon maps of the selected reserves, letting grid intersections determine the location of 1 ha TSPs. In the field the TSPs were established by compass and chain survey from obvious geographic features (Blackett, 1989 in FIP, 1989a; 17). Within each TSP the following measurements and observations were made (FIP, 1989b; App. 1):

---

9 Defined as 1.3 metre above ground level.

10 Eventually the FIP operated with five vegetation zones as it was found useful to divide the moist semi-deciduous zone (MS) into a North Western moist semi-deciduous zone (MSNW) and a South Eastern moist semi-deciduous zone (MSSE). The other three zones are the wet evergreen (WE), the moist evergreen (ME) and the dry semi-deciduous zone (DS), sometimes referred to as the dry semi-deciduous forest zone (DSFZ).

- On the entire plot diameters for all trees having a diameter above 30 cm DBH or 30 cm above buttresses were measured;
- the local species name for all trees was noted, or if unknown a specimen number was given and a field specimen collected for subsequent herbarium identification;
- form and quality codes ranging from 1 to 5 (1 being the best) were given to each tree above 30 cm DBH (to allow for subsequent estimates of utilisable volumes);
- crown classification ranging from 1 to 5 (1 being unshaded) was given to each tree above 30 cm;
- on a subplot of 0.1 ha all trees above 10 cm DBH were measured;
- on a subplot of 0.1 ha all trees above 5 cm DBH were measured;
- relevant observations for each plot such as slope, position, presence or absence of logging activity, burning, rivers, roads, etc.

A diameter-bole volume equation was developed through detailed volume measurements of 3488 trees (Wong, 1990; 8), and volume stand tables generated for trees having form and quality codes 1 or 2 only. This means that TSP volume data only includes *bole* volumes of "good" trees for industrial use (Wong, 1989 in FIP, 1989a; 28). Furthermore, estimates were made of the *utilisable* volumes by subtracting the volume of defective sections (bumps, rot, sweeps, etc.) from the bole volume. The mean defect per tree was 8% of the bole volume. Roughly this percentage lost due to defects remained constant for all trees irrespective of size. This is not reflected in the stand tables, but it should be borne in mind when interpreting the results.

Finally, FIP defined the *commercial* tree volume as the volume under bark of cross cut logs prepared by the timber company minus the volume of defective sections[11]. This should approximate the LMC volumes recorded by FPIB. On average the *commercial* tree volumes were found to be 10% below the TSP bole volumes. If it is assumed that the bark accounts for 2% of the bole volume this meant that the commercial volume would be very similar to the FIP utilisable volume less bark (Wong, 1989 in FIP, 1989a; 29).

However, upon further analysis it was concluded that a statistically significant relation between bole volumes and commercial volumes could not be established. This was probably due to the subjective and highly variable partitioning of boles into logs by the various timber companies (Wong, 1990; 9). Accordingly, the FIP volume measurements did not provide statistically sufficient basis for establishing a mathematical relation between the growing stock and increment (expressed in FIP bole volumes) and the annual extraction of logs (expressed in LMC commercial volumes).

To arrive at national estimates for the growing stock, mean TSP volumes per ha minus the 95% confidence interval were simply multiplied by the area of production working circles within the forest reserves in the high forest zone. The area figures used by FIP are detailed in table 5.1.

---

[11] All 3488 measured trees were felled and utilised normally by timber companies. Measurements for the bole volume were done on the felled but still not cross cut tree, and measurements of the commercial volume were done after cross cutting into logs by the timber company in case.

Table 5.1 **Areas used by FIP to calculate national estimates of the growing stock within forest reserves (ha)**

Region	Protection working circle	Production working circle	Total area of forest reserves
Eastern	83,300	87,200	170,500
Central	27,100	122,800	149,900
Western	94,700	513,400	608,100
Ashanti	124,000	234,000	358,000
Brong-Ahafo	103,200	201,600	304,800
TOTAL	**432,300**	**1,159,000**	**1,591,300**

Source: Ghartey, (1989 in FIP, 1989a; 43-46)

The areas of protection and production working circles were obtained from the regional forestry offices (Wong 1989 in FIP, 1989a; 31).

5.3 Results of the FIP Static Timber Inventory

Unlike the growing stock analysis made by Silviconsult (1985), the FIP static inventory results were not published by vegetation zone, but by administrative region. This, together with the FIP classification of species, makes a direct comparison with the 1985 growing stock estimates impossible, except for the national volumes. Yet, as noted above the FIP class 1 covers almost all of Silviconsult's class A and B species. Table 5.2 summarises the 1989 FIP static inventory results.

Table 5.2 **FIP estimates in 1989 of bole volume within forest reserves by administrative region and species class (1,000,000 m³)**

Region	Class 1			Class 2			Class 3			TOTAL		
	30-70 cm	> 70 cm	Total	30-70 cm	> 70 cm	Total	30-70 cm	> 70 cm	Total	30-70 cm	> 70 cm	Total
Eastern	3.60	6.10	9.70	1.80	1.00	2.80	0.70	0.00	0.70	6.10	7.10	13.20
Central	7.40	9.10	16.50	2.90	1.10	4.00	1.50	0.00	1.50	11.80	10.20	22.00
Western	24.40	41.90	66.30	11.30	5.30	16.60	5.10	0.50	5.60	40.80	47.70	88.50
Ashanti	10.60	18.80	29.40	3.90	1.80	5.70	1.20	0.00	1.20	15.70	20.60	36.30
Brong-Ahafo	8.50	13.80	22.30	3.50	1.90	5.40	0.90	0.00	0.90	12.90	15.70	28.60
TOTAL	**54.50**	**89.70**	**144.20**	**23.40**	**11.10**	**34.50**	**9.40**	**0.50**	**9.90**	**87.30**	**101.30**	**188.60**
Vol/ha (m³)	47.02	77.39	124.42	20.19	9.58	29.77	8.11	0.43	8.54	75.32	87.40	162.73

Notes: The volumes stated are the lower values within the 95% confidence interval around the estimated mean volumes. Estimates are based on the inventory of 1332 TSPs in 43 reserves.
Appendix 2 details the mean volumes by species and diameter class.
Volumes per ha refer to the total area of production working circles (1,159,000 ha).

Source: FIP (1989a)

As table 5.2 shows, the 1989 FIP national estimate of the total growing stock within production working circles is 188.60 million m^3 (162.73 m^3/ha) which is approximately 30 million m^3 higher than the 1985 estimate made by Silviconsult (1985). Also, the estimated 89.70 million m^3 (77.39 m^3/ha) of FIP class 1 species above 70 cm DBH[12] is considerably above the 55.7 million m^3 of Silviconsult's group A and B species above felling limit taken together (see table 4.1). Accordingly, the positive message from the 1989 FIP results was that Ghana was well stocked with species which had been exported at least once since 1973.

However, the 1989 FIP results also indicated that there was reason for concern. As table 5.3 shows, the growing stock above 70 cm DBH of the ten commercially most desired species was approximately 37% of the growing stock above 70 cm DBH of FIP class 1 species, but in 1988 these species accounted for 77% of the total harvest as recorded by the LMC totals.

Table 5.3 **Ten most harvested species compared with their estimated standing volumes (m^3)**

Trade name	Extracted volume in 1988	Volume above 70 cm DBH	Volume above 110 cm DBH
Wawa	366,064	19,911,620	8,194,130
Odum	172,983	2,793,190	1,483,520
Mahogany	66,877	2,283,230	904,020
Kyere	49,556	2,167,330	417,240
Emire	42,242	764,940	428,830
Sapele	41,135	1,495,110	672,220
Makore	39,642	915,610	741,760
Asanfina	35,736	1,564,650	509,960
Edinam	33,167	1,367,620	486,780
Utile	31,891	695,400	382,470
SUB TOTALS (above 10 species)	**879,293**	**33,958,700**	**14,220,930**
TOTALS (FIP class 1 species)	**1,136,737**	**91,317,610**	**35,140,880**
SUB TOTALS AS % OF TOTALS	**77.35**	**37.19**	**40.47**

Note: This table refers mean volumes and not the lower values within the 95% confidence interval around the estimated means.
Sources: FIP (1989b) and LMC totals.

Only Wawa, which made up 21.8% of the FIP class 1 growing stock above 70 cm DBH, appeared abundant. The other most desired species each made up between 0.8 and 3.1% of the FIP class 1 growing stock above 70 cm DBH (see table 5.2). This concern was supported by the FIP dynamic inventory results.

12 FIP (1989a) uses 70 cm DBH to indicate the standing volume above felling limit. Actually, this is an over-estimate since felling limits were and still are species specific, with a number of species having felling limits of 90 and 110 cm DBH. In 1989, 70 cm DBH was the minimum felling limit.

5.4 The FIP Dynamic Inventory

The dynamic inventory attempted to produce timber growth and yield estimates for the forest reserves within the high forest zone. To this end measurements and analysis of data from permanent sample plots (PSPs) were undertaken. On this basis a simple growth table for 14 species and a computer based simulation of the growth of all species which have ever been exported[13] were produced (Alder, 1989 in FIP, 1989a; 47-49).

Data from the PSPs only allowed for calculating simple growth tables for 14 species. The annual growth per ha for each of the 14 species is calculated as the volume that will grow into a diameter class[14] from the class immediately below. It is calculated as the volume of the preceding diameter class times the average proportion of surviving trees and the average percentage that will move up into the next class (Alder, 1989 in FIP, 1989a; 47). Accordingly, growth of trees which remain within diameter classes is ignored.

The AAC for each of the 14 species was calculated as the volume moving from the diameter class immediately below the felling limit into the diameter class above the felling limit. The time this takes was estimated by dividing the annual diameter growth within the particular diameter classes into 20 cm (the diameter class width). National AACs were calculated by assuming that the productive forest area is 1.3 million ha (Alder, 1989 in FIP, 1989a; 4). Why the 14 AAC estimates are based on a productive forest area of 1.3 million ha while the growing stock estimates were based on an area of approximately 1.16 million ha (the total area of production working circles) is not explained. The time it takes to grow from the diameter class immediately below felling limit into the diameter class above felling limit, and AACs together with the rate of extraction in 1988 for the 14 species, are shown in table 5.4.

---

13 It is not clear whether this coincides entirely with the FIP class 1. It must, however, be assumed that only a few species additional to those in FIP class 1 are included.

14 Diameter classes of 20 cm are used.

Table 5.4 **Estimated AACs and resource lifetime for 14 species**

Species (trade names)	Felling limit (DBH)	Time of passage from diam. class below f. limit to the felling limit	Resource above felling limit (m^3)	Estimated AAC (m^3/year)	Rate of extraction in 1988 (m^3/year)	Estimated life of resource (years)
Odum	110 cm	34 years	1,408,000	28,650	172,983	10
Edinam	110 cm	45 years	468,000	7,155	33,167	18
Sapele	110 cm	26 years	720,000	13,496	41,135	25
Utile	110 cm	39 years	465,000	8,081	31,891	20
Mahogany	110 cm	20 years	692,000	31,488	66,877	20
Afrormosia	110 cm	not available	0	0	7,190	0
Wawa	70 cm	29 years	26,356,000	135,779	366,064	114
Guarea	70 cm	39 years	524,000	4,592	10,972	882
Dahoma	70 cm	24 years	5,254,000	75,569	14,915	***
Kyenkyen	70 cm	30 years	3,726,000	33,331	14,801	***
Hyedua	70 cm	107 years	154,000	1,966	10,620	18
Mansonia	70 cm	72 years	695,000	2,753	5,830	226
Danta	70 cm	76 years	1,254,000	10,098	24,787	85
Avodire	70 cm	31 years	2,365,000	13,5488	269	***

Notes: Appendix 1 provides a list of species trade names with the corresponding botanical names and vice versa.
Extraction rates are take directly from the LMC totals of FPIB.
*** indicates that the AAC exceeds the rate of extraction.

Source: Alder (1989) (in FIP, 1989a; 49-52)

Even though the AACs and lifetime estimates in table 5.4 can only be taken as indicative (they still remain the only species specific growth estimates available) the message was very clear. Some species (which also tend to be the most valuable) appeared grossly over-exploited while the extraction of other species could be expanded.

The 14 species in table 5.4 were not entirely representative for all species harvested, so they could not be used to estimate the AAC for all species in all forest reserves within the high forest zone, the national AAC. To this end a computer programme, GHAFOSIM was applied on all FIP class 1 species. The programme combines PSP data on single-tree increment and mortality rates of 24 different species with TSP data on the growing stock within diameter classes and crown classification (the latter used to predict mortality rates) (Alder, 1990). When, in an attempt to approximate a future commercially driven harvest of species, GHAFOSIM was run on only 30 of the 57 FIP class 1 species, this produced a number of interesting results[15] (Alder, 1989 in FIP, 1989a; 48 and Alder, 1990):

---

15 Alder (1990) uses, but does not specify, the term "redwoods" together with 16 other specified species. "Redwoods" is normally used for the commercially most valuable species which have reddish/brown wood. Here this group is considered to contain 14 species.

- If the extraction rate remains at the 1988 level of approximately 1 m^3/ha/yr, this can be sustained for about 100 years or more, but the stock of commercially most desired species will become progressively depleted;
- the 1989 distribution into size classes may only be maintained if a 100 year felling cycle[16] is implemented;
- with a 15 year felling cycle the 1989 stock of mature (big) trees will become depleted, but in the long term yields of approximately 3.4 m^3/ha/yr (a national AAC of approximately 4.0 million m^3) may be sustained;
- with a 25 year felling cycle the 1989 stock of mature (big) trees will become depleted, but in the long term yields of approximately 2.2 m^3/ha/yr (a national AAC of approximately 2.5 million m^3) may be sustained;
- with a 40 year felling cycle the 1989 stock of mature (big) trees will become depleted, but in the long term yields of approximately 1.3 m^3/ha/yr (a national AAC of approximately 1.5 million m^3) may be sustained.

Alder (1990) argues that while felling cycles shorter than 40 years appear to result in higher annual yields, due to the fact that trees are generally removed closer to their point of maximum average annual growth and before natural mortality, the 40 year felling cycle is to be preferred because:

- The uncertainties in yield forecasting are considerable due to data constraints and the fact that GHAFOSIM was still mostly a theory based growth model for whole stands which had not been verified in practice (this is still the case). Therefore, a conservatively set AAC was advisable
- the AAC under the 40 year felling cycle (1.5 million m^3) approximated the annual demand by the industry
- the 40 felling cycle would, compared with shorter felling cycles, minimise the area for harvesting operations to be controlled annually by the Forestry Department (1.16 million ha divided by 40 years gives an annual area of 29,000 ha for harvest)
- the geographical concentration of felling activities would allow for a tighter in-field control of loggers and thereby enhance the possibility of requiring and encouraging the industry to market the abundant but lesser used species more energetically.

Compared with Silviconsult's estimated AAC of approximately 1.1 million m^3 for its group A and B species (see section 4.2), the FIP estimate of 1.5 million m^3 for its class 1 species under a 40 year felling cycle is a little more optimistic. Although the GHAFOSIM estimates of the AAC may have been considered the best available at the time, the warnings about the uncertain nature of this AAC essentially remained the same as in 1985 (Alder, 1990; 53-55):

- PSP growth data was insufficient to allow for accurate predictions of future yield under various management regimes;
- data on the growth of whole stands was (and still is) totally absent;

---

16 The required minimum period of time between any two felling operations within a given area. A 25 year felling cycle allows the residual stand a 25 year recovery period before the next felling operation.

- growth data on a number of individual species were inadequate or absent;
- the knowledge about mortality rates and how they may be influenced by logging and other factors was (and still is) insufficient.

In fact, upon questioning Alder (1989 in FIP 1989a; 52) admits that the precision of the GHAFOSIM simulations of growth and the AAC is about ±50%. Accordingly, Silviconsult's AAC lies within the confidence limits of GHAFOSIM's (FIP's) estimated AAC.

5.5 Other Technical Outputs of the FIP

An issue of concern was the felling cycle. The traditionally used 25 year felling cycle was replaced by a 15 year felling cycle in 1970, with the intention to enable a more rapid removal of old (sometimes called over-mature) trees such that they could be utilised before death and such that a higher volume increment could be achieved for the residual stand (Silviconsult, 1985; 51). During the 1980s, the 25 year felling cycle was re-introduced, but was it sustainable?

Nolan (1989 in FIP, 1989a; 55) calculates an estimated felling cycle of 64 years by dividing a total AAC of 1.4 million m^3 into the FIP class 1 growing stock above 70 cm DBH of 89.7 million m^3 (assuming that this total AAC is correct and that the stock above 70 cm DBH equals the total increment over a felling cycle). Although this was a very rough and scientifically questionable estimate it emphasised, together with the computer simulations in GHAFOSIM and the passage of time for trees to grow from the diameter class immediately below the felling limit into the diameter class above felling limit, that a felling cycle of 25 years was highly unlikely to be sustainable. In fact, Silviconsult came to a similar result in 1985 and recommended a 40 year felling cycle (Silviconsult, 1985; 88).

As mentioned above, the TSPs, sub-plots of 0.05, 0.1 and 0.2 ha were laid out to record all trees of DBHs higher than 5, 10 and 30 cm, respectively. This would indicate how well the forest was regenerating in the different vegetation zones.

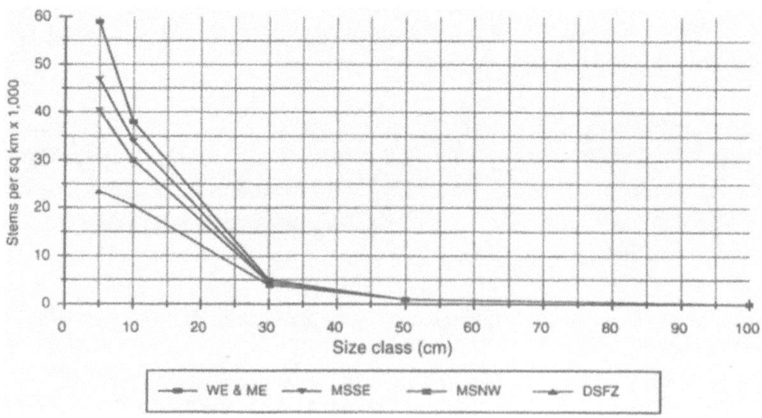

Figure 5.1 **Stem numbers by size classes and vegetation zones**
Source: FIP (1989a; 36)

As figure 5.1 shows, the frequency of stems below 30 cm DBH within the dry semi-deciduous zone is markedly lower than for the other zones. When this observation was combined with the percentage of inventoried area which had been burned (figure 5.2), a pattern emerged.

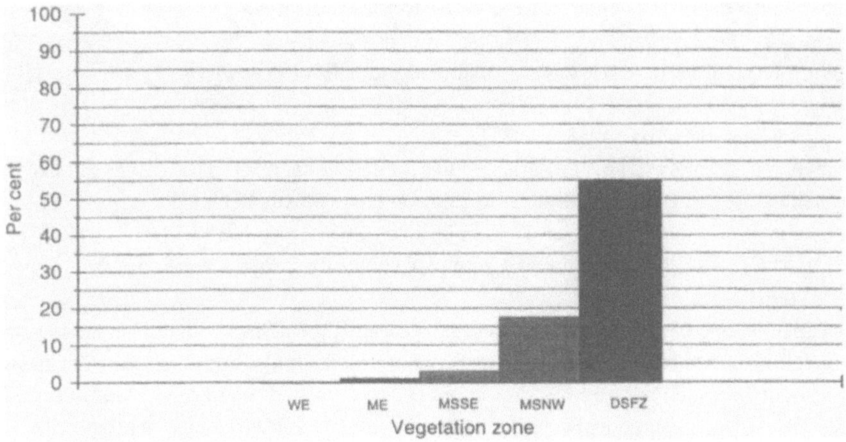

Figure 5.2 **Percentage of burned areas by vegetation zones**
Source: FIP (1989a; 37)

It appeared that the forest in the dry semi-deciduous zone was losing its ability to regenerate, and forest fires were likely to be the direct course.

In general, it was concluded that forest fires were the greatest threat to Ghana's forests and that once burned forests are likely to burn again due to the increased level of (i) annual weeds (especially *Eupatorium*), (ii) windflow at ground level and (iii) volume of dead wood in burned compared to unburned forests. Therefore, it should be matter of urgency to investigate the seemingly logical hypothesis that logged forests are more prone to catch fire than unlogged forests, especially in the dry areas (Hawthorne, 1989 in FIP, 1989a; 12).

5.6 Critique of the FIP

A major point of criticism against the FIP is that the objectives of the project were far from clear and consistent (Flint and Hardcastle, 1992; 39). More specifically, FIP appeared to have no formal project framework. This allowed for three versions of the project objectives.

First, as stated in the terms of reference for the evaluation mission in 1992, the objectives of FIP were (Flint and Hardcastle, 1992; 38):

- The primary objective was to undertake an inventory of 500,000 ha of forest reserves to provide details of the size of the growing stock and the increment of the forest, thereby enabling a more accurate assessment of the annual allowable cut to be calculated. The data would enable a

sustained yield policy to be formulated so that timber could be extracted from the forest while at the same time the floristic and environment values would be maintained
- the second objective was to establish an inventory unit capable of running continuous forest inventory with the ability to provide data on area, volume, species, size-classes and regeneration consistent with the medium and long-term objectives of the Ghana Forestry Department
- the third objective was to prepare a field guide to the forest trees.

Second, the objectives of the inventory were, according to FIP (1989b; App. 1), to:

- Fix the prescribed cut at national level;
- prepare inventory reports and harvesting schedules for individual reserves or groups of reserves;
- report on the status of tree regeneration in relation to forest disturbance.

Third, according to ODA (1987) the immediate objectives of the project were to:

- Carry out an inventory of selected forest reserves totalling 500,000 ha, to conduct mensurational studies on felled trees to produce improved indigenous volume tables and to measure Permanent Sample Plots to provide data on the increment of the forest for use in the planning model;
- prepare a long-term management plan for forest reserves under various management assumptions;
- train local staff both in Ghana and abroad so that they can take on the forest inventory as soon as possible. Skills required included tree identification, measuring with different instruments, sample design and layout, team organisation, air photo interpretation and data processing;
- prepare a practical key to the forest trees of Ghana for use by technical field staff and to write a key with a view to publishing it;
- develop and document all the necessary data processing systems required to carry out the objectives in the future.

In particular, the confusion over objectives seems to be whether the inventory results were intended to be used at management unit level, while it was pretty clear that the results should be used at the national level to guide a sustained yield policy through an estimate of the growing stock and an estimated AAC.

While the forest botanical objective of making a field guide to forest trees in Ghana (Hawthorne, 1990) was fully met, there is reason to criticise a number of issues concerning the inventory results[17]. These include:

- The static inventory design;
- the AAC estimates;
- the (missing) links between AAC estimates, actual harvest and the management of timber harvest inside forest reserves.

5.6.1 Critique of the Static Inventory Design

Basically, the policy and management objectives of the static inventory should have been clarified before the sampling exercise was designed and implemented. It is, for example, not a logical objective to

---

17 Objectives of the FIP other than those specifically concerned with forest management and policy are not considered in detail in this study.

sample 500,000 ha of forest with a 0.25% intensity. There should exist underlying objectives of the inventory which justify the extent and methodology applied.

In general, any inventory based on statistical sampling should include the following steps:

- Define objectives;
- determine population variance and structure, either through pilot sampling or historical information;
- define accuracy required;
- rationalise input costs against accuracy;
- check preliminary results against the objectives and accuracy requirements and, if required, redefine procedures.

As the project was primarily concerned with the sustainability of commercial timber harvest it may be criticised that the project did not focus particularly on big trees within the FIP class 1. The sampling was done proportional to area rather than considering financial value. However, the most important weakness of the sample design lies in:

- The lacking clarity as to which accuracy was acceptable for estimates produced at different geographical levels;
- the definition and areas of production and protection forest;
- the fact that only forest reserves were considered.

The required accuracy of volume estimates appears to have been ±20% on volume estimates, measured by the sampling error defined 95% confidence interval (Flint and Hardcastle, 1992; 50). However, as table 5.5 shows this level of accuracy was more than achieved for some estimates and far from achieved for other estimates.

Table 5.5 **Statistical accuracy of FIP inventory volume estimates for class 1 species above 110 cm DBH by level of analysis**

Level	95% Confidence limits as % of the estimated volumes (SE%)
National	6
Vegetation zone	9 - 22
FMU[1] or reserve	13 - 60
Individual species (national level)	23 - 48

Notes:[1] Forest management unit. Depending on the size of forest reserves a FMU covers one or several reserves.
Based on 1332 one ha plots in 43 reserves (32 FMUs)
Source: Flint and Hardcastle (1992; 18)

At the national level, the sampling design produced very accurate volume estimates for FIP class 1 species above 110 cm DBH as a group. Even within vegetation zones the level of accuracy for the group of FIP class 1 species above 110 cm DBH was acceptable. However, it is only in 19 of the 32 inventoried FMUs that the SE% for all FIP class 1 species equals or lies below 20% (Flint and Hardcastle, 1992; 18). For individual FIP class 1 species above 110 cm DBH, the national estimates did not achieve the desired level of accuracy.

Measured against the (various) objectives of the FIP it is clear that at the national level an accuracy of 6% on the estimated mean value for the growing stock of FIP class 1 species above 110 cm DBH is more than satisfactory[18]. However, to achieve an accuracy of 20% for the 95% confidence interval on the estimated mean volume of all FIP class 1 species above 110 cm DBH it would have sufficed to measure approximately 120 rather than 1,332 plots as was done (Flint and Hardcastle, 1992; 19). At FMU level, and for individual species at the national level, the volume estimates did not, in all cases, achieve the desired level of accuracy.

Accordingly, the FIP static inventory results did only produce statistically satisfactory results at national and vegetation zone levels for FIP class 1 species above 110 cm DBH. Practically, all other data on the volume of the FIP class 1 growing stock were associated with confidence intervals wider than 20% on the estimated means.

A source of error which was practically ignored by the FIP was the area of production working circles. Wong (1989 in FIP, 1989a; 25) only mentions the potential source of area related errors in relation to consistency between forest reserves and their map representations, and concludes that the error from this source is likely to be small. However, the area of production and protection working circles, which were obtained from the regional forest offices, were not investigated by FIP with respect to accuracy and relevance. This invited two sources of error:

- The 1.16 million ha of production forest could be an over-estimate if the forest condition within reserves, due to fire damage, illegal farms, logging damage, etc., suggested that the actual productive area was smaller, and/or if the stated 0.43 million ha of protection forest was too small to meet national objectives of environmental protection;
- the random identification of plots on forest maps must have placed TSPs within obvious protection working circles (steep slopes, watersheds, etc.). If the average stocking of timber in these areas differs significantly from obvious production working circles, this would bias the national estimates accordingly.

The above two points relate to the fact (and problem) that the FIP did not stratify the inventory according to identified areas within forest reserves which were actually suitable for timber production. Ideally, the FIP should have started out by reviewing and perhaps redefining the criteria for (i) protection working circles and (ii) areas which are too poorly stocked to justify any harvest within the coming 100 years or so. Then such areas should have been identified within all forest reserves and TSPs should only have been laid out in areas suitable for logging so as to estimate the growing stock of timber available for harvest in the future.

---

[18] For FIP class 1 species above 70 cm DBH the accuracy on the estimated mean volume is down to 3% (Flint and Hardcastle, 1992; 94)

It may be argued that the FIP did not have the information required to identify areas for permanent protection and areas for medium/long-term protection before logging can be allowed. The forest policy and forest management prescriptions did at the time not provide an adequate basis for defining which areas should be protected, and areal photos or satellite images did either not exist or were unavailable at the right quality to geographically locate protection areas.

However, old maps and management plans might have provided a guide for identifying some areas which would qualify for protection. More importantly, the forest quality parameters recorded on TSP sub-plots, together with an identification of TSPs where the stock of FIP class 1 species were below a certain minimum, could have provided valuable information about the area suitable for logging, and hence the growing stock available for future harvest.

Finally, in 1994 the TSP inventory data was analysed according to the recorded parameters of forest quality, topographic features, and level of growing stock (FIMP, 1994). The result was that, depending on the criteria for forest protection, the estimated area of production forest was between 0.92 and 0.76 million ha, (see section 6.1). A similar analysis could, in 1989, have yielded similar information. Since FIP failed to do so, the area of production forest was over-estimated by at least 0.24 million ha or approximately 20%.

As it was the increasing timber harvest which primarily produced the need for a forest inventory, it is surprising that the FIP never analysed where the extracted wood came from. It was well known that timber was extracted both within and outside forest reserves, but at which ratios was pure guesswork. The reality was that over 50% of the annual harvest came from off-reserve areas (IIED, 1993; 15). With this in mind it is obvious that the inventory should have covered both on- and off-reserve areas in the high forest zone. The reason being that when a substantial proportion of the timber harvest comes from off-reserve areas it is highly relevant to estimate how long this can continue. Furthermore, the forest reserves will be the only source to satisfy the industry's demand for logs after the off-reserve timber resources have been logged out. At such point in time it is crucial that the Forestry Department will be able to control an increased demand for timber on-reserves. In addition, the substantial off-reserve harvest makes it relevant to consider ways of halting or reversing a degradation of the timber resource in these areas.

In summary it may be concluded that:

- The static inventory results as presented in 1989 appeared statistically very accurate, but actually the growing stock available for harvest was over-estimated, due to the non-investigated and, therefore, over-estimated area of production forest[19]
- the inventory results were not accurate enough to be used in management plans for FMUs
- the relevance of the inventory in relation to the timber harvest was limited because off-reserve areas were ignored.

---

19 It is not possible to calculate how much the growing stock was over-estimated in 1989 by simply multiplying the estimated mean volumes per ha by a reduced area of production forest, since the average stock per ha varies according to different definitions of production forest.

5.6.2 Critique of the AAC Estimates

A fundamental problem with national AACs is that, due to data constraints, they are notoriously difficult to calculate on a sound scientific basis. At the same time they are treated as highly important figures at the policy level. Therefore, FIP's objective of producing a national AAC appears ill considered, or it should at least have been modified when the limitations of the PSP data were realised.

Of course the difficulty in calculating the AAC is not a valid reason to refrain from attempting to do so. After all, any kind of forest management planning will have to make do with the data available. The problem arises when policy decisions are based on a calculated national AAC without due consideration of the fact that this is an estimate with confidence limits of ±50%, as the case was for Ghana in 1989 (see section 5.4). Hence, the AAC of 1.3 m^3/ha /year under a 40 year felling cycle for FIP class 1 species is not an accurate estimate.

In combination with the over-estimated area of production forest there was obviously a great danger (with a likelihood of more than 50%) that the national AAC of 1.5 million m^3 could be far from sustainable and damaging to the forest resource if this was actually followed. However, to assess the actual harvest against the calculated AAC requires that the actual on-reserve harvest is known. This was not the case in 1989.

5.6.3 The Missing Links Between AAC Estimates, Actual Harvest and Principles of Forest Management

The justification for calculating AACs is naturally to compare these with the actual harvest. Before this can take place two things must be fulfilled:

- A system to record the harvest must be in place and it must produce reliable data (recorded harvest);
- a relation between recorded harvest and actual harvest must be established.

Neither of the above two points were fulfilled in 1989. There was a system in place to record extracted logs and it produced reasonably reliable data although it is likely that it under-estimates actual extraction (the LMC system managed by FPIB). However, the LMC data did not distinguish between logs coming from forest reserves and logs coming from off-reserve areas. Only the total volume of log extracted by species was recorded. As mentioned, it took till 1993 before the harvest on- and off-reserves was subject to a thorough investigation. This revealed that harvest off-reserves accounted for more than 50% of the total harvest (IIED, 1993;15).

When, as the case is in Ghana, the AAC refers to bole volume felled and the recorded harvest refers to the volume of logs extracted, it is problematic to compare the two without a valid relation between them. It is, in other words, necessary to know the level of logging waste and damage on residual stands to calculate an accurate mean ratio between felled volume and extracted volume. Such information was not and still is not available in Ghana. In fact, FIP came to the conclusion that is was not possible to establish a statistically valid relation between bole volumes and the volume of logs produced from the very same trees (Wong, 1989 in FIP, 1989a; App.1).

Management of timber harvest inside forest reserves is based on a 100% stock enumeration within compartments of approximately 120 ha. Within a compartment all trees above 50 cm DBH are identified, their location recorded, their diameter measured and a stock survey number is carved into the bark. A "stock map" of the compartment indicating the position, species, diameter and stock survey number of each tree is then produced together with a "stock table" where stock survey numbers are arranged by species and diameter class. On this basis the trees allocated for felling are selected and a yield map and yield table for the compartment showing only these trees is produced.

Previously, all trees above felling limit were allocated for felling, but in 1990, as a result of the FIP results, the first version of the "Interim Yield Allocation Formula"[20] was applied in order to reduce the number of trees allocated for felling in each compartment. However, the origin of the formula is obscure. Even though the Interim Yield Allocation Formula in its various versions may be considered a valuable and simple tool to control the on-reserve harvest, it was not directly derived from FIP data (Flint and Hardcastle, 1992; 21).

Surprisingly, after its introduction the Interim Yield Allocation Formula was not readily applied on TSP data to see how the result corresponded with previous estimates of the national AAC. The formula is sensitive to the length of the felling cycle and to numbers (and hence volumes) of trees within diameter classes, so applied at the national volume estimates of FIP class 1 species it should produce a pretty reliable estimate of the *allowed harvest* under the formula (leaving out for the moment the then over-estimated area of production forest).

Appendix 2 details the result of applying the Normal Interim Yield Allocation Formula ($Z = 0.5Y + 0.2X$) on the 1989 TSP volume data assuming a felling cycle of 40 years and approximately 1.16 million ha of production forest. For FIP class 1 species taken together it is seen that the annual allowed cut for all forest reserves should be approximately 0.94 million m^3 or 0.81 m^3/ha/year.

This is only 62% of GHAFOSIM's estimated AAC of 1.5 million m^3 (1.3 m^3/ha/year) under a 40 year felling cycle, and it is below the 1988 recorded harvest of 1.14 million m^3. Although the recorded harvest cannot, for the above mentioned reasons, be compared directly with calculated allowable or allowed cuts, this was actually done. The rough and ready policy analysis was that recorded harvest (wrongly interpreted as equal to actual harvest) was within the limits of the AAC. Had the annual allowed harvest been calculated according to the Interim Yield Allocation Formula, then the policy analysis would have been less optimistic and, despite its questionable interpretation of recorded harvest, closer to the truth as subsequent analyses have indicated.

5.7 Policy Impact of the FIP

Despite the limited success of the project in terms of producing relevant growing stock estimates, and sound estimates of the timber growth and AAC, its impact at policy level appears to have been quite substantial. Cause and effect relations in policy matters are obviously not possible to establish with a 100% certainty (or with meaningful confidence intervals for that matter). Policy decisions are always

---

[20] For each species within the compartment the first version of the Interim Yield Allocation Formula calculates the yield as; $Z = 0.4Y + 0.2X$, where Y is the number of trees above felling limit, X is the number of trees in the diameter class immediately below felling limit and Z is the number of trees above felling limit which are allocated for felling (Ghartey, 1990). However, the formula generally applied (called the normal formula) calculates the yield as; $Z = 0.5Y + 0.2X$ (FD, undated ; Kakrada and Maginnis, 1992 ; FIMP, 1995b).

influenced by a multitude of factors that sometimes produce the most surprising and unforceable results. However, the FIP evaluation summarised the project impact as follows (Flint and Hardcastle, 1992; 21):

Table 5.6 **FIP Contribution to forest policy and management reforms**

Change	FIP contribution
Redefinition of Protection Working Circles.	Botanical work started under FIP.
New felling formula (the Interim Yield Allocation Formula).	Justified by FIP national species availability results, but actual equation not directly derived from FIP data.
Changes in girth limits.	Not informed by FIP data.
Extended felling cycle.	Case for longer felling cycle than 15 or 25 years strengthened, but 40 year cycle not proven.
Additional species for which a special felling permit is required.	Justified by FIP national species availability data.
Increased royalty rates (revised in 1989)[1].	General case for increasing the royalties strengthened by data on national species availability.
New export levies[1].	Justified by FIP national species availability data.

Note: [1]Economic incentives for improved forest management are discussed in chapter 11.
Source: Adopted from Flint and Hardcastle (1992; 21)

Considering the scientific quality of the data produced by FIP, the alleged policy impact as described in table 5.6 is indeed impressing. In particular, it is somewhat surprising that the normally very articulate timber industry lobby in Ghana never questioned the validity of AAC estimates.

6 National Forest Inventory Continued (1989-97)

In 1988, a UK£ 3.7 million agreement was signed between ODA and the Government of Ghana to essentially continue the FIP under a new name, the Forest Inventory and Management Project (FIMP). This project was designed to work within the framework of the wider World Bank, Danida, ODA and Government of Ghana sponsored UK£ 20.2 million forestry sector programme, the Forest Resource Management Project (FRMP), which commenced in 1988 (FRMP, 1988; 1).

Unlike its predecessor the FIMP has a project framework. According to this the immediate objectives and planned outputs in relation to forest management and inventory were (FIMP, 1988): Immediate inventory and management objectives:

- A sustained yield from forest reserves through a 50 year harvesting plan using a 25 year felling cycle which applies (a) allowable cut (b) area control (c) minimum girth;
- improvement in the level of efficiency in the management of the forest reserve;
- conservation of the protection working circles, identification of reserves of special floristic interest.

Planned inventory and management outputs:

- Inventory of 0.6 million ha of forest reserve to complete the inventory of all reserves commenced under the ODA's present Forest Inventory Project;
- mapping and inventory of the blocks of manageable forest situated outside the gazetted reserves. This will involve the coverage of some 6.4 million ha of land;
- inventory and detailed description of all forest plantations within the tropical high forest zone. Believed to cover some 30,000 ha;
- preparation of working plans for the effective management of the forest resources, both with respect to the high forest and the plantations.
- rehabilitation of the forest plantations.

From a cost effectiveness point of view it is surprising that it was possible in 1988 to justify the need for a so-called completion of the inventory to cover all forest reserves, since comprehensive data at the national level was just about to be produced by the FIP. From the outset of the FIP a larger grid of perhaps three by three km covering all forest reserves would probably have been a more sensible approach (Flint and Hardcastle, 1992; 67).

Nonetheless, the remaining forest reserves were inventoried such that during the period 1988-94 a total of 3,101 TSPs of one ha were sampled from 128 reserves (FIMP, 1995a). Furthermore, the botanical work initiated under the FIP yielded impressing and very comprehensive data on the condition of forest reserves which included recommendations on forest protection in Ghana.

The PSP programme was also continued, but as of November 1996 the data collected had not been used in an attempt to update forest growth and yield calculations. In any case it is believed that the PSP programme should run for another 20-30 years to give definitive data on high forest growth, although it should be possible to derive valuable information from the current set of data (Alder, 1996).

This chapter reviews and summarises the analyses of TSP data and forest protection strategies which have been undertaken during the FIMP. This work seems to have produced the best documented and most comprehensive national level estimates of:

- The production area within forest reserves;
- the growing stock by species and diameter class within forest reserves;
- the sustainable harvest potential in forest reserves;
- the timber resource outside forest reserves.

6.1 Production and Protection Forest Within Forest Reserves

In addition to the continued timber inventory a botanical survey was carried out under the FIMP. The aims of the botanical survey were to (Hawthorne and Musah, 1993; 65):

- Build a clearer picture of "what forest, vascular plant lives where" in the forest zone of Ghana and thereby to pinpoint parts of Ghana's forested landscape of greater bioquality;
- collect useful background information on the forest with respect to past management, threats to the forest, local opinion, etc. which might have some bearing on the protective management of the forest.

The latter aim was subordinate to the former but could presumably be pursued at little extra cost when the botanical survey team visited forests to survey plant populations. In addition to the above two aims, the botanical survey provided an excellent training ground for botanists at all levels.

The botanical survey was based on subjectively laid samples (between 1 and 20 in each forest reserve) intended to give the best possible picture of species representation within reserves. Accordingly, the survey deliberately made sure to record the species composition in swamps, logged areas, ridge tops, steep slopes, river sides, etc. (Hawthorne and Musah, 1993; 66-69).

6.1.1 Condition Scores for Forest Reserves

Hawthorne and Musah (1993) apply a condition score between one and six for each forest reserve. The scores are based on observation, conversations with forest staff working in the area, the inventory results and Spot satellite imagery in descending order of importance (Hawthorne and Musah, 1993; 11). Table 6.1 shows the definition of each score.

Table 6.1 **Condition scores used to summarise forest reserves**

Score	Definition
1	EXCELLENT with few signs (< 2%) of human disturbance (logging/ farms) of fire damage, with a good canopy and virgin or late secondary forest throughout.
2	GOOD with < 10% heavily disturbed. Logging damage restricted or light and well dispersed. Fire damage none or peripherical.
3	SLIGHTLY DEGRADED: Obviously disturbed or degraded and usually patchy, but with good forest predominant. Max. 25% with serious scars and poor regeneration; max. 50% slightly disturbed with broken upper canopy.
4	MOSTLY DEGRADED: Obviously disturbed and patchy, but with bad forest predominant; 25-50% serious scars but with max. 75% heavily disrupted canopy. Or forest lightly burned throughout.
5	VERY POOR: Forest with coherent canopy <25% (more than 75% disturbed), or more than half of the forest with serious scars and poor or no forest regeneration; or almost all heavily burnt with conspicuous *Eupatorium* and other pioneers throughout. Not, however, qualifying as condition 6.
6	NO SIGNIFICANT FOREST LEFT: Almost all deforested with savanna, plantation or farm, etc.; <2% good forest; or 2-5% very disturbed forest left; or 5-10% left in extremely poor condition, eg. as scattered trees or riverine fragments. Remnants with little chance of surviving 10 years.

Source: Hawthorne and Musah (1993; 11)

Each forest reserve has been analysed according to the condition scores (Hawthorne and Musah, 1993; App. C). The result is summarised in table 6.2 which provides a national level picture of the forest condition within reserves.

Table 6.2 **Areas of forest reserves by condition**

Condition	Area in km²	Per cent of total area
1	346	1.9
2	2,485	14.0
3	6,190	34.9
4	3,618	20.4
5	2,626	14.8
6	2,480	14.0
TOTAL	**17,745**	**100.0**

Note: The total area differs from that in table 5.1 because it includes reserves in the SM and SO zones and reserves with no forest left in the other vegetation zones.
Source: Based on (Hawthorne and Musah, 1993; 13).

The most important information to be drawn from table 6.2 is that 9,021 km^2 or about half of the gazetted forest reserves are in reasonable condition (scores 1-3) with only 16% of the total reserved area having the lightest or no disturbance in recent history (scores 1-2). The reason being that large areas, particularly in the semi-deciduous zone, have suffered excessively from fire and logging, although a number of reserves in this zone carried no forest at the time of reservation (Hawthorne and Musah, 1993; 13).

6.1.2 Previous Protection of Forest Resources

Forest conservation in Ghana through the gazettment of reserves was very influenced and guided by the environmental and protective functions of forests. However, upon a review of the history of forest protection in Ghana Hawthorne and Musah (1993) conclude that:

- Protection working circles within forest reserves were introduced in the working plans of the 1950s-1960s, but they lacked a clear overall definition. The establishment of protection working circles and prescriptions for their management appear to have rested entirely with the district and regional forestry officers;
- practically unloggable reserves were often defined as protection reserves;
- notwithstanding past definition as protection reserves or protection working circles, many of these areas have de facto been subject to extremely heavy logging, taungya farming, mining, etc.

The overall description of forest protection in Ghana is a failure of management. To a large extent, the concepts have degenerated and upon the slightest pressure the protective legislation or management prescriptions have crumbled. Furthermore, a theoretical framework for protection has been non-exacting, unclear or very often non-existent (Hawthorne and Musah, 1993; 20).

Although forest plant communities are not the sensitive ecosystems that many environmentalists claim, it is very likely that most of the forests in Ghana would become seriously degraded or die if current patterns of utilisation do not change (Hawthorne and Musah, 1993; 42). The types of exploitation to be protected against most forcefully are those that affect directly the main structural elements of the forest, the canopy trees, soils and water courses. Serious adverse effects on the forest as a whole arising from exploitation of non-canopy forest produce is hard to envisage, although the populations of the species being exploited may well suffer (Hawthorne and Musah, 1993; 21). Accordingly, Hawthorne and Musah (1993) developed a new framework for forest protection.

6.1.3 A New National Framework for Forest Protection

Against current trends in nature conservation, Hawthorne and Musah (1993; 7) argue that protective regulation and needs should be decided at the national level and enforced downwards. Local needs for conservation should then be superimposed on the national requirements, not replace them.

This, apparently controversial, approach to forest and nature conservation may be justified by the fact that needs for conservation may, in many cases, only be realised at the broadest scale (at national level or higher). Very rare plants and animals are, for example, often abundant over small areas (Hawthorne and Musah, 1993; 7).

Based on the botanical survey and the analysis of current measures for forest protection Hawthorne and Musah (1993) recommend a new framework for forest protection. This framework is divided into two concepts, fine-grained and large-grained protection.

Fine-grained protection concerns management of single trees and small patches of forest. It should generally be interpreted as management rules that apply to all forest uses (logging in particular) in all production areas, ie. outside larger areas which are set aside entirely for protection. Fine-grained protection should according to Hawthorne and Musah (1993; v-vii) include:

- An upper limit on the number of trees allocated for felling per unit area (about 1.5 trees per ha);
- yield allocation procedures which take into account the ecology of certain over-exploited species, non-pioneer species and species with less widely-broadcast seeds and regeneration (special felling permits and reduced yield for certain species);
- protection of desirable phenotypes as seed-trees in yield allocation;
- attention towards the protection of sensitive landscape areas, particularly localised steep slopes and streams that do not show on 1:50,000 maps;
- attention to forest condition when the yield is allocated such that harvest in fire-damaged and fire-prone forest is reduced;
- attention to the spatial distribution of other trees at national level and in compartments of forest near to the one being considered when yield is allocated;
- total protection of very rare species (eg. *Talbotiella gentii* which is only found in Ghana);
- a total stop of converting viable natural forest into plantations within reserves;
- control of unavoidable mining and quarrying such that areas for these purposes are kept as small as practicable, and progressing activity is made subject to adequate restoration of already worked areas.

Large-grained forest protection will usually be at least the size of a compartment (approximately 120 ha). Protection in these areas need only apply to trees above 5 cm DBH and other canopy plants and rare species. Exploitation of NTFPs is, therefore, not necessarily excluded in these areas. Large-grained protection is recommended to include the following (Hawthorne and Musah, 1993; vii-ix):

- Southern dry forests in the SM and SO forest types. The remaining forests often fall outside forest reserves, but in general they represent a threatened forest type and serve as refuges for a large number of rare species;
- hill sanctuaries with slopes mostly steeper than 15 degrees;
- fire protection blocks fringing the savanna zone or other very burned and fire-prone areas;
- provenance protection areas for community samples and economic "genebanks" where timber trees are completely protected from exploitation;
- convalescence protection areas in severely damaged forest reserves where the forest should be allowed to recover from past over-exploitation through the enforcement of a particularly long felling cycle (significantly longer than the 40 year cycle applied in well stocked forest);
- special biological protected areas with either an unusual abundance of rare species or of an unusual community type;
- sacred forests and forest areas which serve religious and spiritual purposes;
- other environmentally protective blocks like extensive swamps.

It is obvious that, if implemented, the above recommendations on forest protection would affect the area of production forest and the allowed level of harvest within protection forests. However, Hawthorne and Musah (1993) do not attempt to explicitly estimate the effects of forest protection on timber harvest, although they state that a lot of the areas which are recommended to be protected from logging are pretty inaccessible anyway.

6.1.4 Scenarios of Forest Protection

The survey by Hawthorne and Musah (1993) included reserves and forest areas which were not inventoried for timber under the FIMP, but they did not make explicit use of the TSP data in order to estimate the cumulative area within forest reserves which they thought should be protected. Yet, in 1994, the recommendations of Hawthorne and Musah (1993) were operationalised (FIMP, 1994a).

The approach was, on the basis of Hawthorne and Musah (1993), to develop a number of forest protection scenarios for the inventoried forest reserves and then apply the TSP data set, assuming that one plot represents 400 ha in accordance with the 0.25% sampling intensity. In other words, the scenarios represent different levels of large-grained forest protection. The categories of protection areas used to develop the scenarios were defined as follows (FIMP, 1994a):

- Hill sanctuaries: Areas that generally have slopes greater than 30% on 1:62,500 (inch to mile) scale topographic sheets. By translating the maps into TSPs this gives a total area of approximately 210,400 ha;
- swamp sanctuaries: Areas where the TSP sub-plot codes indicate more than 60% swamp in the dry season. This gave 45 plots in 27 reserves, or a total area of 18,000 ha;
- special biological protection areas: Areas with a high density of rare species. To a certain extent these areas fall under hill or swamp sanctuaries as defined above, but additional areas in six forest reserves are included;
- shelter belts: These areas/reserves are not specifically addressed by Hawthorne and Musah (1993), but they were originally gazetted with a view to stabilise the local climate. The shelter belts are in general only a few kilometres wide and several tens of kilometres long

which makes them vulnerable to degradation, especially by fire. In total the area of shelter belts comes to 22,000 ha;

- intact forest sanctuaries: Areas which, in addition to provenance protection areas (see below), contain intact forest. Intact forest is defined by an exceptionally high basal area for the vegetation zone and a low density of pioneer species. In total this area amounts to 2,400 ha, although concrete investigations on the ground may justify a smaller area;
- provenance protection: Areas carrying a high density of high value and heavily exploited timber species. The areas are identified on flat lands in all vegetation zones (not in swamps or hills) which represent the best forest and because certain high value species do not occur in swamps and on hills. In total this area amounts to 8,400 ha, although concrete investigations on the ground may justify a smaller area;
- fire protection: Areas, particularly in the DS zone, which form the border of reserves. The area for fire protection was identified through TSPs that fall within 1,000 m of the boundary of the reserves in the DS zone, assuming that a minimum of 1,000 m width is required for fire protection;
- convalescence: Poorly stocked areas which have a basal area between 5 and 15 m^2 per ha. In total this amounts to 122,000 ha;
- conversion: Areas with less than 5 m^3 per ha, ie. areas where extensive planting is required to establish a forest.

In addition, the continued inventory under the FIMP omitted a number of reserves, covering an area of 270,000 ha, which were known to contain insignificant areas of natural forest or if they were too small to be considered for logging. Accordingly, a total of 128 reserves covering 1,293,600 ha were inventoried through 3101 TSPs of one ha.

Three scenarios were considered. A minimum scenario (scenario 1) according to which only hill sanctuaries, swamp sanctuaries and conversion areas (where logging is difficult or impossible anyway) are protected. A compromise scenario (scenario 2) where, in addition to scenario 1, special biological protection areas, shelter belts and convalescence areas are protected. A sustainable/ideal scenario (scenario 3) where all categories of protection areas are included. Table 6.3 summarises the area impact of the three scenarios.

Table 6.3 Summary of scenarios for forest protection

Category	Area (ha)	Scenario 1	Scenario 2	Scenario 3
Hill sanctuaries	210,400	Yes	Yes	Yes
Swamp sanctuaries	18,000	Yes	Yes	Yes
Special biological protection areas	46,800		Yes	Yes
Shelter belts	22,000		Yes	Yes
Intact forest sanctuaries	2,400			Yes
Provenance protection	8,400			Partial
Fire protection	66,800			DS zone
Convalescence (5 - 15 m^2/ha)	122,000		Yes	Yes
Conversion	150,800	Yes	Yes	Yes
Total area of "good" forest protected (ha)[1]		**194,400**	**241,600**	**292,400**
Residual area of "productive" forest (ha)[2]		**921,200**	**791,200**	**762,400**

Notes: [1]"Good" forest is taken as that which has a basal area greater than 15 m^2/ha. Areas for each protection category do not add up to the total area under protection because many of the categories overlap each other.
[2]"Productive" forest is the residual after taking out protection and does not exclude poorly stocked areas.
Source: FIMP (1994a).

In practice, the exact implications in terms of areas to be protected under the three scenarios would to a large extent have to be determined on the basis of locality specific planning. However, for national level assessments of forest protection strategies, the application of TSP data, as described in FIMP (1994a), certainly provides the best available information.

It is worth noticing that all three scenarios operate with areas of production forest which are significantly smaller than the areas assumed by Silviconsult (1985) and FIP (1989a). Table 6.4 summarises how the estimated area of production forest within forest reserves in the high forest zone has "developed" since 1985.

Table 6.4 **Estimated production area within reserves (1985-94)**

Source	Year	Area (ha)
Silviconsult (1985)	1985	1,182,730
FIP (1989a)	1989	1,159,000
FIMP (1994a) Scenario 1	1994	921,200
FIMP (1994a) Scenario 2	1994	791,200
FIMP (1994a) Scenario 3	1994	762,400

Sources: See table.

The development shown in table 6.4 represents a quite remarkable "learning curve". In a follow-up analysis under the FIMP, scenario 3 is considered to represent the only large-grained protection strategy which, given the data available, can qualify for the term "sustainable" (FIMP, 1995b). Accordingly, the estimated area of production forest (area suitable for timber harvest) has undergone a dramatic reduction by almost 400,000 ha or 34% during the FIP and FIMP.

The total area of gazetted forest reserves (including the reserves not inventoried) in the high forest zone amounts to 1,634,100 ha (FIMP, 1995b). Table 6.5 arranges this area according to the sustainable scenario for forest use (scenario 3).

Table 6.5 **Sustainable use scenario of forest reserves in Southern Ghana**

Category of use	Area (ha)
Timber production	762,400
Permanent protection	352,500
Convalescence	122,000
Conversion	127,200
Not inventoried (conversion)	270,000
TOTAL	**1,634,100**

Source: FIMP (1995b)

As table 6.5 shows, the sustainable use scenario allocates 762,400 ha for timber production. A total of 352,500 ha is set aside for permanent protection to maintain environmental stability and to conserve botanical stability[21]. Convalescence areas cover an additional 122,000 ha, which can be rehabilitated and eventually (in the long term) become production forest again. The remaining area of 397,200 ha includes conversion forest and reserves not inventoried which, in both cases, can neither be harvested in the short term nor rehabilitated like convalescence areas because they carry no or very few trees (FIMP, 1995b).

6.2 Growing Stock within Forest Reserves

As indicated above it is not so much the total growing stock within forest reserves which should be of interest, but rather the on-reserve growing stock on areas which are suitable for logging and the on-reserve forest condition (including estimates of the growing stock) in areas which should be protected from logging.

However, such information has only been indirectly available in the form of AAC estimates based on the stocking of areas suitable for logging (see below). What has been made available are the mean volumes per km^2 based on the inventory of 3101 plots of one ha in 128 reserves. Table 6.6 summarises the result (appendix 3 provides a species specific table for all FIP class 1 species).

21 Hill sanctuaries, swamp sanctuaries, shelter belts, special biological protection areas, intact forest sanctuaries, provenance protection areas and fire protection areas (FIMP, 1995b).

Table 6.6 **FIMP estimates in 1995 of mean bole volumes within forest reserves by species class (m^3/100 ha)**

FIP Class	D-Classes 30-49	50-69	70-89	90-109	110-129	>130	TOTAL	TOT>70	TOT>110
1	**2,268**	**2,678**	**2,642**	**1,171**	**1,197**	**1,320**	**11,822**	**6,877**	**2,517**
2	1,090	827	489	207	85	33	2,731	813	118
3	791	302	102	36	10	11	1,252	159	22
TOTAL	**4,245**	**3,882**	**3,281**	**1,978**	**1,298**	**1,368**	**16,052**	**7,924**	**2,666**

Notes: The volumes stated are mean values based on 3101 TSPs of one ha in 128 reserves.
Due to rounding TOTAL volumes do not always stem.
Source: FIMP volume tables.

As one would expect, the volume estimates in table 6.6 do not differ much from the FIP estimates of 1989 (cf. table 5.2). The total volume per ha of all species were in 1989 estimated to 162.73 m^3. The inventory of an additional 1,769 TSPs in 85 reserves has changed this estimate to 160.52 m^3/ha. For the FIP class 1 species the estimated total volume has changed from 124.42 m^3/ha in 1989 to 118.22 m^3/ha in 1995. The volume above 70 cm DBH for FIP class 1 species has changed from 77.39 m^3/ha in 1989 to 68.77 m^3/ha in 1995. Indeed, an expensive exercise to obtain these fairly marginal refinements of volume estimates.

Yet, one of the justifications for continuing the static inventory after 1989 apparently was to produce estimates that were sufficiently accurate at FMU level. Although comprehensive direct data on this issue has not been published it is possible to make a rough estimate of the 1995 statistical accuracy of volume estimates on the basis of (i) the confidence limits shown in table 5.5 and (ii) the increased number of TSPs inventoried.

If it is assumed that the sampling variances are similar for the FIP and FIMP static inventory data (reasonably realistic), and that the inventories were based on random sampling (more problematic), then the expected confidence interval for estimates based on the FIMP data ($SE\%_{FIMP}$) should be the confidence interval for FIP estimates ($SE\%_{FIP}$) times the square root of TSPs under the FIP (1332) divided by the square root of TSPs under the FIMP (3101). Expressed mathematically:

$$SE\%_{FIMP} \approx SE\%_{FIP} \frac{\sqrt{1332}}{\sqrt{3101}}$$

Although the above assumptions do not hold entirely true, the calculation is likely to produce fairly reliable figures for the accuracy of the FIMP volume estimates. Table 6.7 presents the results of applying the above formula on the figures in table 5.5.

Table 6.7 **Rough estimates of the statistical accuracy of FIMP inventory volume estimates for class 1 species above 110 cm DBH by level of analysis**

Level	95% Confidence limits as % of the estimated volumes (SE%)
National	4
Vegetation zone	6 - 14
FMU[1] or reserve	9 - 39
Individual species (national level)	15 - 31

Notes: The figures are based on assumptions described in the text above the table
[1] Forest management unit. Depending on the size of forest reserves a FMU covers one or several reserves.
Source: Based on Flint and Hardcastle (1992; 18).

As table 6.7 shows the probable accuracy for volume estimates at the national and vegetation zone levels is very high. At FMU level the figures in table 6.7 are not likely to be entirely correct, since the inventory under the FIMP made a deliberate effort to reduce the sampling error for volume estimates within FMU. For individual species at the national level the level of accuracy is presumably close to the figures in table 6.7. Accordingly, national level volume based calculations at species level should produce reliable estimates, especially when estimates concern groups of species (cf. below on AAC estimates).

6.3 The Sustainable Harvest Potential in Forest Reserves

The harvest potential on forest reserves is naturally closely linked to the area of production forest. However, the drastically reduced area of production forest from 1,159,000 ha to 762,000 ha does not result in a proportionally reduced AAC, because the growing stock per ha is higher on the smaller area.

The AAC from reserves does not only depend on the productive area. Fine-grained protection within production forest must also be taken into account. In practice, the recommendations of Hawthorne and Musah (1993) are being implemented under two headings:

- Yield regulation (according to species rarity and vegetation zone) ;
- protection of areas from logging in production forest.

6.3.1 Classification of Species According to Rarity

Yield regulation at species level must be based on the rarity of species. In relation to timber exploitation, economic extinction of timber species in Ghana is far more likely than biological extinction since the marginal cost of harvesting will become prohibitively high long before the very last trees are felled. However, over-exploitation of a given species is likely to focus particularly on the financially most desirable phenotypes (straight boles, few branches, etc.). This will reduce genetic variance and the species will increasingly be represented by trees that are undesirable for timber production.

As part of the computer based Forest Reserves Graphical Information Exhibitor (FROGGIE) which was developed on the basis of the botanical survey, Hawthorne (1993a & 1993b) developed the "Reddish star" terminology. This was done in an attempt to categorise commercial timber species according to the risk of genetic variance (quality) being reduced as a result of exploitation. Reddish stars were developed into three groups (Hawthorne, 1993b; 31-32):

Pink Stars: Significantly exploited but not yet so as to cause concern for their economic future. Pink species will probably become more exploited.

Red Stars: Current rates of exploitation present a significant danger of economic damage over, say, a ten year period.

Scarlet Stars: Under more imminent threat than Red Stars.

In addition to the Reddish Stars, Hawthorne (1993b; 15-19) developed the following Star categories to reflect conservation priorities for species that are not commercially exploited:

Green Stars: Species of no particular conservation concern. That is, species which are fairly common both in the forests of Ghana and surrounding countries; or at a wider geographical scale and not heavily exploited in Ghana.

Blue Stars: Species that are not rare or under economic pressure but less common than Green stars. In general, they are either fairly common in all Ghana's forests and less common in surrounding countries; or restricted to few zones in Ghana while being common in surrounding countries.

Gold Stars: Species that are common in only a few zones in Ghana; or not common in Ghana but sparsely represented in several zones of near surrounding countries; or widespread in Ghana but hardly found outside of Ghana.

Black Stars: Common in only a few zones in Ghana and not found elsewhere; or not common in Ghana but sparsely represented in several zones in near surrounding countries; or not common in Ghana and restricted to a few zones in near surrounding countries.

In order to allocate timber species according to the above classes Hawthorne (1993b; App. B) presents a table including the number of trees by species and diameter class found in 2624 TSPs as well as the LMC totals (officially recorded harvest) for 1989. On this basis he determines the degree of exploitation by estimating the hypothetical area of "average" forest needed to sustain the 1989 recorded harvest. This is simply done for each species by dividing the recorded 1989 harvest by the average population density above 90 and 60 cm DBH. The actual criteria used by Hawthorne (1993b) to allocate species into star classes are not entirely clear. By analysing Hawthorne's data FIMP (1994b) concludes that Scarlet Stars are generally those which require more than 500 km^2 of forest. Red Stars generally require between 100 and 500 km^2 of forest. Pink, Blue and Green Stars were allocated more subjectively and depended on the size of the population as well as the degree of exploitation.

FIMP (1994b) modifies the star ratings of Hawthorne (1993b) by using the recorded average harvest of species during the period 1989-93 and a calculated annual national felling quota for each species. The annual quota is calculated as the national (on-reserve) stocking above minimum felling diameter divided by the applied 40 year felling cycle[22]. Star ratings are then defined as follows (FIMP, 1994b):

Pink Stars:	Species for which the recorded average annual harvest 1989-94 is less than 50% of the calculated annual quota.
Red Stars:	Species for which the recorded average annual harvest 1989-94 is between 50 and 200% of the calculated annual quota.
Scarlet Stars:	Species for which the recorded average annual harvest 1989-94 is more than 200% of the calculated annual quota.

A list of species with previous designations, Hawthorne's (1993b) and FIMP's (1994b) star rating is found in appendix 1[23].

6.3.2 Yield Regulation

On-reserve yield regulation is currently based on a 40 year felling cycle, species specific minimum felling diameters and a yield formula (the Interim Yield Allocation Formula). Within compartments of approximately 120 ha all trees above 50 cm DBH are identified according to species, located on a map of the compartment (stock map), measured and enumerated with a stock survey number carved in the bark (the whole process referred to as a 100% stock survey). The yield formula is applied on each species to calculate the number of trees to be allocated for felling. Trees allocated for felling are selected on the stock map and identified by their stock survey number. The compartment yield is the total number of trees including all species allocated for felling.

Fine-grained forest protection is pursued by applying the "normal" and the "reduced" yield formula according to species rarity and fire-proneness.

The normal yield formula ($Z = 0.2X + 0.5Y$) is applied on Red and Pink Star species as defined by FIMP (1994b) in the WE, ME, MSSE and MSNW vegetation zones.

The reduced yield formula ($Z = 0.2X + 0.25Y$) is applied on Scarlet Star species as defined by FIMP (1994b) in all vegetation zones and on all species in the DS zone, which is generally more fire-prone than other vegetation zones.

In the above formulae Z is the number of trees allocated for felling, Y is the number of trees above minimum felling diameter and X is the number of trees in the diameter class (20 cm) immediately below felling limit. In cases where Y is less than Z the yield is reduced to Y-2, allowing for a retention of minimum two seed trees for each species per compartment.

---

22 This calculation assumes (i) that off-reserve harvest compensates for the on-reserve areas which cannot or may not be harvested, (ii) that this use of on-reserve TSP data represents the best estimate of total national stocking and (iii) that the exploitable stock can be replenished during the 40 year felling cycle (FIMP, 1994b).

23 In the appendix Green and one Blue Star species occur (see section 6.3.1 for definitions).

Whether the yield formulae are sustainable, in the sense that they will maintain or enhance the harvest potential (trees allocated for felling) over subsequent felling cycles of 40 years, is still an open question. The formulae will over-allocate the yield if X is not large enough to generate the present number of Y for the next felling cycle, taking mortality into account. Assessed against Alder's (1992) method for calculating sustainable yield in mixed tropical forests the yield formulae generally produce sustainable yields for groups of species, but assessed against Vanclay's (1993) suggested method of calculating critical de Liocourt's coefficients[24], the formulae are generally not sustainable (FIMP, 1995b).

This suggests that the formulae will be modified when new knowledge on forest growth in Ghana arises. In the interim, regulation of the yield according to the formulae appears pragmatic and sound, not least because they are easy to comprehend and should be possible to implement in practice.

6.3.3 Fine-grained Protection in Production Forest

Fine-grained protection in timber production areas is stipulated in the Logging Manual (FD, 1992), which restricts logging on slopes between 21% and 29% and bans it on slopes greater than 30%. Furthermore, logging is prohibited in a 50 m wide zone on each side of streams occurring on 1:50,000 scale topographic maps and in 25 m wide zones on each side of "minor" streams. Hawthorne and Musah (1993) recommend that areas of swamp are also included in fine-grained protection, and it is the intention of Forestry Department to follow this advise (FIMP, 1995b).

Through analysis of the TSP sub-plot data which record slope, swamp and watercourse, the area impact of fine-grained protection is estimated to take out approximately 14% of the productive forest. Accordingly, when fine-grained protection is applied the total on-reserve area of forest suitable for logging is reduced from 762,400 ha to 645,080 ha. Under the 40 year felling cycle this yields an annual coupe of 16,127 ha.

However, the impact on species availability is not proportional to the area since the stocking of commercial species is relatively low in riverine and swamp areas. The protected 14% of the area only accounts for 8.7% of the Scarlet, 10.7% of the Red and 9.9% of the Pink Star species (FIMP, 1995b).

6.3.3 The AAC in Forest Reserves

By applying the yield formulae on TSP data for areas defined as suitable for logging, the AAC for forest reserves has been estimated by FIMP (1995b). Table 6.8 summarises the result (appendix 4 provides a species specific list of the AACs).

---

24 The de Liocourt's coefficient, Q is the ratio between the numbers of trees in two adjacent size classes expressed as a ratio of the smaller size class (Q>1 is generally the case). Critical Qs may be calculated as $Q = (1 - 0.5S)/(DG - 0.2S)$, where S is the 40-year probability of survival, G is the probability of surviving and growing one or more classes, D is the probability of surviving the harvest undamaged, 0.5 and 0.2 come from the normal yield formula. PSP data suggests values for S of 0.658, G of 0.498 and D of 0.8. This gives a critical Q of 2.5 for the normal yield formula and 1.9 for the reduced yield formula. Calculated Qs (Y/X) for groups of species are, in general, below the critical Qs (FIMP, 1995b).

Table 6.8 **Estimated AACs in the forest reserves of Southern Ghana**

Species Group	AAC under large-grained protection only (m^3)	AAC under large- and fine-grained protection (m^3)
Scarlet Stars	132,437	120,817
Red Stars	224,195	201,353
Pink Stars	534,877	479,518
TOTAL	**891,509**	**801,688**
Total production area (ha)	762,400	645,080
Annual coupe (ha)	19,060	16,127
Average harvest (m^3/ha/yr)	1.17	1.24
Average harvest on coupe (m^3/ha)	46.77	49.71

Source: Based on FIMP (1995b).

Given the high level of accuracy for species specific estimates of standing volumes, the figures in table 6.8 must be considered very reliable. Accordingly, the sustainable harvest should be between 1.17 m^3/ha/year and 1.24 m^3/ha/year without and with fine-grained protection, respectively. Interestingly these estimates do not differ much from those of Silviconsult (1985) and Alder (1989). As a matter of fact, they are practically similar.

Yet, the estimated total AAC of approximately 850,000 m^3 is significantly less than the previous estimate of 1.5 million m^3 by Alder (1989). The reason being an ill informed and over-optimistic assessment of the area of production forest in 1989. As mentioned in chapter 5 the FIP could, in principle, have used TSP data in much the same way as FIMP (1995b) in order to estimate the on-reserve area of unproductive forest and the on-reserve area of productive forest which should be protected. A less optimistic AAC would have been the result.

Even though an AAC of 850,000 m^3 from forest reserves must be considered a very reliable estimate, the relation between recorded harvest and the AAC has still not been entirely clarified. The effect of logging damage on the residual stand should, in a crude form, be catered for in the yield formulae. Yet, the average amount of logging waste and possible under-reporting of volumes extracted from forest reserves are still unquantified factors.

6.4 Off-Reserve Inventory

As mentioned at the beginning of this chapter the FIMP had among its objectives to undertake mapping and inventory of the blocks of manageable forest situated outside the gazetted reserves (FIMP, 1988). However, this part of the project was not implemented until 1996-97. Apparently there were two reasons for this:

- Off-reserve forest and timber resources did not fall under the jurisdiction and management responsibility of the Forestry Department until 1996;
- the off-reserve timber harvest was not believed to be as significant as turned out to be the case when the issue was investigated.

The late implementation of the off-reserve timber inventory is explained by the fact that until 1996 the Forestry Department's management responsibility was restricted to forest reserves (FIMP, 1997a). Since FIMP is implemented within the framework of the Forestry Department it is possible that the project could only justify an off-reserve inventory when the Forestry Department became official management authority for off-reserve timber and forest resources. However, this appears a professionally questionable justification for postponing the off-reserve inventory which was planned and included in the FIMP project document (FIMP, 1988) at a point in time where the Forestry Department did not hold any management responsibility whatsoever for off-reserve forest and timber resources.

By analysing the annual felling returns for forest reserves (submitted by forest districts to Forestry Department Headquarters), IIED (1993; App. 1) managed to estimate the on-reserve harvest by species[25]. Off-reserve harvest by species could then be estimated by subtracting on-reserve harvest from LMC totals.

This exercise produced quite remarkable results. It documented that during the period 1986-91 off-reserve harvest accounted for well over 50% of the total harvest (IIED, 1993; 15) (see chapter 7). Furthermore, Hawthorne and Musha (1993) had documented that many off-reserve forest resources have a high frequency of rare tree species. In addition, off-reserve forest and trees provide environmental services and NTFPs, and serve cultural and spiritual purposes.

The social aspects of managing off-reserve forest and timber resources are, therefore, manyfold and diverse. Yet, a crucial factor is the high level of commercial timber harvest off-reserves. Since the timber industry depends heavily on this dwindling resource there is reason to be concerned over what will happen if and when forest reserves become the only source to satisfy the demand from industrial timber. Will the Forestry Department be able to withstand the pressure, or will the walls come tumbling down if financial interests of timber companies clash with the principles of sustainable forest management?

Accordingly, the documentation produced by IIED (1993) and Hawthorne and Musah (1993) were good reasons for considering options for managing off-reserve forest and timber resources with a view to sustain at least some of them and slow down off-reserve deforestation. An off-reserve timber inventory is the obvious basis for turning such considerations into adequate management planning.

6.4.1 Design of the Off-Reserve Inventory

The inventory covers off-reserve areas in the high forest zone with the exception of towns, villages, savanna grassland and large scale agricultural areas (rubber plantations, oil palm plantations, commercial farms). A total of 5,482,178 ha was covered by the inventory.

Originally it was the plan to operate with a sampling intensity of 500 plots, but due to logistical constraints it was decided to reduce the total number of plots to 200. The plot size was two ha (1000 by 20 m) and the 200 plots were allocated approximately according to the area coverage of vegetation zones. The following stratification was applied (FIMP, 1997a):

---

[25] The calculations of IIED (1993) assume on the basis of Kakrada and Maginnis (1992) that Forestry Department felling returns under-estimate actual harvest by 15%.

- 50 randomly located plots within each of the MSSE, the MSNW and the DS vegetation zones;
- 50 plots randomly located within the area covered by the WE and ME vegetation zones.

Each plot was subdivided into 20 sub-plots (50 by 20 m) and the following recorded on each sub-plot:

- Dominant land-use or vegetation type (six categories)
- site factors (five categories)
- trees above 10 cm DBH identified for species and diameter measured
- condition of each tree is determined (nine categories)
- cause of death is recorded for dead trees (six categories).

6.4.2 Results of the Off-Reserve Inventory

Table 6.9 summarises the estimated off-reserve standing volume by diameter class and star group.

Table 6.9 **Estimated off-reserve standing volumes (millions m³)**

Species Group				D-class (cm)				
	10-29	30-49	50-69	70-89	90-109	>110	**Total**	>70	>F-LIMIT
Scarlet stars	2.727	5.796	9.778	10.346	8.756	8.611	**46.014**	27.713	11.004
Red stars	3.564	7.617	12.577	13.426	8.824	28.785	**74.793**	51.035	38.627
Pink stars	15.431	24.449	33.518	36.931	20.951	16.101	**147.381**	73.983	52.049
TOTAL	**21.722**	**37.862**	**55.873**	**60.703**	**38.531**	**53.497**	**268.188**	**152.731**	**101.68**

Based on: 189 plots of 2 ha.
An estimated off-reserve area of 5,482,178 ha.
Mean volumes within diameter classes (Including the volume of poor quality trees).
Source: FIMP (1997b).

It is seen from table 6.9 that the off-reserve timber resource is still significant. For all species above felling limit the standing volume is just over 100 million m^3 . However, the table also documents that Scarlet Star species account for only 11% of the standing volume above felling limit.

The accuracy of data presented in table 6.9 is indicated by the standard error (95% confidence interval) as per cent of the estimated mean number of trees per ha within species group. This is presented in table 6.10.

Table 6.10 **Standard errors in per cent on the mean number of off-reserve trees per ha**

	D-class							
Species group	10-29 (SE%)	30-49 (SE%)	50-69 (SE%)	70-89 (SE%)	90-109 (SE%)	>110 (SE%)	**Total** (SE%)	>F-LIMIT (SE%)
Scarlet stars	12	13	13	16	21	27	**11**	25
Red stars	11	11	11	15	21	17	**10**	14
Pink stars	8	9	10	11	17	26	**8**	13

Based on: 189 plots of 2 ha.
Source: FIMP (1997b).

As table 6.10 shows, the confidence intervals for mean numbers of trees within diameter classes meet or are close to the general target of 20%. Presumably, the volume estimates presented in table 6.9 are based on (i) the number of trees per ha, (ii) diameter-volume equations from the FIMP on-reserve inventory and (iii) the estimated off-reserve area of 5,482,178 ha. Accordingly, the volume estimates must be associated with a somewhat higher standard error (a wider confidence interval) than the average numbers of trees per ha. It is not possible to calculate the exact size of confidence intervals for volume estimates, but considering the experience from on-reserve inventories it seems likely that the estimated off-reserve area carrying timber and forest resources will change if and when this issue is made subject to further investigations.

By relating the estimated off-reserve resource to the demand for industrial timber, the expected lifetime of the resource may be calculated under different scenarios. Tables 6.11 and 6.12 summarise 12 scenarios where reliable minimum estimates[26] of standing volumes of good trees above felling limit and above 70 cm DBH within star groups are made subject to an annual harvest of either 500,000 or 1,000,000 m^3.

26 The lower level of the 95% confidence interval.

Table 6.11 **Off-reserve standing volume above felling limit and estimated lifetime of resource under annual harvests of 500,000 and 1,000,000 m³**

Species Group	**Standing vol. above f.limit RME of good trees[1] (million m³)**	**Lifetime of resource if annual harvest is 500,000 m³ (Years)[2]**	**Lifetime of resource if annual harvest is 1,000,000 m³ (Years)[2]**
Scarlet stars	7.477	15	7
Scarlet + Red stars	37.586	75	38
Scarlet + Red + 50% of Pink stars	58.073	116	58

Notes: [1]Reliable minimum estimates (lower value of the 95% confidence interval) for trees which are suitable for timber production.
[2]Assumes no re-growth, ie. a mining of the resource.
Source: FIMP (1997b).

Table 6.12 **Off-reserve standing volume above 70 cm DBH and estimated lifetime of resource under annual harvests of 500,000 and 1,000,000 m³**

Species Group	**Standing vol. above f.limit RME of good tre-es[1] (million m³)**	**Lifetime of resource if annual harvest is 500,000 m³ (Years)[2]**	**Lifetime of resource if annual harvest is 1,000,000 m³ (Years)[2]**
Scarlet stars	19.863	40	20
Scarlet + Red stars	58.352	117	58
Scarlet + Red + 50% of Pink stars	86.696	173	87

Notes: [1]Reliable minimum estimates (lower value of the 95% confidence interval) for trees which are suitable for timber production.
[2]Assumes no re-growth, ie. a mining of the resource.
Source: FIMP (1997b).

The "mining" scenarios of tables 6.11 and 6.12 show that if only the Scarlet Star species are harvested off-reserves, this financially most valuable resource will be logged out within a foreseeable future. On the other hand, the total size of the off-reserve timber resource suggests that is not too late to implement some kind of off-reserve management planning with the objectives to maintain:

- A certain off-reserve tree cover to cater for social and spiritual needs, provide environmental protection and preserve biological values;
- a sustained supply of timber from off-reserve areas in the interest of the timber industry and the sustainable management of forest reserves.

However, time may be running out faster than the tables 6.11 and 6.12 indicate. If the substantial off-reserve harvest is concentrated on Scarlet Star species, and if the off-reserve area where timber harvest is financially feasible turns out to be significantly smaller than the estimated 5,4482,178 ha, then the pressure on forest reserves for timber might increase tremendously within the next five or ten years. This issue is further analysed in chapter 7.

In addition to industrial timber harvest, the rural population contributes to off-reserve deforestation. Clearings for farmland are bound to continue and people are likely to get rid of trees on farmland unless they have clear positive financial incentives to grow and protect trees on the land they cultivate. This is analysed further in chapter 13.

7 Analysis of the Timber Harvest

As mentioned in the previous chapters it is necessary to distinguish between on- and off-reserve timber harvest in order to assess the sustainability of timber harvest in the high forest zone. This chapter attempts to combine the on- and off-reserve inventory results with the available information about the on- and off-reserve timber harvest.

7.1 The Log Measurement Certificate System

Since 1988 the Forest Product Inspection Bureau (FPIB) has published so-called Log Measurement Certificate totals (LMC totals) which detail the recorded annual timber extraction by species and by timber company (the latter used to bill timber royalties). Timber trucks are stopped at road check-points throughout the high forest zone and the following information about the load of logs is recorded:

- Tree species;
- under bark volume of logs;
- number of logs per tree;
- stock survey number (if the log comes from a forest reserve);
- name of forest reserve (if the log comes from a forest reserve); otherwise the log must be marked with the letters "OFR" to indicate it comes from an of-reserve area;
- locality mark identifying the origin of the log in terms of administrative district and land owning stool (local community);
- the timber company's property mark.

Until 1993, the LMC system did not distinguish between logs from on- and off-reserves. However, for the period 1993-95 LMC data for on- and off-reserve timber extraction, respectively, are not reliable. It simply took three years to go from just recording the total extraction to record on- and off-reserve extraction specifically.

7.1.1 The Quality of the LMC data

In relation to growth estimates and estimates of the standing volume two issues must be considered when using the LMC data:

- The relation between bole volume and volume of extracted logs from felled trees;
- the effectiveness of road check-points in monitoring timber extraction.

The first issue has been assessed in section 5.2.1 and basically the question remains unanswered since it has so far proved impossible to establish a statistically acceptable relation between bole and log volumes. Yet, it is clear that the volume of extracted logs is lower than the bole volume of trees from which they origin. On average this difference is approximately 10% (FIP, 1989a; 29). In addition, allowance should be made for the volume of trees which are felled but not extracted due to defects (rot, hollowness or cracks). Unfortunately, this is another issue which has not been subject to quantitative assessment.

Is the LMC system effective? How many logs pass unrecorded by the FPIB road check-points? Again, these are important questions which cannot be answered given the information available. Since royalties are generally billed on the basis of LMC data, timber companies have a financial incentive to avoid the

LMC system. Another reason to avoid check-points is when logs from illegally felled trees are transported. This may be under-sized trees or trees from areas where the harvesting activity lacks a legal basis, eg. on other companies' or unallocated concessions.

Both reasons to avoid the LMC system are known to apply in practice. Timber trucks do sometimes not stop at FPIB road check-points, logs are illegally transported during hours when the FPIB road check-points are not manned (public holidays, week-ends and after 6 p.m. during working days) and trees are illegally felled on other companies' concessions. We have personally witnessed all three offences. It is also generally realised within the Ministry of Lands and Forestry that these illegal activities do take place.

Nonetheless, the LMC system has never been subject to a comprehensive official assessment. In view of the time and resources spent on timber inventories and the objectives of doing so it appears illogic and very surprising that the quality and relevance of harvesting records has attracted so little interest.

One study by the Forestry Research Institute of Ghana (FORIG) revealed that many logs from under-sized trees pass through the FPIB check-points. Of 4,023 logs examined at two major FPIB check-points (Kumasi and Takoradi) approximately 45% were found to stem from under-sized trees, and logs from under-sized trees were found for 34 out of the 52 species that were included in the study (FORIG, 1994; 5-8). Furthermore, the study documented a large number of irregularities in log markings which made it impossible to determine the origin of logs and the company responsible for the felling (which is not necessarily the same as the company transporting the logs) (FORIG, 1994; 21-22). These results will be further analysed in relation to procedures for managing the timber resource of which the minimum felling diameter is one (see section 11.2).

Unfortunately, the study of FORIG (1994) did not consider to which extent logs pass unrecorded through FPIB check-points. However, the obvious financial incentive to do so, together with the number of documented other irregularities, suggest that this is likely to happen when research teams and the like are not around to detect this.

Despite the uncertainties and obvious weaknesses of the LMC system it must be concluded that LMC totals form the most comprehensive information available about:

- The annual extraction of timber by species;
- the extraction of timber from on- and off-reserves (since 1996).

However, for comparisons with inventory data it should be remembered that LMC figures undoubtedly under-estimate the *actual* harvest. Judging by the available (but inadequate) information, actual harvested volume is likely to be at least 10% above the recorded LMC volume. Whether the correct figure is significantly higher (as one might suspect) is an open question. To avoid the introduction of questionable and arbitrary assessments, the LMC figures in the tables below and the tables presented in appendices to this study are those officially recorded. Implications of the quality of LMC figures are discussed under the interpretation of data.

7.2 Felling Recorded by the Forestry Department

The Forestry Department has a system which records on-reserve felling. District forest officers (DFOs) are supposed to keep records of on-reserve felling within their district and submit felling reports (sometimes referred to as felling returns) to Forestry Department Headquarters.

7.2.1 The Quality of Forestry Department Felling Returns

Kakrada and Maginnis (1992) report that the felling returns which forest guards report to their district forest offices are often based on information from the concessionaire bush manager. Accordingly, it is likely that these felling returns understate the actual harvest, since timber royalties may be billed on the basis of felling returns instead of LMC figures.

Furthermore, Kakrada and Maginnis (1992) report that district forest officers were often unaware of the scale of this problem, and it was found that under-reporting was in a range of five to ten per cent. However, IIED (1993; app. 1) assumes that the under-reporting of felling returns is 15%. Data are too meagre to determine the correct degree of under-reporting in felling returns. In any case this is not so important because the most interesting information to be drawn from the data is historical trends in on- and off-reserve harvest at the level of species groups.

7.3 This Study's Use of Available Data on Timber Harvest

This study adopts the methodology of IIED (1993) and estimates off-reserve harvest for the period 1973-91 by subtracting recorded on-reserve harvest plus 15% from LMC totals. For 1992, on- and off-reserve data are from FIMP (1995b). For 1993, LMC data are used directly although they must be considered unreliable.

For the period 1994-96, the LMC totals specify extracted volumes from unidentified sources. In 1994 a total of approximately 1,470,000 m^3 was recorded as coming from unidentified sources. In 1995 this figure was down to approximately 253,000 m^3, and in 1996 the LMC totals include only approximately 60,000 m^3 from unidentified sources (see appendix 5). As an estimate this study allocates the volumes from unidentified sources to on- and off-reserve extractions by the ratios of 1992. This is only done to indicate the development in on- and off-reserve harvest for the entire period 1973-96.

For specific analysis of harvest and estimated AACs, LMC data from 1996 will be used. These data are, despite the problems discussed in section 7.1.1, assumed to be the best available.

7.4 On- and Off-reserve Harvest

The total recorded timber extraction from on- and off-reserves is summarised in the figures 7.1 and 7.2 (appendix 6 provides the data in tabular form).

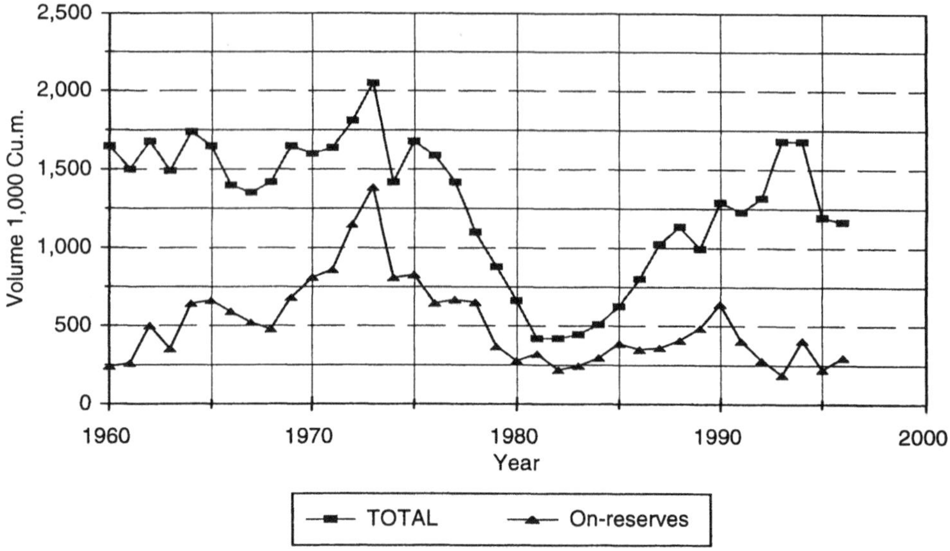

Figure 7.1 **Total and on-reserve recorded timber extraction 1960-96**

Sources:
IIED (1993)
FIMP (1995b)
FPIB LMC Totals (1986-96)

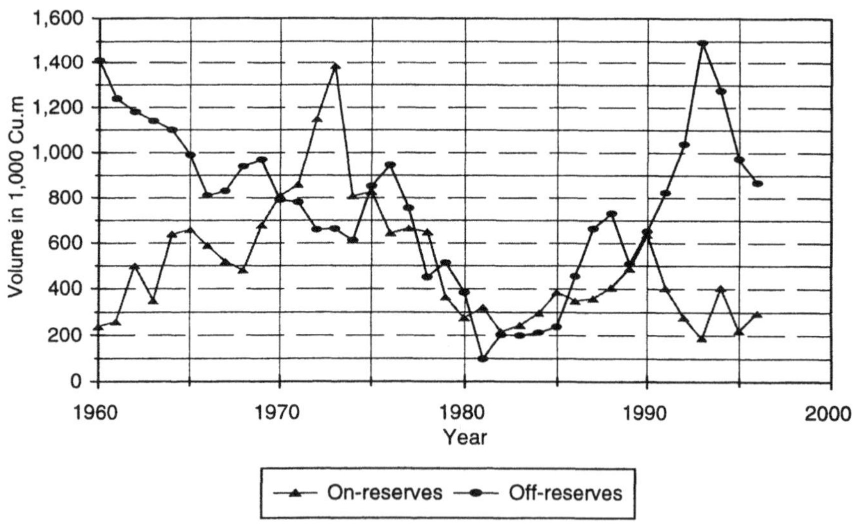

Figure 7.2 **On- and off-reserve recorded timber extraction 1960-96**

Sources:
IIED (1993)
FIMP (1995b)
FPIB LMC Totals (1986-96)

The curves in figures 7.1 and 7.2 clearly show that off-reserve timber extraction has always accounted for a substantial share of the total recorded timber extraction in Ghana. More often than not the recorded extraction of timber off-reserves is higher than the recorded on-reserve timber extraction, and during periods where off-reserve extraction is below on-reserve extraction the difference between the two is, with the exception of 1971-75, very small.

From 1971 onwards there is a tendency for decreasing on-reserve extraction to be compensated by increasing off-reserve extraction. This has been taken as an indicator of the Forestry Department being more successful in controlling on-reserve harvest, which should make timber companies use off-reserve resources as a "buffer" to balance variations in the availibility of on-reserve timber resources (IIED, 1993; 16).

However, this interpretation of the timber extraction curves does not cover the entire picture. The log export market has, for good financial reasons, had a tremendous impact on the felling activity in Ghana. This was last documented in 1991-94. Following rumours about a coming log export ban on Wawa, this species was harvested and exported at increasing rates until 1993 when the ban became reality. Through a series of partial log export bans Malaysia significantly reduced its log exports during the period 1993-95 (ITTO, 1996; 21). This sent Far East buyers to West and Central Africa in the search for logs to offset the declining supply from Malaysia. The effect in Ghana was that log export contracts on substantial volumes were offered for species that previously accounted for a very small fraction of the annual harvest and log export. In particular, logs of Ceiba and Kyenkyen became prominent export commodities over-night.

This sudden change in overseas log demand led timber companies in Ghana to focus on off-reserve timber resources, presumably because the harvest control on-reserves was more strict. Yet, the fall in on-reserve timber extraction during the period 1991-93 is probably not only a result of strictly enforced harvest controls. Rather it is the result of many timber companies allocating resources to participate in the log export driven off-reserve harvest galore. During the period 1991-94 many companies simply closed their sawmilling activities and devoted their entire capacity to the export of logs (see sections 10.2 and 11.1.3 for a more detailed discussion of forest fee policies and the resulting financial incentives to export logs rather than processed wood). In 1995 Ghana issued a ban on all log exports to control the situation.

7.4.1 Total Recorded Extraction by Species Groups

Detailed data for timber extraction by species is only available for 1986 onwards. Figure 7.3 depicts the recorded total timber extraction during the period 1986-96 for species within star groups (cf. section 6.3.1).

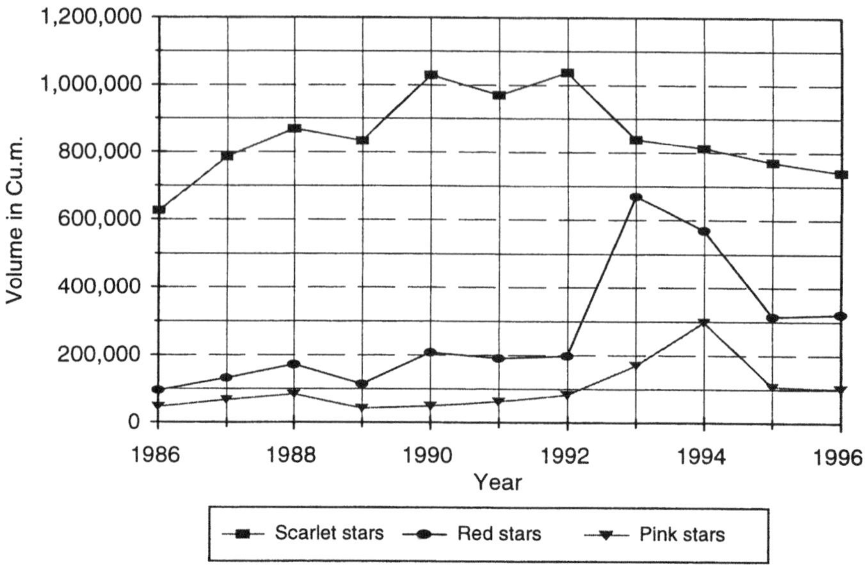

Figure 7.3 **Total recorded timber extraction by species group 1986-96**
Source:
FPIB LMC Totals (1986-96)

Figure 7.3 clearly shows that the mainstay of timber extraction is made up by a substantial and fairly stable extraction of Scarlet Star species. This group comprises 15 species (in terms of trade names, 18 in terms of botanical species). Log export was banned for nine of these species in 1979. In 1988 and 1993, respectively, log exports were banned for two more species and in 1995 log export was banned for the remaining four (see appendix 5). Accordingly, the timber industry in Ghana must have well established markets for processed Scarlet Star species, which makes it highly dependent on a constant supply of these species. This explains the constant and high rate of extraction compared with Red and Pink Star species. The falling trend since 1992 in Scarlet Star extraction is, therefore, likely to reflect a decreasing (off-reserve) availability of these species combined with a possible stricter on-reserve harvest control. Appendix 5 gives the details behind figure 7.3.

This dependency of the timber industry on Scarlet Star species is a crucial forest policy problem in Ghana today. The reason being that AAC estimates show that the current extraction level of Scarlet Star species cannot be sustained in the future (see section 7.5).

The peak in extraction of Red Star species during 1993-94 reflects the sharp rise in extraction of Ceiba and Kyenkyen for log exports. The 1993-94 peak in extraction of Pink Star species is mainly due to rise and fall in extraction of Ofram and Otie, which like Ceiba and Kyenkyen were part of the log export boom to meet the demand from Far East buyers. The extraction level of Red and Pink Star species in 1995-96 does mark an increase in domestic processing of these species compared to the level in 1992, which may be the first sign that the timber industry has initiated a process of diversifying its log demand in order to reduce the heavy dependence on Scarlet Star species. However, more radical changes in the species composition are required for extraction to match the AAC (see section 7.5).

7.5 On-Reserve Harvest and AAC

As mentioned above it is only the 1996 LMC totals which provide a reasonably reliable picture of the on- and off-reserve extraction by species. Table 7.1 summarises, for species groups, the 1996 recorded on-reserve extraction together with the estimated on-reserve AAC under (i) large-grained protection only and (ii) fine- and large-grained protection. Appendix 4 details the recorded extraction and AAC volumes by species.

Table 7.1 **Estimated on-reserve AACs compared with the 1996 recorded on-reserve extraction by species groups**

STAR GROUP	AAC total volume[1] (m^3)	1996 on-reserve extraction (m^3)	1996 extraction as % of AAC	AAC total volume[2] (m^3)	1996 on-reserve extraction (m^3)	1996 extraction as % of AAC
Scarlet	132,437	183,869	139	120,817	183,869	152
Red	224,195	66,470	30	201,353	66,470	33
Pink	534,877	33,420	6	479,518	33,420	7
TOTAL	**891,509**	**284,932**	**32**	**801,688**	**284,932**	**36**

Notes: [1]Under large-grained protection only (762,400 ha of production forest).
[2] Under fine- and large-grained protection (645,080 ha of production forest).
Sources: FIMP (1995b) and FPIB LMC totals (1996) (see appendix 4 for details).

The general message from table 7.1 is clear. Scarlet Star species are over-harvested, whereas the harvest of Red and Pink Star species could be increased substantially. Yet, important details are revealed by a closer analysis of the star groups.

7.5.1 Scarlet Star Species

According to the estimated AACs, ten out of the 15 Scarlet Star species are seriously over-harvested. Moreover, these ten species accounted for 96% of the recorded Scarlet Star species extraction in 1996. In this connection, it should be remembered that recorded extraction under-estimates actual harvest by at least 10%. Furthermore, table 7.1 does not account for the approximately 60,000 m^3 recorded as coming from unidentified sources in 1996.

Accordingly, the current on-reserve harvest of Scarlet Star species is almost certainly unsustainable. Measured against the fine- and large-grained protection scenarios, the annual harvest of the ten commercially most desired species should be reduced by at least 75,000 m^3 compared with the 1996 extraction. The harvest of less desirable Scarlet Star species could only be increased by approximately 15,000 m^3 annually (considering that actual harvest exceeds the recorded extraction).

7.5.2 Red Star Species

Red Star species are generally extracted at rates below the estimated AACs - although unrecorded logging waste and under-reporting of actual log extraction by the LMC system might paint a somewhat rosy picture of the realities. However, not all Red Star species could be harvested at rates exceeding the 1996 level. Candollei was recorded to be extracted at 247% of the estimated AAC, while Guarea and Niangon are reported to be extracted at levels close to their AACs. In fact, these species should probably belong to the Scarlet rather than the Red Star category (see section 7.5.4).

Much of the under-utilised harvest potential within the Red Stars is concentrated on only three species, Ceiba, Kyenkyen and Dahoma. The annual harvest of Ceiba could possibly be increased by some 60,000 m^3, while the annual harvest of Kyenkyen and Dahoma might be increased by approximately 25,000 m^3 and 8,000 m^3, respectively. The remaining annual harvest potential of approximately 100,000 m^3 is divided fairly evenly between ten different species.

7.5.3 Pink Star Species

Pink Star species represent a substantial but largely un-utilised harvest potential of 480,000 to 530,000 m^3 per year depending on which level of forest protection the AAC is related to. Yet, the industrial demand for Pink Star species has always been very limited (with the exception of Ofram and Otie during the log export boom in 1993-94).

Counted by trade names, 15 of the 32 Pink Star species were not subject to any harvest in 1996. Danta, Otie, Sterculia, Ofram and Avodire are the only Pink Star species which were recorded to be extracted at rates above 2,000 m^3 in 1996. These five species accounted for approximately 27,000 m^3 or 82% of the total recorded Pink Star extraction. The un-used harvest potential within the group of Pink Star species is concentrated on Celtis, Denya, Ananta, Danta, Asoma, Esia, Ofram and Avodire. These eight species represent an AAC of approximately 330,000 m^3 (depending on the level of forest protection). However, in 1996 the recorded extraction of these species was only approximately 29,000 m^3, which illustrates the currently low market value of this abundant resource.

7.5.4 The Star Group System and Related AACs Revisited

It will be recalled that the star group classification used by FIMP is based on the information available in 1994, which was (i) recorded extraction from on- and off-reserves assessed against (ii) on-reserve inventory data only (FIMP, 1994b). Accordingly, the then developed star group classification rested on fairly rough assumptions about the connection between total extraction and the resource base on- and off-reserves.

Today the scientifically well founded on-reserve AACs may, together with on-reserve extraction data, be considered a better foundation for star classification of tree species. This is specifically so because the reduced yield formula is only applied on-reserves with the specific objective to reduce felling of species threatened by commercial harvest, ie. the Scarlet Star species.

On the basis of species specific on-reserve AACs and the recorded on-reserve extraction by species it may, since recorded extraction notoriously under-estimates actual harvest, be relevant to define star categories as follows:

- Scarlet stars: More than 90% exploitation.
- Red stars: Between 50 and 90% exploitation.
- Pink stars: Less than 50% exploitation.

These criteria would shift a few species from the Scarlet and Red Star groups to the group of Pink Star species and some species would go from the Scarlet to the Red Star group and vice versa (see appendix 4). Yet, these changes can only be taken as indicative because the AAC calculations are based on the reduced yield formula for Scarlet Star species according to the star classification of 1994. Available data does not permit re-calculation of AACs and re-allocation of species into star groups on the basis of recorded 1996 on-reserve extraction rates. The result would have been a refinement of the current AAC estimates by species. AACs for current Scarlet Star species which rightfully belong in the Red or Pink Star group would be increased, and vice versa for Red Star species which belong in the Scarlet Star group (no Pink Star species would change group).

However, the star group system and AACs should obviously not rely on a single year's extraction of timber on-reserves, but rather on average figures for, eg. the latest three years. Furthermore, the Scarlet Star group AAC is not likely to change much since the currently ten over-harvested species will probably remain Scarlet Stars under the above proposed criteria. Today these species account for 87% of the Scarlet Star group AAC.

Since star groups are used to calculate on-reserve yields, the system should allow for species to change star groups such that harvest controls may be timely applied when market forces increase or decrease the harvest pressure on individual species. An example might be Candollei which is currently placed in the Red Star group while the 1996 extraction level suggests that the species is seriously over-harvested.

7.5.5 Policy Implications of the Extraction Rates and AACs

It may be concluded that the total on-reserve harvest is well within the limits on the total on-reserve AAC, even when the uncertainties of using recorded extraction as a proxy for actual harvest and log extraction from unspecified sources are taken into consideration. However, according to the AAC estimates approximately 70,000 m^3 or 25% of the 1996 recorded timber extraction represents a notorious over-exploitation, predominantly of Scarlet Star species.

This is where the crux of the matter lies. In relation to the resource base, market forces create an excess demand for certain species (in general the commercially most valuable ones) which will lead to their economic (not botanical) extinction if nothing is done about it. The policy challenge is to:

- Protect over-harvested species against the demand pressure through control-of-supply measures;
- divert demand away from over-utilised and towards under-utilised species.

It is fortunate for Ghana that the latter option actually exists. The substantial resource within forest reserves of under-utilised Red and Pink Star species forms a security of raw material supply *and* a marketing challenge for the timber industry. The over-exploitation of certain species within forest reserves forms a management challenge to the Forestry Department, which should set things straight in the long-term interest of the nation (including the timber industry).

7.6 Off-Reserve Timber Extraction and the Resource Base

As for on-reserves, the most reliable off-reserve extraction figures are those of 1996. Tables 7.2 and 7.3 summarise the off-reserve recorded extraction for star groups together with the off-reserve inventory results of standing trees above felling limit and above 70 cm DBH (see appendix 5 for details).

Table 7.2 **Off-reserve standing timber resource above felling limit and estimated lifetime of resource if the 1996 recorded extraction is continued**

Species Group	Standing vol. > f.limit RME of good trees[1] (m³)	1996 recorded extraction[2] (m³)	Lifetime of resource[3] (Years)
Scarlet Stars	7,477,000	515,349	15
Red Stars	30,109,000	240,633	125
50% of Pink Stars	20,487,000	65,068	315
Scarlet + Red Stars	37,586,000	755,982	50
Scarlet + Red + 50% of Pink Stars	58,073,000	821,050	71

Notes: [1]Reliable minimum estimates (lower value of 95% confidence interval) for trees of FIP quality codes 1 and 2 (good trees).
[2]Includes recorded *extraction* of all Pink Star species (not only 50%).
[3]Assumes that the resource is effectively mined, ie. growth of standing timber is cancelled out by felling of trees for other purposes than timber (firewood, clearing of land for agriculture, etc.).
Sources: FIMP (1997b) and FPIB LMC totals (1996).

Table 7.3 **Off-reserve standing timber resource above 70 cm DBH and estimated lifetime of resource if the 1996 recorded extraction is continued**

Species Group	Standing vol. > 70 cm RME of good trees[1] (m³)	1996 recorded extraction[2] (m³)	Lifetime of resource[3] (Years)
Scarlet Stars	19,863,000	515,349	39
Red Stars	38,489,000	240,633	160
50% of Pink Stars	28,344,000	65,068	436
Scarlet + Red Stars	58,352,000	755,982	77
Scarlet + Red + 50% of Pink Stars	86,696,000	821,050	106

Notes: [1]Reliable minimum estimates (lower value of 95% confidence interval) for trees of FIP quality codes 1 and 2 (good trees).
[2]Includes recorded *extraction* of all Pink Star species (not only 50%).
[3]Assumes that the resource is effectively mined, ie. growth of standing timber is cancelled out by felling of trees for other purposes than timber (firewood, clearing of land for agriculture, etc.).
Sources: FIMP (1997b) and FPIB LMC totals (1996).

Table 7.2 assumes that (i) the total 1996 recorded off-reserve extraction of 821,050 m^3 is continued with the 1996 ratios of Scarlet, Red and Pink Star species, (ii) that only trees above felling limit are harvested, and (iii) that only 50% of the available Pink Star species are of commercial interest. Table 7.3

is similar to table 7.2 except that it assumes a general off-reserve felling limit of 70 cm DBH, which is probably fairly realistic at least for the Scarlet and Red Star species.

7.6.1 Off-Reserve Extraction of Scarlet Star Species

Scarlet Star species are clearly under serious threat of economic extinction. This group of species represents a significantly lower growing stock compared to Red and Pink Star groups, while the recorded off-reserve extraction of Scarlet Star species accounted for 63% of the total 1996 recorded off-reserve extraction.

Inventory data does not yet allow for an evaluation of the off-reserve extraction by species, but judging from the extraction figures certain species are under a more immediate threat than others. Together Wawa, Odum, Mahogany, Kyire and Asanfina accounted for 464,082 m^3 or 90% of the recorded extraction of Scarlet Star species in 1996 (cf. appendix 5). To some extent this obviously reflects the availability of species off-reserves, but market demand certainly also plays an important role in the selection of trees for harvest off-reserves.

Considering that the recorded extraction under-states actual harvest by at least ten per cent and that the timber industry depends heavily on Scarlet Star species, tables 7.2 and 7.3 predict major changes in the timber industry and/or major pressure on the forest reserves within the near future. Despite the uncertainty about estimated lifetimes for Scarlet Star species the tables document that this resource is being logged out.

For some species this could easily happen within the next 15 years and for other species the time span might be 50 years. The point is that it *is* happening, so *now* is the time to take action at the policy and management level if Scarlet Star species both on- and off-reserves are to be protected from an even heavier over-exploitation than is already the case.

7.6.2 Off-Reserve Extraction of Red Star Species

The 1996 recorded off-reserve extraction of Red Star species accounted for 240,633 m^3 or 29% of the total 1996 recorded off-reserve extraction. Even though this is a substantial volume and at least ten per cent below actual harvest, Red Star off-reserve resources should allow for harvest at the present level in at least another 100 years (cf. tables 7.2 and 7.3).

The extraction of Ceiba stands out among the Red Star species. In 1996 it was recorded to be approximately 136,000 m^3 or 57% of the total recorded Red Star extraction which is substantial. However, Ceiba is known to be the most abundant tree species in Ghana, and it is not likely that detailed off-reserve inventory data, when it becomes available, will forecast a soon economic extinction of Ceiba even if the current extraction level is continued.

7.6.3 Off-Reserve Extraction of Pink Star Species

Pink Star species accounted for approximately 65,000 m^3 or only 8% of the total recorded 1996 off-reserve extraction. Of the 32 Pink Star species (by trade name) only Sterculia, Ofram, Danta, Esia, Yaya and Otie were in 1996 reported to be extracted at rates above 1,000 m^3. Together these species accounted for approximately 62,000 m^3 or 96% of the 1996 recorded Pink Star extraction, while the other Pink Star species were either left untouched or extracted at very low rates (cf. appendix 5).

It is clear that the current off-reserve extraction of Pink Star species reflects market demand rather than resource availability. As the case is on-reserves, the stock of Pink Star species represents a resource which the industry might turn to when the commercially more attractive Scarlet and Red Star species become less available, either because the resource is depleted or because new management procedures efficiently control supply.

7.7 Policy Implications of Extraction and Resource Data

For the first time since the mid 1980s when the timber harvest in Ghana began to increase as a result of the Export Rehabilitation Programme it is, as shown above, now possible to assess the annual timber harvest against the resource base both on- and off-reserves. Data is, as discussed in this and earlier chapters, not perfect but by far good enough to draw important conclusions at the policy level.

The timber industry's preference for Scarlet and Red Star species has for many species within these two groups led to severe over-exploitation on-reserves. Moreover, inventory data and recorded extraction have revealed that off-reserves Scarlet Star species will, according to the most optimistic scenario (table 7.3), be logged out within the next 39 years, if the 1996 extraction rate continues.

Accordingly, if the Forestry Department cannot bring the timber exploitation on-reserves under control within the coming 15-30 years, this could easily be the end of many Scarlet and Red Star species as far as raw material for the timber industry is concerned. Botanically they would probably survive as single trees which are too rare and of too poor stem form to be pursued for industrial utilisation.

What the environmental consequences will be of an on-reserve species composition with fewer Scarlet and Red Star species than today is an open question. However, a reduced genetic diversity within and between species in a natural forest is hardly justifiable unless it can somehow be proved socially desirable. Such justification is absent in Ghana and it is certainly not official policy to mine the commercially most valuable timber species for short-term financial profits and hand over whatever is left to future generations.

The substantial resource of Pink Star species both on- and off-reserves provides some hope for a future where the present extraction level may be maintained, but based on a wider selection of species than today. However, this safety net function of the Pink Star species will only materialise if there is a market for these species. The current and historical harvesting pattern shows that this is not yet the case.

It would at this point be naive to expect that the timber industry is voluntarily going to reduce harvest of the currently over-exploited Scarlet and Red Star species and focus its resources on developing markets for abundant Pink Star species. A few sawmills may have the strategic wisdom to do so (we have come across one company which did). Yet, the majority of timber industries naturally pursue short-term profit maximisation when this is an option, ie. when they have access to Scarlet and Red Star species.

Particularly off-reserves and to a limited extent on-reserves, the economic system appears to be one of open access which is leading to the tragedy of the certain Scarlet and Red Star timber species. The only remedy to reverse this trend is policy interventions that combine:

- Economic incentives which favour conservation of timber trees for future use (demand-side measures);
- restrictions on felling activities (supply-side measures).

The latter point is necessary to ensure that long-term social interests of maintaining a diverse on- and off-reserve timber resource get the upper hand when these are in conflict with short-term financial interests of private companies. The options for such policy interventions and results of interventions already applied will be analysed in the following chapters.

8 Timber Exports and the Wood Industry

The export of wood products and the timber industry as such are prominent in Ghana's economy. Estimates suggest that the timber industry accounts for 4-7% of GDP, and that wood exports represent some 18% of exports (Overseas Consultancy Service, 1995; 16-17).

In chapter 3 it was mentioned that part of the 1984 IMF/World Bank Economic Recovery Programme for Ghana included an Export Rehabilitation Programme including timber and wood products. This chapter examines the apparent success of this programme, which until 1994 has resulted in:

- Increasing volumes of wood exports;
- increasing foreign exchange earnings;
- increasing value per cubic metre of wood exports.

From 1994 to 1996 the export revenue and volume have gone down a bit, but the general trend during the period 1986-96 has been increasing export revenue and an increasing volume of processed wood exports. However, the above points do not complete the picture. Clearly, one negative and crucial development in Ghana is the fact that the country's timber resource has been and still is seriously over-exploited. Harvest for export plays a significant role in this respect. Furthermore, the success of further processing before export depends, from a resource conservation and foreign exchange earning point of view, entirely on how efficient the further processing is.

8.1 Development in Wood Exports by Volume

As figure 8.1 shows, the export volume of products has rapidly increased from approximately 90,000 m^3 in 1982 to approximately 960,000 m^3 in 1994, after which the export volume dropped to approximately 360,000 m^3 in 1996. The very high volumes in 1993 and 1994 mainly reflect a sudden "log export boom" of especially Kyenkyen, Ceiba, Yaya, Otie and Ofram.

Until 1994, logs were a major export article closely followed by lumber, which has shown a steadily increasing trend since 1982. The sharp fall in log exports after 1994 is due to the complete log export ban that was, in effect, implemented administratively during 1995 as the Government stopped issuing log export licences.

Sliced veneer and the group "All Others" have also shown a steadily increasing trend although at modest volumes. Appendix 7 provides the details behind figure 8.1.

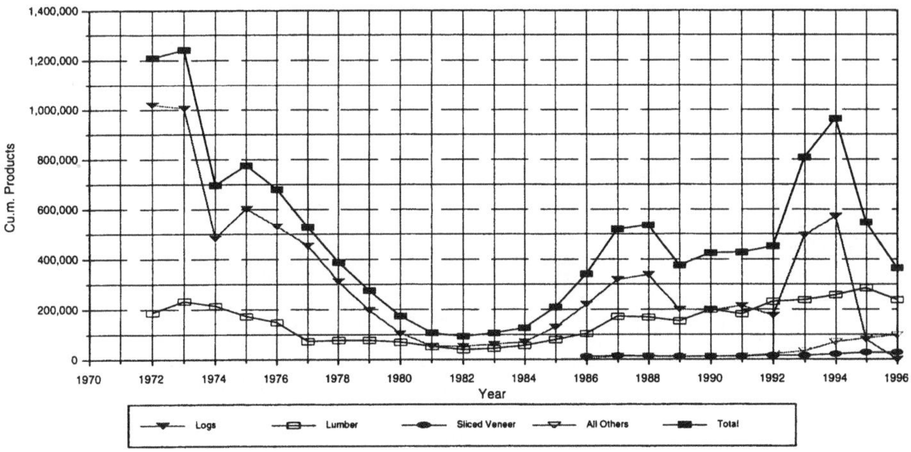

Figure 8.1 **Volume of wood exports by product groups 1972-96**

Sources: General Woods (1992; WD. no. 3 App. 3).
FPIB Export Permits (1986-96).

8.2 Apparent Domestic Consumption

In addition to export demands the timber harvest must also satisfy domestic needs. No direct data on the domestic timber consumption is available, but a rough estimate may be provided by subtracting the export harvest from the recorded extraction (stock change is, in other words, ignored). The export harvest is here defined as the volume of wood exports converted into roundwood equivalents (RWE). This is done by assuming that the recovery rates[27] for lumber, sliced veneer and "all others" are 40%, 30% and 35% respectively (see section 8.3). In estimating the domestic wood consumption it is further assumed that 55% of the residues from the export production are re-processed for domestic use other than fuelwood and charcoal. For the period 1986-96 figure 8.2 shows the result of this exercise (appendix 7 provides the data in detail).

27 The volume of wood products in per cent of the volume of raw logs used for their manufacture.

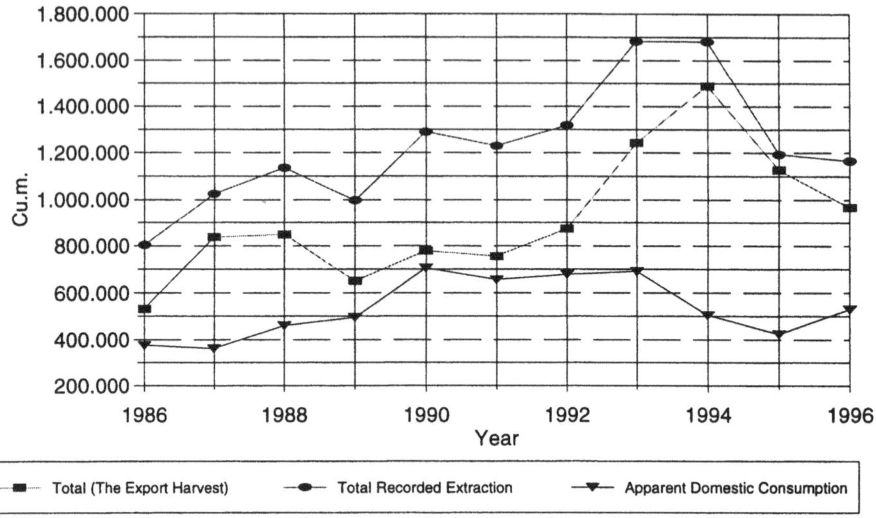

Figure 8.2 **Estimated use of the recorded timber extraction (RWE) 1986-96**
Sources: FPIB Export Permits (1986-96)
FPIB LMC Totals (1986-96)

The falling trend in apparent domestic wood consumption from 1993 to 1996 reflects the export harvest in these years making up between 74 and 94% of the recorded extraction. Yet, it seems unlikely that the domestic consumption should have fallen to the extent indicated by the graph. A more logical explanation would be that a combination of illegal logging and a depletion of stocks of processed wood took place during the period 1994-96. Certainly illegal logging was a serious problem during this period, where a working group to recommend on control of illegal timber harvesting outside forest reserves had to be formed by the Ministry of Forests (FD, 1994a). During the log export boom in 1993-94, exports of processed wood continued to increase (see figure 8.1). This suggests that the wood industry, while stepping up log exports tremendously, may have managed to increase the export of processed wood by depleting stocks. In conclusion, the size of the domestic wood consumption is probably at least 700,000 m^3 in RWE per year, excluding charcoal and fuelwood.

A major domestic market for timber is the construction industry which is expected to continue to expand. The two most important "driving forces" of domestic wood demand are likely to be population and economic growth. Only an increase in the real price of wood would tend to depress the demand. Based on a population growth of 3.3% and real economic growth of 3.8% in 1994, Overseas Consultancy Service (1995; 14) roughly estimates that the domestic wood consumption could grow 3.5% per annum. This would amount to a domestic consumption of nearly one million m^3 by 2005.

Of course, more sophisticated analyses and less indirect calculations than the above could be made. Yet, it seems reasonable to assume that the domestic wood consumption in RWE is going to be in the region of one million m^3 within a foreseeable future. Since the domestic demand/need is likely to be satisfied legally or illegally and the AAC from forest reserves is only between 0.8 and 0.9 million m^3, it is clear that the present level of wood exports (almost one million m^3 RWE in 1996) cannot continue without depleting the resource base unless a significant increase in sustained timber yield from the high forest zone is accomplished. This could be through plantations (off-reserves or on areas inside reserves with no forest) and/or a kind of sustained yield timber production off-reserves (on farmlands). In fact, this appears the most important forest policy problem in Ghana today. The financial and legal perspectives for adding to the on-reserve AAC through plantations and off-reserve sustained yield timber production are further assessed in chapters 12 and 13.

8.3 Export as the Dominant Driving Force for Timber Harvest

Apart from estimating the apparent domestic wood consumption, figure 8.2 also indicates that, during the period 1986-96, the export harvest made up 60-94% of the total recorded timber extraction. Harvest for export has, in other words, been the dominant driving force behind timber exploitation in Ghana.

Naturally, the estimated percentage of harvest for export out of the total recorded extraction depends entirely on the assumed recovery rates and the reliability of data. Yet, the recovery rates used here are probably on the high side. For example, General Woods (1992; WD no. 4; I-15), in an extensive survey covering about 75% of Ghana's timber industries, finds that the recovery rates for export lumber is between 20% and 40%. For sliced veneer the technically feasible recovery rate on modern mills is 35-40%, so the estimated 30% recovery may even be generous given the generally low technical standard of Ghanian veneer mills and keeping in mind that the calculations are based on export quality products only. The recovery rate of 35% used for the group "All others" is also considered generous as this group apart from rotary veneer and boles consists mainly of further processed lumber. Accordingly, the export harvest used here is more like to be an under- than an over-estimate.

The reliability of data on timber extraction has been discussed in section 7.1.1, where it is concluded that official timber extraction data most probably under-estimates the actual harvest by at least 10%. There are no indications of significant discrepancies between the official data on export volumes and actual export volumes during the period 1986-96.

In addition to the above technically based estimations of how important the export harvest is in relation to the total timber harvest, it has clearly been the objective of the Export Rehabilitation Programme, and subsequently the World Bank supported Forest Resource Management Project, to boost the timber industry and foreign exchange earnings from wood exports (FRMP, 1988; 49). It is, therefore, no surprise that the export harvest accounts for a significant if not the major part of the recorded timber extraction.

In 1988, the World Bank naturally emphasised that the AAC would ultimately set the limits for wood export earnings and new jobs created in the wood industry. Yet, at that time the Bank assumed that the AAC was not likely to be less than the, at that time available, estimate of 1.1 million m^3. In fact, the Bank seemed to anticipate that the AAC would prove to be considerably higher, in the range of 2.0 to 2.7 million m^3 (FRMP, 1988; 49-51). This has proved to be completely wrong as the AAC from forest reserves is subsequently estimated with fairly good accuracy to be between 0.8 and 0.9 million m^3.

8.4 Species Composition of the Export Harvest

As shown above the export harvest has, since 1986, generally equalled or exceeded the on-reserve AAC, which in itself should be a cause of concern. However, the species composition of the export harvest reveals that the situation is even more alarming than the total volumes indicate. A minimum estimate of the relative importance of species harvested for wood exports may be calculated by summing up the export harvest of logs, lumber and sliced veneer for each species[28]. This is shown for star groups in figure 8.3, which also includes the total recorded extraction from on- and off-reserves.

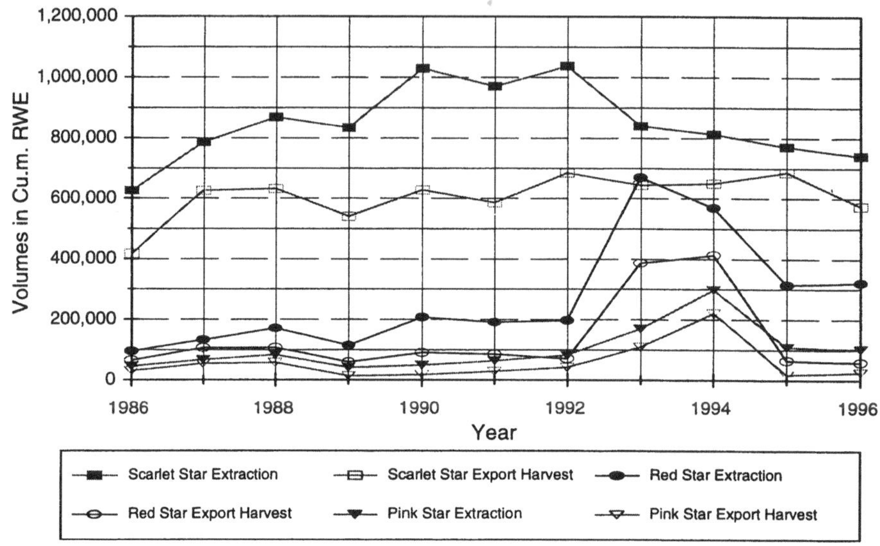

Figure 8.3 Total recorded extraction and minimum export harvest estimates for species groups 1986-96

Sources: FPIB Export Permits (1986-96)
FPIB LMC Totals (1986-96)

It is seen from figure 8.3 that, during the period 1986-96, the export harvest of Scarlet Star species has dominated the picture and fluctuated little around an average of some 600,000 m^3 RWE annually, with a span of 100,000 to 400,000 m^3 up to the recorded extraction. The average export harvest of Red and Pink Star species during this period was approximately 140,000 m^3 and 60,000 m^3 RWE, respectively. However, as figure 8.3 indicates, the average volumes for Red and Pink Star species are influenced by the log export boom in 1993-94 where the export harvest of Red Star species went up to some 400,000 m^3 RWE. For Pink Star species the export harvest was approximately 100,000 m^3 and 220,000 m^3 RWE in 1993 and 1994, respectively (see appendix 7 for details).

---

28 Together these three commodity groups covered, on average, 93% of the export volume during the period 1986-96 (98% in 1986 and 73% in 1996). See appendix 7

8.4.1 The Unsustainable Export Harvest

Since forest reserves so far are the only (de jure if not de facto) secured source of raw logs it is of interest to compare the export harvest with the estimated on-reserve AACs. The result of doing this is, in summary, shown in table 8.1 (details are found in appendix 7).

Table 8.1 **Estimated on-reserve AACs compared with the estimated export harvest of logs, lumber and sliced veneer**

	Large-grained protection only (production forest = 762,400 ha)			Large- and fine-grained protection (production forest = 645,080 ha)		
SPECIES GROUP	AAC total volume (m^3)	Avg. 1986-96 Export harvest in RWE (m^3)	Avg. 1986-96 Export harvest as % of AAC (%)	AAC total volume (m^3)	Avg. 1986-96 Export harvest in RWE (m^3)	Avg. 1986-96 Export harvest as % of AAC (%)
Scarlet Stars	132,437	606,459	458	120,817	606,459	502
Red Stars	224,195	137,389	61	201,353	137,389	68
Pink Stars	534,877	56,906	11	479,518	56,906	12
Others		13,761	-		13,761	-
TOTAL	**891,509**	**814,515**	**91**	**801,688**	**814,515**	**102**

Sources: FIMP (1995b)
FPIB Export Permits (1986-96)

As table 8.1 shows, Scarlet Star species accounted the majority (74%) of the average export harvest during 1986-96. Furthermore, the 1986-96 average export harvest of Scarlet Star species is four to five times the on-reserve AAC for this species group, depending on which level of forest protection the AACs represent. A more detailed look at the Scarlet Star species shows that all species except Albizia, Daniella and Hyedua are harvested for export in volumes that are four to 14 times their on-reserve AACs (see appendix 7).

The average 1986-96 export harvest of Red Star Species represents between 61% and 68% of the on-reserve AAC for this group as a whole. Yet, three species, Afzelia, Candollei, and Emire have been harvested for export in volumes that are two to five times their on-reserve AACs. Pink Star species were all harvested for export at rates that are significantly below their on-reserve AACs.

Bearing in mind that the above export harvest volumes for star groups are minimum estimates, it is clear that the species composition of the wood exports is unfortunate, to put it mildly. The 1996 on-reserve recorded extraction of Scarlet Star species was approximately 180,000 m^3 (see section 7.5), which is 1.4 to 1.5 times the AAC. The 1996 off-reserve recorded extraction of Scarlet Star species was around 520,000 m^3 which, if exploitation continues at the 1996 level, will deplete off-reserve Scarlet Star timber resources over the next 15 to 40 years (cf. tables 7.2 and 7.3). Accordingly, the timber industry is and has at least for the last ten years been mining the Scarlet Star timber resource for export.

Judging from the 1996 extraction records it may be assumed that the depletion of Scarlet Star species has been most intensive off-reserves. The question is what happens when (if) Scarlet Star species are logged out off-reserves while the export demand pull for these species probably remains at a level of some 600,000 m^3 RWE annually. Theoretically and in a perfect world the harvest should, as assumed by the World Bank, be reduced to the AAC. Yet, the value of wood exports suggests that vested financial interests in the current level and species composition of wood exports are going to make a reduction of the Scarlet Star species export rather difficult.

8.5 The Value of Wood Exports

The FOB value of Ghana's total wood exports has, measured in 1996 prices, increased steadily from approximately DM 145 million in 1986 to approximately DM 220 million in 1996 (see figure 8.4). The peak of DM 365 million in 1994 is mainly due to the log export boom which culminated that year (cf. the curve for logs and poles) and fell to zero in 1996 where no logs were exported. The export value of lumber, sliced veneer and the group "All Others" has, as figure 8.4 shows, followed a fairly steadily increasing trend over the period.

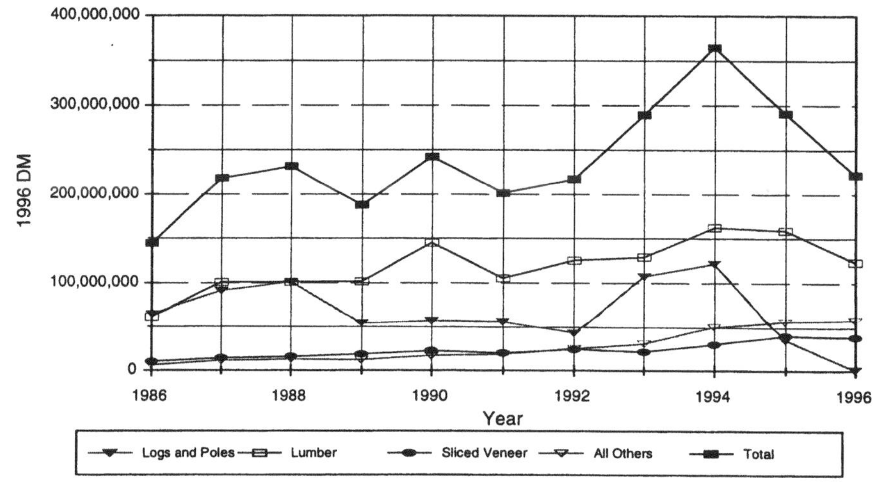

Figure 8.4 **FOB value in 1996 DM of the total wood exports 1986-96**

Sources: FPIB Export Permits (1986-96)
Statistisches Bundesamt Deutschland (1999)

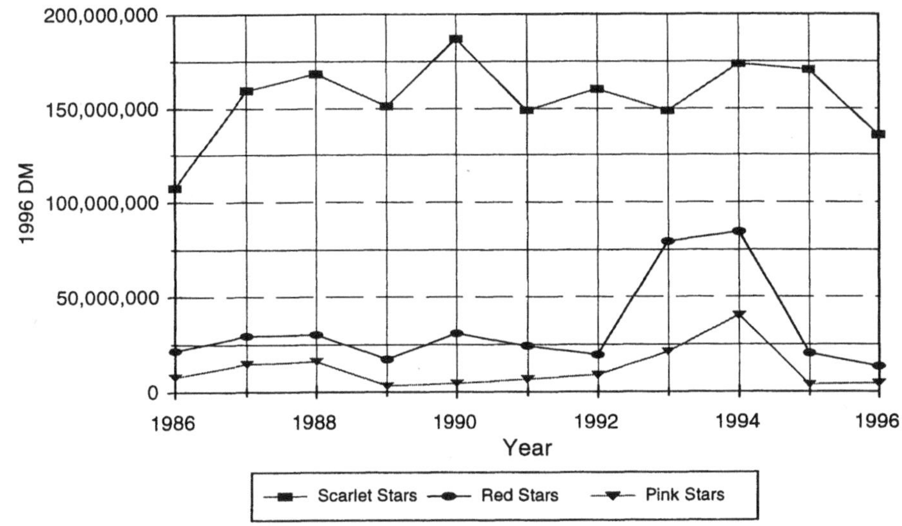

Figure 8.5 **FOB value in 1996 DM of the log, lumber and sliced veneer export by species groups 1986-96**

Sources: FPIB Export Permits (1986-96)
Statistisches Bundesamt Deutschland (1999)

In terms of species composition it is no surprise that the Scarlet Stars dominate like they do in terms of volume. For the export of logs, lumber and sliced veneer only it is seen from figure 8.5 that, measured in 1996 prices, Scarlet Star species have accounted for around DM 150-175 million annually. Red and Pink Star species have fluctuated around DM 25 and 10 million, respectively if the log export boom in 1993-94 is excluded.

If 1996 is taken as a proxy for the composition of wood exports in the near future, ie. with no export of logs and no radical market changes for other products, it is seen from figures 8.4 and 8.5 that the Scarlet Star species will account for more than 70% of the total wood export revenue. Since the export markets are not likely to change taste in terms of demand for certain species as long as these are available, and since Ghana is a relatively small world market supplier, the major export value and notorious over-exploitation of Scarlet Star species clearly poses a serious forest policy problem.

Essentially, the problem is that a significant number of jobs in the wood industry and a significant fraction of the much needed foreign exchange earnings from wood exports are achieved through a mining of the Scarlet Star timber resource. The irony is that the obvious imbalance between the timber resource and the timber export is created through an IMF and World Bank supported deliberate policy to boost the wood export.

8.5.1 Unit Values of Wood Exports

In figure 8.6 the development in FOB export unit value is, in 1996 prices, depicted for the period 1986-96. If one looks at the total FOB export value per m^3 of all products taken together, this has gone up impressively from around DM 420 per m^3 to DM 600 per m^3 during the period. This figure's temporary fall to approximately DM 350 per m^3 in 1993-94 was a result of the log export boom which was concentrated on relative low value Red and Pink Star species.

Another way to analyse the development in export unit value is to base the calculation on the export harvest. When this is done the trend becomes *negative*. Measured in 1996 prices, the value has gone down from DM 270 per m^3 RWE in 1986 to DM 230 per m^3 RWE in 1996 (the detailed data behind figure 8.6 is found in appendix 8). This strongly indicates that the apparent successful development of Ghana's wood export may not be so successful after all.

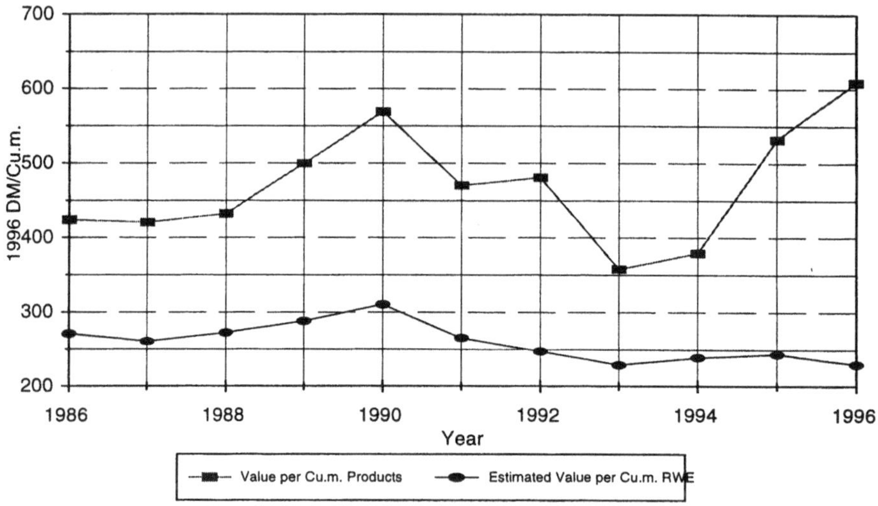

Figure 8.6 **FOB value in 1996 DM per m^3 of products and per RWE of total wood exports 1986-96**

Sources: FPIB Export Permits (1986-96). Statistisches Bundesamt Deutschland (1999)

A closer look at the FOB weighted average unit export value for commodity groups is provided in figure 8.7. The graphs in figures 8.6 and 8.7 show that the overall weighted average unit value of wood exports closely follows that of lumber. The weighted average export unit value of logs has fallen gradually which is also the case for the group "All others". Sliced veneer has as the only commodity group shown a remarkable increase in weighted average unit export value.

As pointed out earlier, lumber is the dominant wood export commodity both in terms of volume and value. A detailed picture of the lumber export value per RWE is provided in figure 8.8. This shows that, during the period 1986-96, the weighed average FOB export value in 1996 prices for all species taken together (which closely follows the Scarlet Star group) has fluctuated around DM 250 per m^3 RWE with a slightly falling trend. The weighted average unit value for Red Star lumber is higher than Scarlet Star lumber because the relative low value species, Wawa in the Scarlet Star group is exported in much higher volumes than any other species.

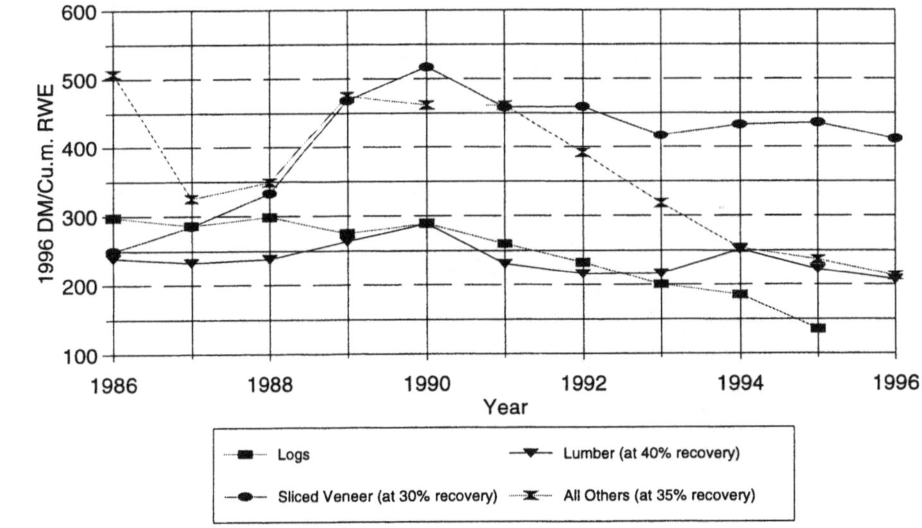

Figure 8.7 **FOB value in 1996 DM per RWE for commodity groups of total wood exports (1986-96)**

Sources: FPIB Export Permits (1986-96)
Statistisches Bundesamt Deutschland (1999)

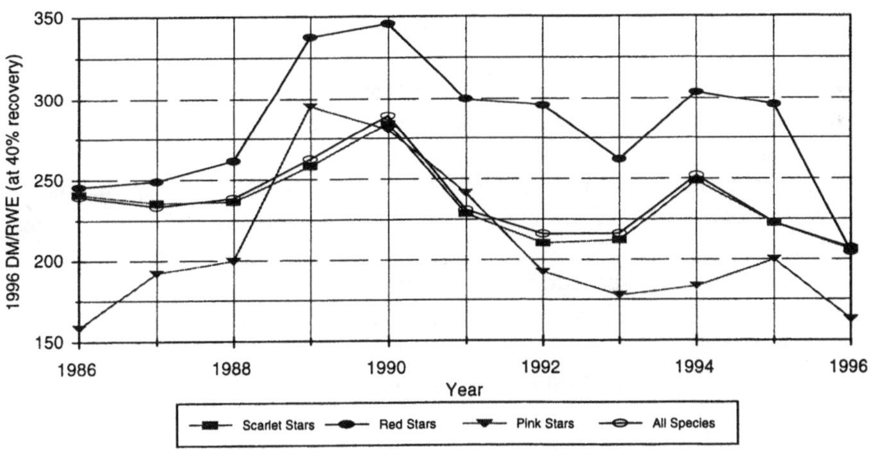

Figure 8.8 **Average FOB value in 1996 DM per RWE of lumber exports by species groups (1986-96)**

Sources: FPIB Export Permits (1986-96). Statistisches Bundesamt Deutschland (1999)

A similar analysis for sliced veneer is shown in figure 8.9. It is seen that the weighted average unit value of all species taken together has, in 1996 prices, gone up significantly from DM 250 per m^3 RWE in 1986 to an impressive DM 520 per m^3 RWE in 1990, after which it has fallen to a fairly stable level around DM 410 per m^3 RWE. Furthermore, Scarlet Star species are dominant although there is much less variation in the unit value among species groups than the case is for lumber.

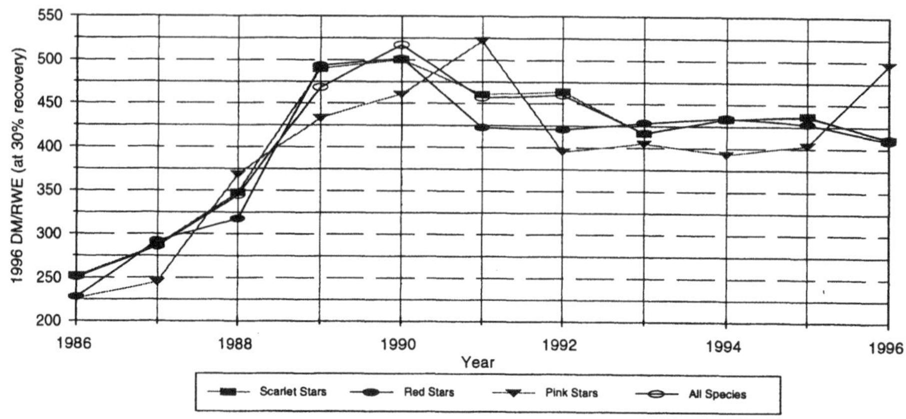

Figure 8.9 **FOB value in 1996 DM per RWE of sliced veneer exports by species groups (1986-96)**

Sources: FPIB Export Permits (1986-96). Statistisches Bundesamt Deutschland (1999)

In conclusion, the unit value per RWE for Ghana's wood export has stayed at a low level and even shown a somewhat falling trend during the period 1986-96. Only sliced veneer clearly stands out as an apparent positive deviation from the general rule, but this commodity is not exported in big volumes. Accordingly, the increasing export of processed wood appears to have had little positive effect on the foreign exchange earnings per m^3 timber felled for export. In fact, the opposite seems to have been the case. This can hardly be termed a successful development. In particular, this is so because the lion's share of the export revenue has been and still is based on an indisputable over-exploitation of Scarlet Star species.

8.6 The Hidden Costs of Value Added Wood Exports

Section 8.5 has demonstrated that the increasing overall unit price of wood exports, when calculated on the basis of product volumes, provides a false image of the economic realities. When the timber resource base and the degree to which it is utilised is considered, the gloss of the wood export is wearing off.

To make things worse, Ghana appears to be foregoing substantial export revenue earnings by promoting the export of so-called value added products. This may be illustrated by comparing actual export revenues with a hypothetical log export of the timber that has gone into manufacturing the exported products (see table 8.2).

Table 8.2 **FOB value of the 1996 lumber and sliced veneer export compared with a hypothetical alternative log export**

PRODUCTS	Unit	SPECIES GROUP Scarlet Stars	Red Stars	Pink Stars	Others	All Species
AIR-DRIED LUMBER:						
Export volume	m^3	106,604	14,256	8,812	10,625	140,298
Export volume in RWE at 40% recovery	m^3 RWE	266,511	35,641	22,030	26,563	350,745
Weighted average actual export revenue	DM/RWE	207	258	157	274	214
Weighted average log export value[1]	DM/RWE	283	318	193	-	-
Actual export revenue	DM	55,297,790	9,183,869	3,461,953	7,283,455	75,227,066
Revenue from hypothetical log export	DM	75,310,107	11,331,941	4,242,508	-	90,884,556
Break even recovery rate[2]	%	54	49	49		48
Export revenue foregone[3]	**DM**	**20,012,317**	**2,148,072**	**780,556**	**-**	**22,940,945**
KILN-DRIED LUMBER:						
Export volume	m^3	82,197	1,161	995	565	84,919
Export volume in RWE at 40% recovery	m^3 RWE	205,493	2,902	2,489	1,413	212,297
Weighted average actual export revenue	DM/RWE	217	334	214	367	220
Weighted average log export value[1]	DM/RWE	242	312	197	-	-
Actual export revenue	DM	44,578,577	969,592	533,638	518,150	46,599,958
Revenue from hypothetical log export	DM	49,694,510	904,502	490,197	-	51,089,209
Break even recovery rate[2]	%	45	37	37		44
Export revenue foregone[3]	**DM**	**5,115,932**	**-65,090**	**-43,440**	**-**	**5,007,402**
SLICED VENEER:						
Export volume	m^3	25,736	1,784	343	19	27,882
Export volume in RWE at 30% recovery	m^3 RWE	85,786	5,945	1,143	64	92,938
Weighted average actual export revenue	DM/RWE	411	409	496	284	411
Weighted average log export value[1]	DM/RWE	584	281	190	-	-
Actual export revenue	DM	35,220,984	2,429,974	567,114	18,163	38,236,234
Revenue from hypothetical log export	DM	50,130,091	1,670,788	217,484	-	52,018,363
Break even recovery rate[2]	%	43	21	12		41
Export revenue foregone[3]	**DM**	**14,909,107**	**-759,186**	**-349,630**	**-**	**13,800,292**
TOTAL ACTUAL EXPORT REVENUE	**DM**	**135,097,351**	**12,583,435**	**4,562,704**	**7,819,769**	**160,063,258**
TOTAL EXPORT VOLUME	**m^3**	**214,537**	**17,201**	**10,150**	**11,210**	**253,098**
TOTAL EXPORT REVENUE FOREGONE	**DM**	**40,037,357**	**1,323,796**	**387,486**	**-**	**41,748,638**
EXPORT REVENUE FOREGONE	**DM/m^3**	**187**	**77**	**38**	**-**	**165**

Notes: [1]Weighted by the volume of the actual export of species within star groups and product category.
[2]The recovery rate at which the export revenue of products per RWE would equal that of the hypothetical log export.
[3]Negative values indicate export revenue gained through processing of logs.
Sources: FPIB Export Permits (1993-94 and 1996) and TEDB (1993)

The weighted average log export values shown in table 8.2 are calculated on the basis of actual log export prices in 1993-94 and 1993 FOB log prices in other West African countries (used for species that were not exported in log form from Ghana during 1993-94). Accordingly, the table compares 1996 lumber and sliced veneer prices with 1993-94 log prices. Nevertheless, the results are considered valid, as the calculated export revenue foregone can hardly be explained by exceptionally high log prices in 1993-94 compared with what might have been the 1996 level. Furthermore, the weighted average "break even" recovery rates (the recovery rates at which the export revenue per RWE of products would equal that of logs) for "All Species" are only slightly below what is technically feasible within the three product categories. Assuming the market is efficient, this supports that the price relations between logs and products in the table are realistic[29].

It is seen from the table that, in 1996, the export revenue foregone by exporting sliced veneer, air-dried and kiln-dried lumber alone was in the area of DM 40 million or 25% of the actual export revenue. In addition, it is clear that the export of air-dried lumber and Scarlet Star species in general (the prime export product and prime species group) account for the majority of export revenue foregone. The better performance of kiln-dried lumber compared to air-dried lumber is perhaps slightly overstated, as a 40% recovery rate is used for both groups (some losses due to defects developing under drying are bound to occur).

An interesting point is that the further processing of Red and Pink Star species to kiln-dried lumber and sliced veneer generally has a positive effect on the export revenue in comparison with a hypothetical log export. Considering the resource base this leaves some hope as to maintaining current export revenues through further processing. Unfortunately, the volumes in these categories are relatively small.

8.7 Structure of the Wood Industry

Before concluding on the export driven timber harvest a brief overview of Ghana's wood industry is relevant. Four issues in this relation have a direct bearing on the overall forest policy problems of managing the high forest resource. These are:

- Installed wood processing capacity;
- employment in the wood industry;
- general structure and ownership;
- size of the wood industry in the national economy.

8.7.1 Wood Processing Capacity

The recorded extraction in 1996 (after the complete log export ban) of approximately 1.2 million m^3 shows that the processing capacity of the wood industry is at least at this level. However, a survey by General Woods (1992, WD. no 4) estimated the processing capacity of the wood industry to be around 2 million m^3 annually. On the basis of a new survey and in the light of the 1995 complete log export ban, Overseas Consultancy Service (1995; 21) estimated that the processing capacity in the wood industry is growing and could reach a capacity of over 3 million m^3 annually in 1997.

---

29 The FOB log prices must reflect overseas buyers' willingness to pay for logs, which in turn is a function of their ability to make profits from converting them into products (here lumber and sliced veneer). Since these products are sold on the same markets as similar products from Ghana, a calculation of "break even" recovery rates on the basis of FOB log and product prices in Ghana/West Africa should, if the market is efficient, be slightly lower than the average recovery rates realised by overseas buyers since the unit transport costs of products from Ghana should be lower than the unit transport costs of products that are based on log imports.

Although the above estimates on processing capacity may be debatable it is unlikely that they are totally off the mark. Accordingly, there can be no doubt that the installed capacity in the wood industry is around twice the on-reserve AAC. Clearly this is an unsustainable situation.

8.7.2 Employment in the Wood Industry

Even though the wood industry appears to be running at approximately 50% of its processing capacity it employs a significant number of people. The forestry sector employs about 75,000 people, and provides a livelihood of an estimated 2 million people (TEDB, 1995; 1). Although the exact number of people whose livelihood depends on the timber industry is inherently difficult to estimate, there can be little doubt that the wood industry is nationally very important both in terms of direct employment and supporting household economies.

8.7.3 General Structure and Ownership of the Wood Industry

Today the wood industry is totally dominated by private enterprises as the former big wholly or partly state owned companies have been either closed or privatised. According to Overseas Consultancy Service (1995; 15), the structure of the industry may be summarised as shown in table 8.3.

Table 8.3 **Structure of the Wood Industry**

Type of Industry	Number of Entities
Logging	250
Saw milling	110
Ply milling	9
Veneering	15
Furniture/Parts	200
Flooring	4
Doors	6
Profile Boards	12
Toys	4

Note: Entities may be involved in more than one activity. Source: Overseas Consultancy Service (1995;15)

In terms of wood exporting companies the picture is slightly different. The 1996 situation is summarised in table 8.4.

Table 8.4 **Structure of Wood Exporters 1996**

Export Commodity	Number of Exporters
Air-dried Lumber	132
Kiln-dried Lumber	52
Plywood	9
Sliced Veneer	16
Rotary Veneer	14
Processed Lumber/ Mouldings	15
Furniture Parts	4
Flooring	4
Profile Boards	5

Note: Exporters often export more than one commodity.
Source: FPIB Export Permits (1996)

Since most exporters are also producers one may, on the basis of tables 8.3 and 8.4, assume that the predicted increase in the number of sawmills holds true. As the high level of export harvest in relation to the total recorded extraction would suggest, it is further seen that, with the exception of producers of furniture parts, a significant fraction of the wood industries are also exporters. Yet, behind the quite high number of wood exporters lies the fact that only a few companies dominate this market. For 1996 this is illustrated in table 8.5

***Table 8.5* Numbers of dominant wood exporters 1996**

Export Commodity	Number of Exporters	Market share by value
Air-dried Lumber	8	43%
Kiln-dried Lumber	6	54%
Sliced Veneer	6	81%
Rotary Veneer	4	74%
Processed Lumber/ Mouldings	3	72%
Furniture Parts	1	98%
Plywood	1	91%
Flooring	2	91%
Profile Boards	2	81%

Source: FPIB Export Permits (1996)

The message from table 8.5 is quite clear. Very few companies almost entirely dominate Ghana's wood export. Actually, the table slightly understates this situation because some companies are among the dominant exporters of more than one commodity. For example, the private companies, Ghana Primewood and John Bitar & Co. LtD. are dominant in both air-dried lumber and sliced veneer.

8.8 Conclusion on Wood Exports

The bottom line currently seems to be that not only does the export of Scarlet Star species deplete this group of species at an alarming speed. The export of these species also represents a considerable amount of foreign exchange revenue foregone because log exports are not allowed. Ironically, the export of Pink and to some degree Red Star species could, from a resource point of view, be increased significantly and, judging from table 8.2, these species would, as opposed to the Scarlet Stars, even represent a net gain in foreign exchange revenue when exported as products rather than logs.

Accordingly, the composition of wood exports is not desirable from a long-term national economic point of view, but obviously it must be profitable for the wood exporting industries of which very few dominate the market. Otherwise the export of wood products would not have continued the way it has during the 10 year period from 1986 to 1996. Furthermore, there is a substantial overcapacity in the wood industry.

There can only be one explanation to this. Lacking incentives for the wood exporting sector to act in the long-term interest of the nation and adjust to the resource base. In other words, market failures. The courses and possible solutions should, therefore, be sought in the domestic market and politically defined regulative mechanisms associated with commercial timber harvest.

9 Forest Fees and Taxes

The previous chapters have reviewed the "technical evidence" sustaining that the timber resource and certain forest areas in Southern Ghana are being degraded, and that the speed of degradation forms an immediate threat of economic extinction for some species. Furthermore, the substantial off-reserve timber harvest and the industry's export driven preference for especially Scarlet and to some degree Red Star species is a ticking bomb under the potentials for implementing sustainable management or at least sustained timber yield management of forest reserves.

In several places it has been indicated that financial incentives must form a major underlying reason for the over-exploitation. Since most of the species which are in highest demand are over-exploited and since the processing capacity of the wood industry appears to be under-utilised, there should be a "seller's market" for logs. In practice, this is not the case because the market does not operate properly. The price of standing timber (royalties and other forest fees) and who gets access to harvest timber is determined administratively rather that through market forces.

This chapter reviews the different forest fees and taxes, including their legislative background and lays the foundation for analysing the financial incentives to harvest and process different timber species. Forest fees and taxes may in this respect be considered "demand-side" measures intended to regulate the timber harvest.

Generation of government revenue to finance forest management, including timber exploitation, is another (perhaps the most) important reason for forest fees and taxes. In addition, the legislation also covers aspects of rights and responsibilities, of which some of the more important ones are mentioned. These issues are further discussed in chapters 11 and 12.

9.1 Overview of Forest Fees and Related Laws

There are and have been a number of official forest fees in Ghana. A list of the different fees, the fee level (where standardised) and the institution responsible for their collection is presented in table 9.1.

Table 9.1 **Current and recent forest fees and institutions collecting them**

Institution: - Fee	Current level	Set by /in consultation with[1]/
Forestry Department (FD):		
- Royalties/stumpage fees	Variable	Min. of Lands and Forestry[2] /after *Ad Hoc* Committee[3]/
- Concession rent	¢ 1,000/ha/year	Min. of Lands and Forestry[2] /after *Ad Hoc* Committee[3]/
- Contract Area rent (on-res.)	¢ 1,200/ha/year	Min. of Lands and Forestry
- Contract Area rent (off-res.)	¢ 1,000/ha/year	Min. of Lands and Forestry
- Silvicultural fee (on-reserves)	¢ 50 /ha/year	Min. of Lands and Forestry[2] /after *Ad Hoc* Committee[3]/
- Silvicultural fee (off-reserves)	¢ 20 /ha/year	Min. of Lands and Forestry[2] /after *Ad Hoc* Committee[3]/
- Compensation payments	Variable	Magisterial Courts
- Fines	Variable	Min. of Lands and Forestry[2]/Timber Lease Committee[4]
- Property mark	¢ 400,000/year	Min. of Lands and Forestry[2]/Timber Lease Committee[4]
- Conveyance Certificates	¢ 10,000/truck load	Forestry Department
- Taungya fees	Variable	Forestry Department: costs of cleaning perimeter of area
- Permits: major for. products	Variable	Royalty rates: Ministry of Lands and Forestry
- Permits: minor for. products	Variable	District Forestry Officers of the Forestry Department
- Concession preparation fee	¢ 500,000	Forestry Department
Forest Products Inspection Bureau (FPIB):		
- Retailer registration fee	¢ 5,000/year	Forest Products Inspection Bureau
- Exporter registration fee	¢ 30,000/year	Forest Products Inspection Bureau
- "Sale" of LMCs	¢ 200/log	Forest Products Inspection Bureau
- General export levy	2% FOB value	FPIB/TEDB to cover institution costs
- Exp. levy on air dried lumber	Variable	Min. of Lands and Forestry[2] /after *Ad Hoc* Committee[3]/
Timber Export Development Board (TEDB):		
- General export levy	1% FOB value	FPIB/TEDB to cover institution costs
Export Promotion Council (EPC):		
- Registration fee	¢ 30,000/year	Export Promotion Council
Ghana Chamber of Commerce (GCC):		
- Export document fee	¢ 5,000/cargo	Ghana Chamber of Commerce
Ghana Shippers Council (GSC):		
- Registration fee	¢ 120,000/year	Ghana Shippers Council

Notes: 1 Where only one institution decides there is no consultation on level of fees.
2 In 1993 the Ministry of Lands and Forestry replaced the Ministry of Lands and Natural Resources.
3 *Ad Hoc* Committees formed by representatives of the Forestry Department, the Timber Export Development Board, the Forest Products Inspection Bureau, the Forestry Commission (FC) and the Association of Ghana Timber Industries (AGTI).
4Timber Lease Committees formed by representatives of the Forestry Department, the Timber Export Development Board, the Forest Products Inspection Bureau, the Forestry Commission (FC) and the Lands Commission (LC).
Sources: Based on: IIED (1993; 92). Updated and corrected in accordance with current legislation.

9.1.1 Timber Royalties and Stumpage Fees

Royalties on trees felled within forest reserves have since 1976, when the system was introduced, been collected by the Forestry Department (Forest Fees Regulations, 1976).

The Lands Commission was responsible for collecting off-reserve timber royalties until 1994 where the Ministry of Lands and Forestry transferred this responsibility to the Forestry Department, in an attempt to streamline the procedures for royalty collection and to improve the collectionrate (MLF, 1994).

This change in collection of timber royalties off-reserves had been recommended for a long time by various consultants. As part of the Forest Resource Management Project it was even agreed between the Government of Ghana and the World Bank that the Forestry Department should have taken responsibility for the off-reserve collection of forest fees by October 1, 1990 (FRMP, 1988; 53). However, during the period 1988-94 the Lands Commission managed to keep royalty collection off-reserves under its jurisdiction, even though it was never able to produce any accounts for collected timber royalties, which obviously raised doubts about its competence and credibility.

On April 4, 1991 the Ministry of Lands and Natural Resources (now Ministry of Lands and Forests) issued a directive which transferred the responsibility for collection of timber royalties off-reserves to the Forestry Department (MLNR, 1991). Yet, with reference to the PNDC Law 42[30], according to which the Lands Commission claimed to be the only legal authority to collect off-reserve timber royalties, and by documenting that the Forestry Department did not re-distribute fees collected off-reserves to the legal beneficiaries (District Assemblies, Stools and Traditional Councils) the Lands Commission managed to fight the ministerial decree with success (LC, 1991).

After the responsibility to collect off-reserve timber royalties was transferred to the Forestry Department a number of complaints, of which some were sent through the Ministry of Local Government (MLG), from district councils over not receiving their legal share of these began to materialise. In response, the Ministry of Lands and Natural Resources redrew its directive of April 4, 1991 and re-instituted the Lands Commission as collector of timber royalties off-reserves as of April 7, 1992 (MLG, 1992 and MLNR, 1992).

When the Ministry of Lands and Forestry was formed under the 1992 Constitution, the Lands Commission, in effect, came under the Ministry's jurisdiction (Constitution 1992; 258, 2). This finally paved the way for the long awaited transfer of off-reserve royalty collection to the Forestry Department (MLF, 1994).

Timber royalties have been defined through legislative instrument (the Forest Fees Regulations), and rates were adjusted through revision of the regulations which has been very time consuming compared with the obvious alternative, annual administrative revisions in accordance with legally defined guidelines. The combination of royalties defined in Cedis (¢) and high inflation rates have eroded their real value between revisions of the rates.

Royalties were species specific and until 1994 they were charged on a per tree basis. This system opened for a differentiation of royalties according to species value and rarity which, in theory, could be used to maximise government revenue and discourage extraction of over-exploited species. The most recent revision of royalties was by the Forest Fees (Amendment) Regulations (1993), which was gazetted on 21 January, 1994. Apart from raising royalty rates the Forest Fees (Amendment) Regulations (1993) stipulated royalties to be charged per cubic metre instead of per tree.

---

30 The PNDC (Provisional National Defence Council) Law 42 (1982) abrogated the 1979 Constitution and essentially functioned as Ghana's constitution from 31 December, 1981 to 7 January, 1993 where the present (democratic) 1992 Constitution came into force.

In December 1998 timber royalties were substituted by stumpage fees (Timber Resources Management Regulations, 1998; 21). A new formula was introduced according to which the stumpage fee for each species is calculated as:

Stumpage fee = Log volume x *Timber price* x *Stumpage rate*

Where:

- Log volume is calculated in m^3.
- Timber price is 35% of the FOB price per m^3 for air-dried lumber.
- Stumpage rate is 20%, 10% or 5% depending on species (see appendix 9).

The timber price will be based on the previous quarter's average price as published by the Forest Products Inspection Bureau. This should prevent stumpage fees from being eroded by inflation.

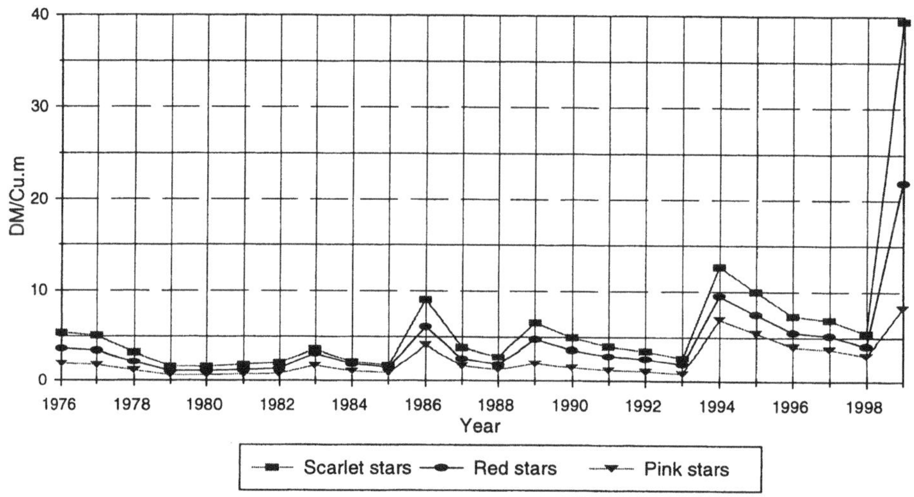

Figure 9.1 **Royalty rates in DM/m^3 during the period 1976-99, average figures for star groups**
Sources:
Forest Fees Regulations (1976)
Forest Fees (Amendment) Regulations (1983)
Forest Fees (Amendment) Regulations (1986)
Forest Fees (Amendment) Regulations (1989)
Forest Fees (Amendment) Regulations (1993)
Timber Resources Management Regulations (1998)

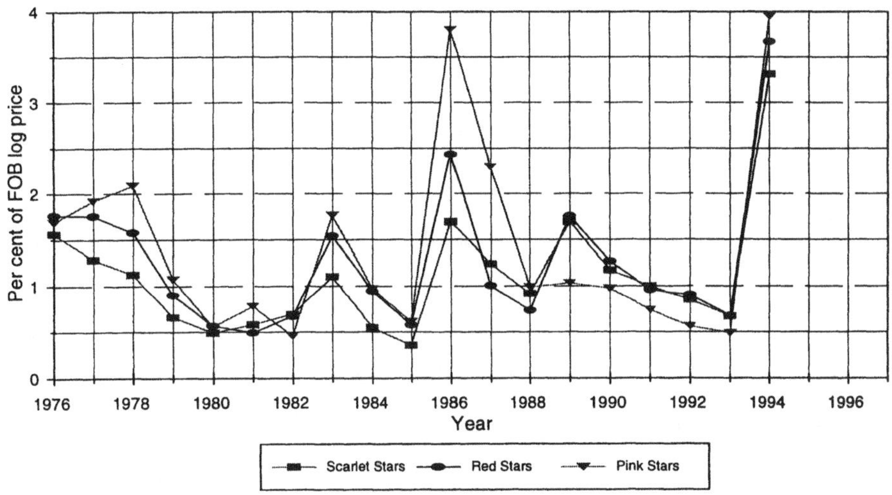

Figure 9.2 **Royalty rates as per cent of FOB log prices 1976-99, average figures for star groups**

Sources:

Forest Fees Regulations (1976)
Forest Fees (Amendment) Regulations (1983)
Forest Fees (Amendment) Regulations (1986)
Forest Fees (Amendment) Regulations (1989)
Forest Fees (Amendment) Regulations (1993)
FAO Forestry Paper 104 (1992)
Forestry Department Annual Reports (1971-85)
FPIB Export Permits (1986-94)
TEDB (1993)

For the period 1976-1999 figures 9.1, 9.2 and 9.3 depict the development of royalty rates per m^3 in German Marks (DM)[31], as per cent of FOB log export prices and as per cent of FOB lumber export prices, respectively. It should be noted that figure 9.2 is based on (i) the available FOB log prices in Ghana and, in cases where figures from Ghana are few or unavailable due to log export bans or poor statistics, (ii) average FOB log prices from West and Central African countries (see appendix 11). In figure 9.3 actual FOB prices and volumes for the total lumber export (air-dried, kiln-dried and overland) have been used for the period 1986-96, after which 1996 prices have been used as a proxy due to data constraints.

31

Standard Forestry Department volumes per tree have been used to calculate royalties per m^3 before 1994 (see appendix 9). As a proxy for real values the current Cedi values of royalties are converted to German Marks (DM) at bank exchange rates.

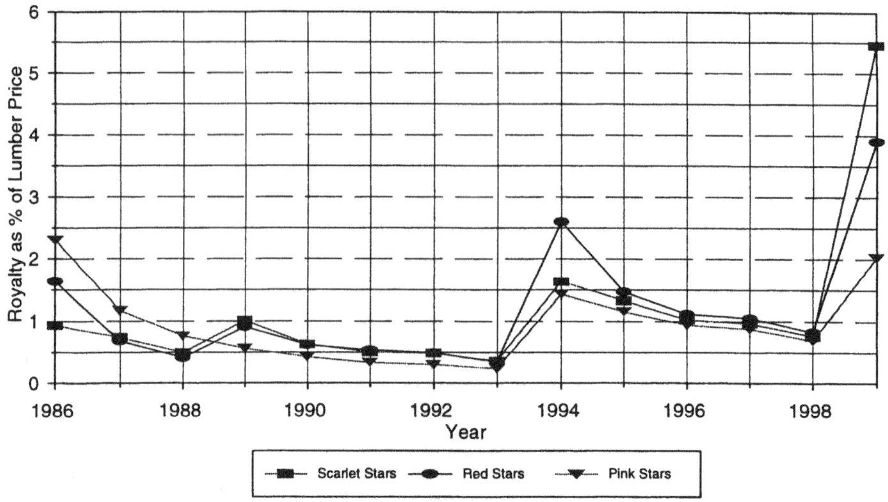

Figure 9.3 **Royalty rates as per cent of FOB lumber prices 1976-99, average figures for star groups**

Sources:
Forest Fees Regulations (1976)
Forest Fees (Amendment) Regulations (1983)
Forest Fees (Amendment) Regulations (1986)
FPIB Export Permits (1986-96)
Forest Fees (Amendment) Regulations (1989)
Forest Fees (Amendment) Regulations (1993)
Timber Resources Management Regulations (1998)

Three important characteristics about the royalty system are illustrated by figures 9.1, 9.2 and 9.3. These are:

- Between revisions of the rates in 1983, 1986, 1989, 1994 and 1999 the real value of royalties has been seriously eroded by inflation;
- the "real" value of royalties as measured in current DM, and particularly in relation to FOB log and lumber prices, has stayed at a very low level until 1999;
- during the period 1986-98 the small difference between royalty rates for Scarlet, Red and Pink Star species (especially when measured against log and lumber export prices) have hardly created effective financial incentives to reduce the harvest of Scarlet Star species and increase the harvest of Pink Star species instead.

The effects of log export bans and costs of domestic processing prior to export of certain species should be included in a comprehensive analysis of figures 9.1 9.2 and 9.3 before the latter two points above can be fully documented. Such analyses are found in chapter 11.

9.1.2 Concession and Contract Area Rent

Concession rent was based on the entire concession area regardless of timber harvest potentials, ie. concession rent was also billed on protected and inaccessible areas within a concession. The rent was introduced in paragraph 2 of the Forest Fees Regulations (1976) and was then ¢ 0.25 (DM 0.55) per ha per year. In 1983 the Forest Fees (Amendment) Regulations (1983) increased the concession rent to ¢ 2 (DM 0.24) per ha per year, and in 1992 it was supposedly increased from ¢ 2 to ¢ 100 (DM 0.71 in 1992) per ha per year (Ahadome, 1992; 2 and FIMP, 1995b; 9). Yet, it is doubtful whether this was ever implemented as the Forest Fee Regulations law was not amended in 1992 and the previous amendment in 1989 did only regulate timber royalties, not the concession rent (Forest Fees (Amendment) Regulations, 1989). In any case, the concession rent has up to 1994 hardly affected the logging profitability since there has been little resistance among concessionaires to pay the rent (FIMP, 1995b; 9).

With effect from 1 March, 1993 the concession rent was increased to ¢ 1,000 (DM 1.74 in 1994) per ha per year (Forest Fees (Amendment) Regulations, 1993). Since then, concessionaires have been increasingly calling for waivers of rent for poorly stocked areas, which has made the Forestry Department consider whether concession rent should be billed on the basis of productive areas only (FIMP, 1995b; 9).

Concession rent has now been substituted by contract area rents of ¢ 1,200 per ha per year within forest reserves and ¢ 1,000 (DM 0.65 in 1999) per ha per year outside forest reserves and (Timber Resources Management Regulations, 1998; 27). All old timber concessions and leases have, in principle, been cancelled as at October, 1998 (Timber Resources Management Act, 1997). It is the duty of the Forestry Department to re-assess all potential timber utilisation areas and on this basis advertise timber utilisation contract areas (Timber Resources Management Regulations, 1998; 1-9)[32]. Accordingly, timber contract areas must in the future be expected to cover mainly productive areas.

9.1.3 Silvicultural Fee

A silvicultural fee was charged both on- and off-reserves. It was legally defined in the Forest Fee Regulations (1976) and based on the area which had been harvested over the last year. The fee was until recently ¢ 50 and ¢ 20 per ha per year on- and off-reserves, respectively, a level defined last time the rates were adjusted (Forest Fees (Amendment) Regulations, 1983). There were at least three problems with this fee:

- It was minimal and probably not even enough to cover the costs of collection;
- it was difficult to calculate as the areas harvested are difficult to check;
- it was difficult to justify as the Forestry Department hardly undertakes any silvicultural work in areas where logging takes place, especially not off-reserves.

IIED (1993; 100) assumed that the Silvicultural Fee was discontinued after 1992. However, this did not happen until December, 1998 when the Timber Resources Management Regulations (1998; 42) revoked the Forest Fees Regulations (1976).

9.1.4 Compensation Payments

Compensation payments is a charge to concessionaires to cover the costs of restoring site damage from timber felling and extraction. The charge is legally backed in the Economic Plants Protection Decree (1979) and concerns the damage on agricultural crops caused by the felling of timber trees. In other words, the Economic Plants Protection Decree (1979) has been central in the legal relationship between

32 See section 12.3.2 for further details.

timber companies holding off-reserve concessions and farmers who cultivate land which carries timber trees.

The Economic Plants Protection Decree (1979) opens two ways for specifying the level of compensation. It may be set by the courts in accordance with specific cases, or the Commissioner (Minister) responsible for cocoa (now Minister of Agriculture) may in liaison with the Commissioner (Minister) for Lands and Natural Resources (now Minister for Lands and Forestry) prescribe fixed rates for specified plants according to the age of such plants.

Unfortunately, the law speaks only of cocoa as specified plants to be protected. The Commissioner (now Minister of Agriculture) may by executive instrument amend the law to cover other plants, but this has never happened. Accordingly, the now Ministers of Agriculture and Lands and Forestry are empowered to:

- Let the Economic Plants Protection Decree (1979) cover all relevant agricultural crops;
- specify compensation rates for crop damages due to timber felling and extraction.

This issue has been and still is crucial for the management of trees on farmland (off-reserve timber resources) because logging companies have in many instances managed to avoid paying adequate compensation for crop damages if any at all (IIED, 1993; 61 and Personal observations). The reason being that individual farmers cannot, in practice, drag logging companies to court and claim their right to compensation.

Since the responsible Ministers have not used their opportunity to amend the Economic Plants Protection Decree (1979) the compensation rates for damaged crops have always been subject for discussion between farmers and timber companies. On several instances during field trips we have personally witnessed (and been dragged into) disputes over compensation payments between representatives of logging companies and farmers.

However, the Timber Resources Management Act (1997; 4 (2d)) now gives the individual farmer a *de jure* right to veto felling of trees on land he/she cultivates, regardless the crops. Accordingly, he/she may, in principle, set the price for allowing a certain tree to be felled. This is an improvement of farmers' rights. The impact on the ground of this rule is still to be discovered (see section 12.3.6 for further analyses).

9.1.5 Fines and Imprisonment

Fines are charged and imprisonment imposed as penalties for unauthorised activities. A number of laws specify different forest offences and fines, but the most important ones in this respect are:

- The Forest Protection Decree (1974);
- the Trees and Timber Decree (1974);
- the Timber Resources Management Act (1997);
- the Timber Resources Management Regulations (1998).

The Forest Protection Decree (1974) was amended in 1986 and changed name to the Forest Protection (Amendment) Law (1986). Within forest reserves it regulates the right to:

- Fell, extract and convey trees and wood products;
- collect, hunt and extract non-timber forest products;
- cultivate land;
- light fire;
- graze livestock;
- obstruct the channel of rivers, streams, canals and creeks;
- use property marks on standing and felled trees.

All the above activities are illegal without written consent from the competent forest authority (the Forestry Department) and offenders may be fined by a maximum of ¢ 10,000 for first time offences and ¢ 20,000 for second or subsequent offences. In 1986 ¢ 10,000 equalled DM 251.57. The 1999 value of ¢ 10,000 is down to DM 6.45, which illustrates how inflation erodes the preventive effect fines may have had. However, if offences relate to the felling or transport of trees, timber or other manufactured forest produce (eg. cane furniture), the offender shall on top of the fine be liable to pay twice the commercial value of each tree or forest produce which is the subject-matter.

An important legal aspect of the Forest Protection Decree (1974; 4) is that, with reference to the Forest Ordinance (Cap. 157) (1927;23), it does not prohibit the exercise of admitted rights established when the forest reserve was constituted. Such rights usually include collection and extraction of NTFPs by the land owners.

The Trees and Timber Decree (1974) was amended in 1994 and changed name to the Trees and Timber (Amendment) Act (1994). Among other issues this law specifies:

- The use of locality and property marks on stumps and trees extracted from on- and off-reserve concessions;
- the right of forest officers to arrest suspected offenders.

With the amendment in 1994 the standard fine for offences was raised from ¢ 5,000 to ¢ 5,000 ,000, which is still a substantial sum of money although inflation has reduced the real value of the fine by approximately 60% since 1994 (¢ 5,000 ,000 equalled DM 8,130 in 1994 and DM 3,226 in 1999). The general offences are fraudulent use of property marks (other companies' marks or expired marks), incorrect marking of stumps and logs and clearing of land for agriculture in protected areas.

The Timber Resources Management Regulations (1998; 41) have maintained the provision of summary conviction and a maximum fine of ¢ 5,000 ,000 or imprisonment for a term not exceeding 12 months, or both for obstructing any authorised person from the Forest Products Inspection Bureau or the Forestry Department in carrying out his/her duties and for possessing or transporting timber/wood without a valid conveyance certificate. For any officer who is found to be involved in such illegal activities the same penalties apply (Timber Resources Management Regulations, 1998; 41 (3)). This is a new provision which attempts to address the fact that previous legislation did not explicitly mention punishment of officials. The 1992 Constitution, however, requires that criminal offences, including administrative ones, and penalties must be established by law (Cirelli, 1997; 27).

Officials charged with management of forest resources who by any act or omission in the performance of their duties facilitate the breach of any provision of the Timber Resources Management Act (1997), or condone/connive with any person in breach of a provision of the Timber Resources Management Act (1997), may on summary conviction be sentenced to imprisonment for not less than six and not more than 12 months *without the option of a fine* (Timber Resources Management Act, 1997; 17 (1)). The same provision applies for people who harvest, haul, offer for sale or stock timber in contravention with the Act (Timber Resources Management Act, 1997; 17 (2)). This is a tightening of previous provisions, intended to effectively discourage serious offences even if fines have been eroded by inflation (Cirelli, 1997; 57).

9.1.6 Property Marks

The Trees and Timber Decree (1974) requires all companies to register a property mark at the Forestry Department Headquarters before they are allowed to fell any trees. The property mark identifies the company and it is endorsed for localities where the concessionaire proposes to operate. According to the Trees and Timber (Amendment) Act (1994), property marks must be renewed every six months (previously it was every year) and the 1994 amendment substantially increased the fee for renewal of the mark from ¢ 1,000 to ¢ 400,000 (DM 650 in 1994 and DM 258 in 1999).

Apart from generating some government revenue, the renewal of property marks serves an important regulatory function. The Chief Conservator of Forests can, at his discretion, refuse to renew or cancel a property mark, which means that the company cannot operate legally. The financially most important aspect of this power is that the Chief Conservator of Forests may refuse to renew property marks of companies that are in arrears with their royalty payments. Concessionaires who are aggrieved by a refusal to register or cancellation of their property mark may complain to the responsible Minister (now the Minister of Lands and Forestry).

However, recent history in Ghana has illustrated that such sensible legislation can only produce sensible results when it is *de facto* supported by the political leaders. During the 1980s and early 1990s, collection of royalty debts subject to property mark renewals was totally defunct. Companies in arrears with royalty payments were allowed to operate even though they had accumulated substantial royalty debts. Pressure was put on forestry officers to let timber companies operate, even though the Forestry Department had formally refused to renew their property marks. It was not uncommon that people claiming to come from "the Castle"[33] turned up at District Offices and told the District Forestry Officer to let timber companies operate without valid property marks. These people never carried their message in writing and rarely identified themselves. However, some District Forest Officers who continued to refuse the operation of companies without valid property marks were called to "the Castle" to explain their actions (Personal communication, several forestry officers).

9.1.7 Conveyance Certificates

Conveyance certificates have, at the request of the Ministry of Lands and Forestry, been instituted by the Forestry Department in 1994. The system of conveyance certificates was primarily developed in response to an accelerating illegal timber harvest (particularly off-reserves) by making the transport of such timber more difficult (FD, 1994a; 29). The legal basis for requiring conveyance certificates was and is the Forest Protection Decree (1974; 7) according to which "the burden of proof that any forest

---

33 The Castle is a commonly used notion for the Danish built fortress, Christiansborg in Accra. Since independence this has been the seat of the Ghanaian president and his nearest staff.

produce has not been taken in contravention with this Decree shall lie upon the person in whose possession it is found".

FD (1995b) stipulated that the procedures to be followed by anyone in relation to the felling of any off-reserve tree should include the following three steps:

- Trees to be felled should be inspected by the applicant (concessionaire or chainsaw operator), a representative of the Forestry Department and community representatives. After this inspection the District Forest Officer (DFO), acting upon the advice of the community representatives on farmers' objection to tree felling, will indicate which trees may be felled.
- A Forestry Department official (in most cases the DFO) must issue a permit to fell trees outside a concession area, to saw logs to boards anywhere, or must indicate and certify which trees may be felled by a concessionaire as part of his logging plan.
- A conveyance certificate is issued for all logs and boards after tree felling and processing (where appropriate). The Conveyance Certificate is needed by anyone transporting wood anywhere, whether it has come from a forest reserve or from outside a forest reserve. This certificate will be issued after an inspection by a Forestry Department agent. The same procedures will apply for on- and off-reserves.

Personal field observations in March, 1997 made it clear that all of the above three recommendations were not followed by the Ministry of Lands and Forestry, or they were not fully implemented by the Forestry Department. The first two points were only partially implemented, whereas conveyance certificates were from the outset required to transport all logs and processed wood because the Forest Products Inspection Bureau was and is not allowed to issue LMCs unless valid conveyance certificates for the logs are presented.

Today the Timber Resources Management Regulations (1998; 24) requires conveyance certificate to transport wood and logs anywhere. The rules on inspection of timber before and after felling off-reserves have been modified according to the new system of timber utilisation contracts and permits (see section 12.3).

The Forestry Department charges a fee of ¢ 10,000 per truck load (DM 6.45 in 1999). As a truck carries 10-15 m^3 logs per load, this fee equals to a maximum of ¢ 1,000/m^3 (DM 0.65/m^3 in 1999) for logs, which must be considered minimal.

9.1.8 Taungya Fees

Taungya fees are payments made to the Forestry Department by groups of farmers for demarcation of areas within forest reserves allocated for taungya. The rates are based on the costs of clearing around the perimeter of the taungya area. Previously, the taungya system was applied in an effort to re-plant degraded on-reserve areas. However, this approach has largely been discontinued because farmers tended to clear the same land over and again or clear non-degraded forest and (presumably in collusion with Forestry Department personnel) use taungya as window dressing (Personal communication, J. Francois, former Chief Conservator of Forests, 1993). Accordingly, taungya payments are insignificant (IIED, 1993; 99).

9.1.9 Permits for Major Forest Products

Major forest products are logs "sold" for special purposes such as carving, making of dugout canoes[34], etc. Fees for felling permits are charged at current royalty/stumpage fee rates in accordance with the Forest Fees Regulations (1976). Until Conveyance Certificates were required (cf. above), permits were subject to widespread abuse (IIED, 1993; 99).

9.1.10 Permits for Minor Forest Products

Fees on permits to collect minor forest products (NTFPs) within forest reserves are charged by District Forest Officers. The rates are, in principle, set on the basis of an annual review of wholesale market prices, but, in reality, there are large variances for the same product from District to District (IIED, 1993; 99).

There has been some doubt about the legal basis for the Forestry Department to charge any fees for permits to collect NTFPs. Section one in The Forest Protection Decree (1974) makes it illegal for anyone to collect and extract such goods from forest reserves without written authority from the Forestry Department. However, the Forest Protection Decree (1974; 1 (4)) states that "nothing in this section shall prohibit the exercise in a Forest Reserve by any person of any right under the Forest Ordinance (Cap. 157) for the time being is, or is treated as an admitted right".

This formulation is copied from the Forest Ordinance (Cap. 157) (1927; 23) which protects existing admitted rights within forest reserves. The burden of proof that any good has not been illegally extracted from a forest reserve lies upon the person in whose possession it is found (Forest Ordinance (Cap. 157), 1927; 31). Furthermore, the Governor (now Minister of Lands and Forestry) is empowered to prescribe fees for any purpose (Forest Ordinance (Cap. 157), 1927; 35)).

Accordingly, the Forestry Department is legally allowed to charge fees for permits to collect NTFPs unless the collector can prove that he/she acts in accordance with an admitted right to freely collect such products within a forest reserve. Such rights and the persons holding them should be described and publicly announced when individual forest reserves are gazetted. For example, when the Sukusuku Forest Reserve was gazetted in 1977, admitted rights of "the Subjects of Sefwi Wiawso Traditional Area" include free collection of NTFPs for domestic use, but collection of NTFPs in larger quantities for sale requires a permit (Land and Concessions Bulletin, 1977; 84-85). Yet, many reserves were constituted more than 50 years ago, which has made it increasingly difficult for individuals to prove their admitted rights (see also section 12.1).

9.1.11 Concession Preparation Fee

In 1992 the Forestry Department introduced a once only fee of ¢ 500,000 (DM 1,789 in 1992 and DM 322 in 1999) to cover the costs of preparing each new concession. The legal basis for this fee is found in the Forest Ordinance (Cap. 157) (1927; 35).

---

[34] Dugout canoes of considerable size (some over 15 m long) are the most common fishing vessels in Ghana. They are found in thousands along the coast where local fishermen launch them from the beach straight into the ocean. Big Wawa trees have highest preference for canoe making.

9.1.12 Registration Fees

Section two of the Forest Products Inspection Bureau Law (1985) lists the functions of the Bureau, which include registration of all enterprises engaged in timber and wood processing industries. Furthermore, subject to written approval by the Secretary (now the Minister of Lands and Forestry), the Bureau may levy such charges as it may think fit on persons in the timber industry to cover the costs of running the Bureau (Forest Products Inspection Bureau Law, 1985; 7).

Local wood product dealers are charged an annual registration fee of ¢ 5,000, non-exporting saw millers pay an annual fee of ¢ 10,000 and all exporters pay ¢ 30,000 per annum (IIED, 1993; 100).

The Ghana Chamber of Commerce, the Council of Export Promotion and the Ghana Shippers Council all receive payments for registration or licences to export (see table 9.1). However, the service provided by these institutions is practically nil. They are rather institutionalised bottlenecks which make the process of exporting unnecessarily cumbersome (Treue and Armstrong-Mensah, 1995).

9.1.13 "Sale" of LMCs

As previously described the Forest Products Inspection Bureau issues LMCs for extracted logs in Ghana. This is done at road check-points and the Bureau charges ¢ 200 per log in accordance with the Forest Products Inspection Bureau Law (1985).

9.1.14 Export Levies to Cover the Costs of Running TEDB and FPIB

The Timber Export Development Board and the Forest Products Inspection Bureau receives 1% and 2% of the FOB value, respectively, of all wood exports to cover the costs running the Bureaus. The Timber Export Development Board Law (1985; 14) and the Forest Products Inspection Bureau Law (1985; 14) provide for these charges.

9.1.15 Export Levies on Logs and Air Dried Processed Timber

The Trees and Timber (Amendment) Act (1994) introduced export levies on logs and air dried processed timber (lumber). The levy on logs varies between 10 and 30% of the FOB price, depending on species, and the levy on air dried lumber varies between 10 and 30% of the FOB price.

Yet, the levy on logs was never applied as, in 1995, the Government put a complete ban on log exports. This was done administratively by putting a halt to the issuing of log export licences. The ban is, in principle, temporary and still not included as part of the forestry legislation, but the forestry sector generally expects it to be permanent. The levies on air-dried lumber were for unexplained reasons not enforced before January, 1997 (FPIB, 1997).

10 The Economics of Timber Exploitation

Standing timber in Ghana is not sold on a free and open market. Until December, 1998 logging companies operated on a concession basis and concessions were allocated by the Government through the Ministry of Lands and Forests. With the Timber Resources Management Act (1997) and the Timber Resources Management Regulations (1998), timber rights are now supposed to be allocated according to technical, financial and social qualifications of timber companies. An expected element of financial bidding for acquiring access to standing timber must be considered abolished by the Government of Ghana (see section 12.3.2 for further details). By far most land, including forest reserves, is community owned (stool land), yet all rights to timber or trees on any land were in 1962 vested in the president in trust for the stools concerned (Concessions Act, 1962; 16, 1-3).

Rights to timber have never been allocated according to a system of competitive financial bidding. The only official payments an applicant for a timber concession had to make related to the costs of processing the application (Timber Leases and Licences (Amendment) Regulations, 1979). This principle has been maintained for the future timber utilisation contracts (Timber Resources Management Regulations, 1998). Accordingly, the right to harvest timber trees had and still has no official price. Yet, the Government has and is officially influencing domestic timber prices through:

- Timber royalties;
- stumpage fees (substituted royalties in December, 1998);
- export levies;
- forest fees other than export levies;
- log export bans.

Unofficially the actual procedures for acquiring timber concessions may, of course, also influence the market. In any case, the domestic timber market is distorted and domestic log prices do not depend on market forces alone. Rather, they grow out of a combination of:

- FOB export prices on wood products;
- export levies/bans;
- forest fees and taxes;
- logging costs;
- costs of further processing;
- informal payments.

The room for informal payments (bribes) is logically determined by the difference between actual willingness to pay for rights to standing timber and the forest fees which must be payed to acquire and exercise such rights. If informal payments do not mop up the difference between forest fees and the actual willingness of a company to pay for rights to standing timber, then the company earns windfall profits (profits that are higher than "normal" profit rates).

If companies in general cannot afford to make informal payments, forest fees must, on average, equal the industry's willingness to pay for rights to standing timber. However, this does not necessarily mean that the timber industry is efficient and that the Government is maximising its revenue from forest fees. If there are restrictions on log exports or, as the case is in Ghana, a complete ban on log exports, low forest fees in combination with the world market prices of processed products may still generate "normal" profits at low recovery rates.

Through a system of forest fees, log export bans and an officially controlled system of allocating rights to standing timber, the Government attempts to:

- Protect and develop the domestic timber industry;
- maximise earnings of foreign currency from wood exports;
- control or promote harvest of different species;
- generate revenue for the Government, local institutions and the land owning communities.

This is quite a cocktail with many conflicting interests. Apart from to the above listed rational reasons for a Government to intervene the timber market, the very nature of political market interventions also opens the possibility of (unofficially) pursuing other and less commendable objectives. To which degree this happens depends on the credibility and accountability of the political and administrative system.

This chapter will document that, until December, 1998, forest fees and taxes have been far below the average willingness of the industry to pay for standing timber. This has been of great benefit to the timber industry. Companies that operated at relatively high conversion rates could earn windfall profits (or finance high unofficial payments) and companies that were utterly inefficient and operated at very low conversion rates were able to earn "normal" profits and stay in business. The costs have been borne by the average Ghanaian taxpayer and the land owning communities because the most valuable timber species have been over-exploited (loss of natural capital). Furthermore, the official forest fees have not always covered the costs of forest management and in any case resulted in small benefits for the land owners (this is further analysed in section 11.1 and chapter 12).

10.1 Stumpage Values

The willingness to pay for rights to standing timber is generally known as the stumpage value of standing timber. In a controlled market where the willingness to pay for standing timber cannot be observed directly, the stumpage value is defined as the selling prices of logs or other wood products less the costs of production, including a "normal" profit on capital (FAO, 1983; 33-48).

The same tree may, therefore, represent different stumpage values depending on which product (lumber, plywood, paper, veneer, etc.) is used as basis for the calculation. Furthermore, different companies manufacturing the same product may, due to differences in efficiency and use of installed processing capacity, have different willingness to pay for getting access to the same standing timber.

10.2 Domestic and FOB Log Prices

The log export from Ghana was first regulated in 1979 where the export in log form of 13 species (counted by trade names) was banned. In 1988, log exports of an additional four species were banned, and in 1993 the log export of Wawa was banned. Following the log export boom in 1993-94, export of logs was completely banned in 1995. The species banned from log exports are shown in table 10.1 and appendix 11. It is seen that the export in log form of most Scarlet and many Red Star species was already banned in 1979 or 1988.

It is, therefore, not possible to make direct comparisons between recent domestic and FOB log prices for many of the most important export species. However, in 1993, during negotiations over revising the timber royalties, TEDB published a comprehensive list of FOB log prices from other West-African countries. Domestic prices of logs delivered at mills are not subject to official monitoring. However, price lists for the most important species in 1993, 1994 and 1996 have, for this study, been created through a combination of interviews and literature. The summary results are shown in table 10.1 and appendix 11 provides the details.

Table 10.1 **Domestic and FOB log prices for star groups and selected species**

SPECIES Trade name	FOB log prices 1993 (DM/m³)	1993 Domestic log prices (DM/m³)	1993 Per cent of 1993 FOB price (%)	1994 Domestic log prices (DM/m³)	1994 Per cent of 1993 FOB price (%)	1996 Domestic log prices (DM/m³)	1996 Per cent of 1993 FOB price (%)
SCARLET STARS							
Asanfina	650	176	27	106	16	103	16
Edinam	300	176	59	130	43		
Sapele	512	244	48	163	32	188	37
Utile	546	244	45	179	33	188	34
Hyedua	550			98	18		
Mahogany	360	176	49	114	32	75	21
Odum	325	217	67	130	40	160	49
Afrormosia	661	325	49	195	30		
Kyere	320	149	47	98	30	85	26
Makore	329	244	74	114	35	94	29
Wawa	230	95	41	57	25	56	24
AVERAGE (all Scarlet Stars)	**381**	**177**	**46**	**114**	**30**	**103**	**27**
RED STARS							
Kyenkyen	181	95	52	49	27	42	23
Ceiba	178	81	46	41	23	42	24
Candollei	300	176	59	114	38		
Niangon	320	190	59	98	30	94	29
Walnut	354	149	42	98	28		
Mansonia	447	149	33	57	13	42	9
Dahoma	176	95	54				
Emire	320	163	51	57	18	47	15
AVERAGE (all Red Stars)	**284**	**123**	**43**	**69**	**24**	**50**	**18**
PINK STARS							
Danta	280	81	29	49	17	42	15
Otie	180	95	53				
Sterculia	196	81	41				
Ofram	220	95	43			66	30
Avodire	280	122	44	81	29		
AVERAGE (all Pink Stars)	**188**	**84**	**45**	**57**	**30**	**54**	**29**

Sources:
FPIB Export Permits (1993) [FOB log prices, 1993]. TEDB (1993) [1993 FOB log prices from West-Africa for banned species from Ghana]. GTMO (1994) [domestic log prices, 1993] Treue and Armstrong-Mensah (1995) [domestic log prices, 1994]. Log sales contracts form 14 companies [domestic log prices 1996]

The table clearly illustrates that domestic log prices are much lower than the 1993 FOB log prices. Measured against the 1993 FOB log prices or in current DM, domestic log prices have also shown a generally falling tendency since 1993. This result is not surprising for species which are subject to the log export ban and it illustrates the substantial indirect subsidy that log export bans are to wood processing companies.

Why the domestic log prices for species that could be exported as logs are also generally much lower than the FOB log prices is less obvious. One reason might be that logs of otherwise exportable species sold at the domestic market are, in general, not of export quality (this may also to some degree be the case for banned species). Another reason could be that companies supplying logs to the domestic market (these tend to be small enterprises) have not been able to penetrate the export market due to lacking organisational skills and/or because their supply capacity per company has been too small to attract overseas log buyers.

There are also indications that the bigger integrated wood industries are suppressing the domestic log prices through cartel-like arrangements (Overseas Consultancy Service, 1995; 21). This observation is supported by the fact that a few companies are dominating the wood export (cf. section 8.7.3) which certainly increases the risk of viable cartel-like arrangements to be formed. Furthermore, the very low 1996 domestic log prices indicate that, contrary to what the resource scarcity would suggest, the log market in Ghana has been a buyer's rather than a seller's market.

Whatever the complete explanation for the difference between the FOB and domestic log prices might be, the magnitude of this difference in itself proves that the domestic log market is heavily distorted. Furthermore, it is logical that the log export bans must account for a major part of this price distortion and that the bans have greatly strengthened the bargaining power of processing companies over logging companies.

10.3 Estimated Stumpage Values based on Log Prices

On the basis of the FOB and domestic log prices given in table 10.1 it is possible to estimate the stumpage value of standing timber. A problem in this respect is that the costs of logging may, in general, only be obtained through interviews. Since such cost figures can be used politically to increase royalty rates, logging companies have an obvious interest in overstating their costs when asked by outsiders.

The costs of logging and transport to mill have been estimated by IIED (1993; app. 14), Bach, Gram and Treue (1995) and Treue and Armstrong-Mensah (1995; 87). Included in the cost estimates are:

- Planning by the logging company;
- road/bridge construction;
- costs of felling (chainsaw) and extraction to bush road (caterpillar, skidder or tractor);
- costs of transport to mill (log truck);
- depreciation costs on capital (caterpillars, graders, skidders, tractors, chainsaws and trucks);
- a 20% "normal" profit on capital.

Average figures for the different estimates of logging costs converted into current DM/m^3 are shown in table 10.2.

Table 10.2 **Available estimated logging costs in current DM/m³**

Variable costs	**Average of available cost factors**
Felling and extraction	21.13
Transport to mill	31.82
Other	5.03
Sub-total	**57.98**
Fixed costs	
Planning	0.11
Inventories	1.16
Roads, bridges, skid trails, culverts, log dumps	2.75
Monitoring/inspection	0.76
Social costs	0.20
Depreciation	1.88
Financial costs	0.13
Sub-total	**6.98**
"Normal" profit on capital	**9.21**
TOTAL COSTS	**74.17**

Sources:
IIED (1993)
Bach, Gram and Treue (1995)
Treue and Armstrong-Mensah (1995)
Ghana Timber Association (1994)

The table covers a rather significant variation among the estimated logging costs. In addition, the DM values from different sources are not adjusted for inflation, but this is considered a minor source of error in relation to the obvious uncertainties attached to this kind of calculations. Accordingly, the estimated logging cost of DM 74.17 per m^3 logs should only be taken as a rough estimate.

For comparison it is of interest to calculate the, for some species, hypothetical 1993 stumpage value on the basis of FOB log prices. This involves some extra costs in addition to the logging costs. Namely, export levies, port handling charges and bank charges. A 1% and 2% export levy on all wood exports is paid to TEDB and FPIB, respectively. Bank and port handling charges are approximately 1.75% and 5.25% of the FOB value, respectively (Treue and Armstrong-Mensah, 1995; 103).

On the basis the above mentioned data table 10.3 summarises the estimated stumpage values in current DM/m^3 for star groups and selected species.

Table 10.3 **Estimated stumpage values of selected species and star groups based on log prices and the costs of logging and export**

SPECIES Trade name	Stumpage value based on 1993 FOB log prices (DM/m³)	Stumpage value based on 1993 domestic log prices (DM/m³)	Stumpage value based on 1994 domestic log prices (DM/m³)	Stumpage value based on 1996 domestic log prices (DM/m³)
SCARLET STARS				
Asanfina	511	102	32	29
Edinam	196	102	56	
Sapele	387	170	88	114
Utile	417	170	105	114
Hyedua	421		23	
Mahogany	250	102	40	1
Odum	218	143	56	85
Afrormosia	521	251	121	
Kyere	214	75	23	10
Makore	222	170	40	20
Wawa	133	21	-17	-18
AVERAGE (All Sc. Stars)	**269**	**103**	**40**	**29**
RED STARS				
Kyenkyen	89	21	-25	-32
Ceiba	86	7	-34	-32
Akasa	241	7	-25	-32
Candollei	196	102	40	
Niangon	214	116	23	20
Walnut	244	75	23	
Mansonia	328	75	-17	-32
Dahoma	84	21		
Emire	214	88	-17	-27
AVERAGE (All Red Stars)	**181**	**49**	**-5**	**-24**
PINK STARS				
Danta	178	7	-25	-32
Otie	88	21		
Sterculia	102	7		
Ofram	124	21		-8
Avodire	178	48	7	
AVERAGE (All Pink Stars)	**95**	**10**	**-17**	**-20**

Sources:
Bach, Gram and Treue (1995). FPIB Export Permits (1993). Ghana Timber Association (1994)
GTMO (1994). IIED (1993). TEDB (1993).Treue and Armstrong-Mensah (1995)

Not surprisingly table 10.3 shows that the stumpage values based on 1993 FOB log prices are significantly higher than the stumpage values based on domestic log prices. The table also indicates that the logging business has become less and less profitable apparently leading to negative stumpage values for a number of species in 1994 and 1996. This suggests that the estimated logging costs are on the high side, but the generally falling domestic log prices also suggest that the log buyers (processing industries) have increasingly succeeded in negotiating log prices downwards. The complete ban on log exports in 1995 has naturally helped the log buyers in doing so.

Another issue which strongly suggests the logging costs in table 10.2 are higher than the actual costs is that the estimated stumpage values for many species are either negative or do not leave much or any room for timber royalties and other forest fees, which have actually been paid. The negative stumpage values for Wawa in 1994 and 1996 are particularly surprising and unrealistic since the recorded extraction of Wawa was approximately 365,000 m^3 and 437,000m^3, or 22% and 37% of the total recorded extraction in 1994 and 1996, respectively.

10.4 Estimated Stumpage Values based on FOB Lumber prices

Since saw mills can export their products it is reasonable to assume that stumpage values calculated on the basis of FOB lumber prices might be higher than the ones derived from domestic log prices. However, the lumber processing costs must also be based on information obtained from saw mills and like the case is for logging companies they too have a political interest in making their processing costs appear as high as possible.

In fact, it seems that the saw mills which have been interviewed about the costs of processing and exporting lumber have answered strategically. The results of different saw milling cost surveys are summarised in table 10.4.

Table 10.4 **Production costs for air-dried export lumber in current DM/m^3**

	Year					
	1991	1991	1993	1994	1994	Average
	A	B	C	D	E	
Cost factor						
Transport to Port	-	-	14	25	3	14
Port Handling	-	-	3	-	5	4
Fixed Costs (overheads)	105	103	67	39	49	73
Depreciation	15	23	22	23	7	18
Interest	24	38	12	9	11	19
Bank Charges	-	-	12	11	1	8
Operational Costs	164	21	298	342	458	256
TOTAL COSTS	**307**	**185**	**428**	**449**	**534**	**392**
Wawa lumber FOB price	350	350	359	426	426	382

Sources:
A: General Woods (1992; WD no. 5, App. 4) [1991 average costs for 11 mills].
B: IIED (1993; App. 14) [1991 average costs for 11 mills].
C: Bach, Gram and Treue (1995) [1993 costs for one medium sized mill].
D: Bach, Gram and Treue (1995) [1994 costs for one large mill].
E: Treue and Armstrong-Mensah (1995;96) [1994 costs for one large mill].
FPIB Export Permits (1991-94)

It is seen from table 10.4 that, except for the IIED (1993) survey (B), the costs of producing air-dried lumber are claimed to be rather high. A simple test of the data reliability is to compare it with actual FOB lumber prices of Wawa, which, during the period 1991-94, accounted for 45-54% of the lumber export by volume. The table shows that the FOB lumber price on Wawa is significantly below the claimed lumber production costs from 1993 and 1994, whereas the 1991 cost estimates allow for some profit and costs of logs. However, the 1991 domestic log price was approximately DM 25/m^3 (appendix 12). Assuming a 40% lumber recovery rate this equals to a log cost of some DM 63 per m^3 of produced lumber.

Accordingly, the costs presented in table 10.4 are likely to be a good deal higher than the actual costs of lumber production. The FOB lumber prices for Wawa in table 10.4 are even on the high side when used in this context, since they include an unknown proportion of kiln-dried lumber (the data available does not allow for a separation of kiln- and air-dried lumber).

10.4.1 General Stumpage Value based on Wawa Lumber

It makes little sense to estimate stumpage values on the basis of FOB lumber prices when the available data on lumber production costs are of doubtful quality. However, based on the 1996 domestic log price of Wawa and the 1996 FOB prices of air- and kiln-dried Wawa lumber, it is possible to estimate the maximum lumber production costs (including profit on capital). This exercise is summarised in table 10.5.

Table 10.5 **Estimated average lumber production costs based on 1996 FOB lumber and domestic log prices**

Assumed recovery rate: 40.00%

	(DM/m^3 lumber)	(DM/m^3 RWE)
Air-dried Wawa:		
FOB Lumber price	388.24	155.29
Export levy of air-dried Wawa (10% of FOB)	38.82	15.53
Levies, bank and other export charges[1]	23.29	9.32
Mill to port transport costs	25.00	10.00
Domestic log price (delivered at mill)	140.85	56.34
Maximum lumber production costs	**160.27**	**64.11**
Kiln-dried Wawa:		
FOB Lumber price	476.12	190.45
Levies, bank and other export charges[1]	28.57	11.43
Mill to port transport costs	25.00	10.00
Kiln-drying costs	149.00	59.60
Domestic log price (delivered at mill)	140.85	56.34
Maximum lumber production costs	**132.71**	**53.08**

Note:
[1] 6% of FOB (Treue and Armstrong-Mensah, 1995).
Sources:
FPIB Export Permits (1996).
Treue and Armstrong-Mensah (1995) [cost estimates].
Log sale contracts from 14 companies [1996, domestic log prices].

Judging from the estimates in table 10.5, the lumber production costs should in 1996 prices roughly be between DM 130 and 160 per m^3 lumber (between DM 53 and 64 per m^3 RWE at a 40% recovery rate). If DM 64 per m^3 RWE is taken as a proxy for the general costs of producing lumber, this may be used to estimate stumpage values based on FOB prices for air-dried lumber. This is summarised in table 10.6 for selected species and star groups.

Table 10.6 **Stumpage values in current DM/m³ for star groups and selected species derived from FOB prices for air-dried lumber, estimated costs of processing and domestic log prices**

SPECIES Trade name	1996 FOB price on air-dried lumber[1] (DM/RWE)	FPIB, TEDB, bank and other charges (6% of FOB) (DM/RWE)	Export levy on air-dried lumber (DM/RWE)	Mill to port transport costs (DM/RWE)	Estimated lumber production costs (DM/RWE)	Average domestic log prices 1994 & 96 (DM/m^3)	**Derived stumpage value (DM/m^3)**
SCARLET STARS							
Albizia	175	10		10	64	43	**47**
Asanfina	341	20		10	64	104	**142**
Edinam	212	13	21	10	64	130	**-26**
Sapele	297	18	30	10	64	175	**-0**
Utile	360	22	36	10	64	183	**45**
Hyedua	347	21	52	10	64	98	**102**
Mahogany	259	16	26	10	64	94	**49**
Odum	247	15	37	10	64	145	**-24**
Kusia	184	11		10	64	46	**53**
Makore	293	18	29	10	64	104	**68**
Wawa	155	9	16	10	64	57	**-0**
AVERAGE (all Sc. Stars)	**261**					**113**	**32**
RED STARS							
Afzelia/Papao	274	16		10	64	98	**86**
Kyenkyen	58	3		10	64	46	**-66**
Bonsamdua/Ayan	224	13		10	64	43	**93**
Candollei	233	14		10	64	114	**31**
Guarea	252	15		10	64	57	**106**
Niangon	300	18		10	64	96	**112**
Walnut	233	14		10	64	98	**47**
Mansonia	303	18		10	64	50	**161**
Emire	259	16		10	64	52	**118**
AVERAGE (all Red Stars)	**228**					**62**	**77**
PINK STARS							
Celtis/Essa	207	12		10	64	41	**80**
Potrodum	152	9		10	64	57	**11**
Danta	209	13		10	64	46	**77**
Ofram	146	9		10	64	66	**-3**
Avodire	287	17		10	64	81	**114**
AVERAGE (all Pink Stars)	**59**					**58**	**56**

Note:
[1] Assuming a 40% recovery rate.

Sources:
FPIB Export Permits (1996) [FOB lumber prices].
Trees and Timber (Amendment) Act (1994) [export levies].Treue and Armstrong-Mensah (1995) [cost estimates].Log sales contracts from 14 companies [domestic log prices, 1996].

Naturally, the derived stumpage values in table 10.6 must be on the low side since the calculation presupposes that the stumpage value of Wawa is zero. Considering that Wawa by volume is both the prime export and most harvested species, its actual stumpage value to both saw millers and loggers is hardly zero. Nevertheless, the stumpage values in table 10.6 are likely to be better estimates for the actual willingness to pay for standing timber than the 1996 stumpage values derived in table 10.3.

10.5 Conclusion on Derived Stumpage Values

The most valuable information to be drawn from the above calculations of stumpage values is that the available data on logs and lumber processing costs must be rather misleading. Technically this is a meagre result, but politically it carries some weight. It shows that, until December, 1998, the policy on timber taxation (revision of royalty rates) must have been based on false premises or, alternatively, no serious attempt has been made to base the forest fees and timber taxation on what the timber industry presumably was willing to pay for rights and access to standing timber. Neither of these explanations are very flattering.

It this connection it is worthwhile to mention that both the study of General Woods (1992) and the one of IIED (1993) had recommendations on forest fees and taxation among their primary objectives. In addition, both of these studies have been used by the Forest Resource Management Project, which has policy reform at the top of its agenda (FRMP, 1988; iv).

The revision and substantial increases of timber royalties, which the new stumpage fee system represents, shows that at least some forest policy makers in Ghana have also been aware that the willingness of the timber industry to pay for standing timber was probably higher than the various estimated stumpage values indicated, and definitely higher than the pre-1999 forest fees and taxes. That timber royalties were not increased substantially at an earlier stage probably indicates how difficult it can sometimes been to make politicians face the economic, social and environmental realities of timber exploitation.

10.6 Observed Actual Willingness to Pay for Standing Timber

One of the appraisal studies for the Forest Resource Management Project refers to an observed willingness to pay for standing timber. It was noticed that wood processing industries were paying 50% of the domestic log price for timber they extracted from other companies' concessions (World Bank, 1986, Working Paper V; 26). However, it appears that the economic and political significance of this observation was for a long time not being (allowed to be) pursued in any detail. At least it is not explicitly referred to in the final Forest Resource Management Programme document (FRMP, 1988).

Interestingly, the issue was later followed-up by the Ghanaian Government, and timber royalties that equalled the actual willingness to pay have, in fact, been enforced for a short period during the early 1990s. This was when the National Investigation Committee (NIC) looked into various "Malpractices in the Timber Sector" and, among a lot of other things, found that, through a sub-letting system, 64 out of 438 concessions were operated illegally[35] by companies other than the concession holders (NIC Report, 1991; Part A. No. II). That this was going on could, however, not have come as a great surprise as Silviconsult already pointed it out in 1985 (Silviconsult, 1985; 48).

---

35 No sub-letting of concessions was valid without the written consent of the Minister (Concessions Act, 1962; 13), and no timber lease is transferable (Timber Lease and Licence Regulations, 1962; 5,3).

Concerning the commercial value of standing timber, the NIC Report (1991) found that the sub-letting system was based on species specific prices per m^3 of logs extracted from the concessions. Not surprisingly, these prices were simply 50% of the domestic log prices and it was the original concession holder who paid the royalties.

In 1991, most of the companies which operated on sub-let concessions were allowed to continue on temporary leases because they employed a lot of people and contributed to the wood export. Some of these companies did not have their own concessions and many of the original concession holders did, in fact, not have any or sufficient logging equipment to operate their concessions. Under the temporary timber leases payments for extracting timber from sub-led concessions (50% of the domestic log prices) were continued in the place of royalties. Revenue from this special source was paid into the Government's "Special Forestry Improvement Account"[36]. Fines and various voluntary payments, resulting from the investigation of the timber industry, were also paid into this account. By 1991, a total of DM 25,551,248 had been paid into the Special Forestry Improvement Account and fines totalling DM 32,403,752 and Cedis 72,550,000 (DM 296,558 in 1991) were issued (NIC Report, 1991).

Following the NIC investigation into the timber industry it was the intention that the principles af allocating timber concessions should be completely overhauled. However, no substantial progress on this issue was made until 1998 (see section 12.3 for further details). Meanwhile, the timber industry needed raw logs and the illegal sub-letting of concessions continued. This has, during the collection of data for this study, been confirmed in numerous interviews with timber companies. The timber sector referred to the sub-letting system as the "fifty-fifty arrangement", because the concession holder got 50% of the domestic log price from the company which operates his/her concession.

The available data of observed willingness to pay for standing timber are for selected species and star groups summarised in table 10.7

By comparing the data in table 10.7 with the estimated stumpage values in table 10.3 it is seen that for Scarlet Star species, excluding Wawa, the difference between observed willingness to pay and the domestic log price based stumpage value estimates is not so big. This is, however, the case for Red and Pink Star species where table 10.7 indicates a much higher commercial value than does table 10.3.

36 It seems that the existence of this account is only semi-official. The MLF has not been willing to inform about the current balance of the account and how money from it has been used, except that only the Minister can operate the account and the money must be used for "special" forestry related purposes.

Table 10.7 **Available observed willingness to pay for standing timber in current DM/m³ for selected species and star groups**

SPECIES **Trade name**	**1986** (DM/m³)	**1991** (DM/m³)	**1993** (DM/m³)	**1994** (DM/m³)	**1996** (DM/m³)
SCARLET STARS					
Asanfina	20	25	88	53	52
Edinam	38	25	88	65	
Sapele	38	33	122	81	94
Utile	38	33	122	89	94
Hyedua	14	16		49	
Mahogany	38	25	88	57	38
Odum	24	16	108	65	80
Afrormosia		33	163	98	
Kyere	14		75	49	42
Makore	38	25	122	57	47
Wawa	17	12	47	28	28
AVERAGE (all Sc. Stars)	**25**	**24**	**92**	**59**	**55**
RED STARS					
Afzelia/Papao	14	12	88	49	
Kyenkyen		12	47	24	21
Ceiba			41	20	21
Akasa			41	24	21
Candollei	35	25	88	57	
Guarea	20	16	75	28	28
Niangon	14		95	49	47
Walnut			75	49	
Mansonia	14	25	75	28	21
Dahoma	20	12	47		
Emire	14	16	81	28	23
AVERAGE (all Red Stars)	**19**	**17**	**60**	**33**	**25**
PINK STARS					
Danta	20	16	41	24	21
Otie		12	47		
Ofram	20	16	47		33
Avodire	14	12	61	41	
AVERAGE (all Pink Stars)	**16**	**14**	**42**	**28**	**27**

Sources:
World Bank (1986, Working Paper V; 26) [1986, willingness to pay].
Temporary felling permit of Feb. 11, 1991 to MIM Timber [1991, willingness to pay].
GTMO (1994) [domestic log prices, 1993].
Treue and Armstrong-Mensah (1995) [domestic log prices, 1994].
Log sales contracts from 14 companies [domestic log prices, 1994].

10.6.1 Validity of the Observed Willingness to Pay for Standing Timber

Before applying 50% of the domestic log price as a measure for the "real" commercial value of standing timber it is reasonable to consider the justification of doing so. The obvious strong point of using an actual market price is that it directly reflects the timber industry's willingness to pay for standing timber, since the concession holders have paid the timber royalties and other taxes. Accordingly, this value measure combines all industrial uses of timber, and it is not influenced by questionable estimates of the cost and profitability of logging and further processing.

A weakness is that an unknown volume, but clearly not all standing timber, has been "sold" at 50% of the domestic log price. The observed price could, therefore, reflect the marginal rather than the average willingness to pay for standing timber and may, if used on the total recorded timber extraction, over-estimate the real commercial value of the resource. In other words, companies which have operated their own as well as others' concessions and bought some logs at the domestic market may, on average, have paid less than 50% of the domestic log price per m^3 of log intake. Yet, the actual price of acquiring a timber concession has hardly been limited to the costs of processing the application.

It should also be considered that, since 1979, the domestic log prices have been heavily distorted through log export bans. After the complete log export ban in 1995 domestic log prices may even have been further suppressed through cartel-like arrangements between the big processing companies. Accordingly, the domestic log prices have probably been below the biggest and most efficient mills' actual willingness to pay.

In conclusion, 50% of the domestic log price is considered the most reliable current estimate of the "real" commercial value of standing timber. Theoretically, this may over-estimate the commercial value of all timber extracted in the past. Yet, the falling trend in domestic log prices since 1993 and indications of cartel-like arrangements among big log buyers suggest that the actual willingness to pay for logs could be higher than what the domestic log prices imply.

Furthermore, from a national economic, resource conservation and industrial development point of view, domestic log prices that are far below the FOB log prices cannot be desirable in the longer term, because they must reflect either an inefficient domestic wood industry or an inefficient domestic log market or (most likely) a combination of the two. Accordingly, in the present situation where domestic log prices appear to be not even half the FOB log prices (cf. table 10.1) it would, from a political point of view, not be rational to accept 50% of domestic log prices as a long-term maximum commercial value of standing timber. Rather, it ought to be regarded as an interim minimum value that, through appropriate political interventions, should be gradually increased.

10.7 Willingness to Pay Compared with Forest Taxes and Fees

Based on the above considerations about the observed willingness to pay for standing timber it is possible to estimate the financial room for informal payments and/or windfall profits in the timber sector. The forest fees and taxes to be compared with the observed willingness to pay include, in order of magnitude (see appendix 12 for details):

- Timber royalties/stumpage fees;
- concession rent/contract area rent;
- property mark renewal;
- conveyance certificates;
- silvicultural fees.

The difference between the observed willingness to pay and official forest fees and taxes is, for star groups, illustrated in figures 10.1, 10.2 and 10.3 (the data behind the figures is presented in appendix 12). It is seen that for 1993 this difference is exceptionally high due to the low level of forest fees and the apparently very high domestic log prices. Otherwise, during the period 1986-98 the difference roughly varies between DM 10 and 40 per m^3, depending on year and species group. Apparently, the new stumpage fees have closed the gap between forest fees and the willingness to pay for Scarlet and Red Star species[37]. This shows a remarkable change in the political commitment to tax timber exploitation. The data available for Pink Star species is somewhat meagre, so the gap between forest fees and the willingness to pay in 1999 may not be of the magnitude indicated by figure 10.3. It does, however, make some sense if forest taxes on Pink Star species are kept at a relatively low level, as this would promote the utilisation of these species.

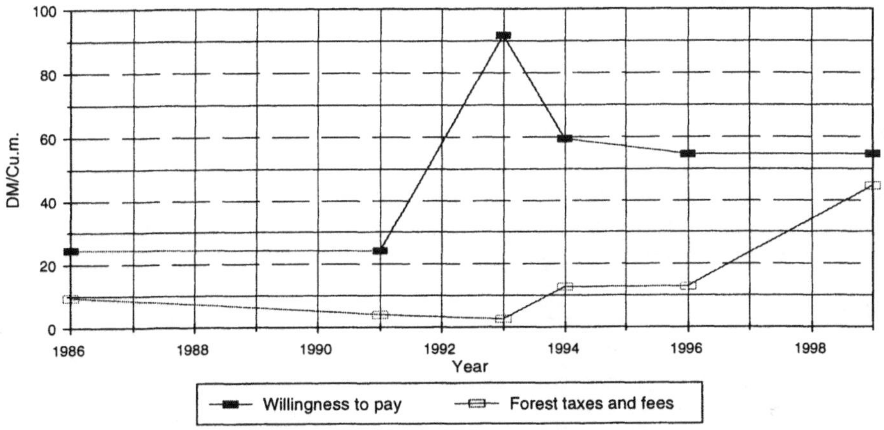

Figure 10.1 **Observed willingness to pay for standing timber compared with official forest fees and taxes for Scarlet Stars (averages)**

[37] The observed 1996 willingness to pay for standing timber has, due to data constraints, been used as a proxy for 1997-99.

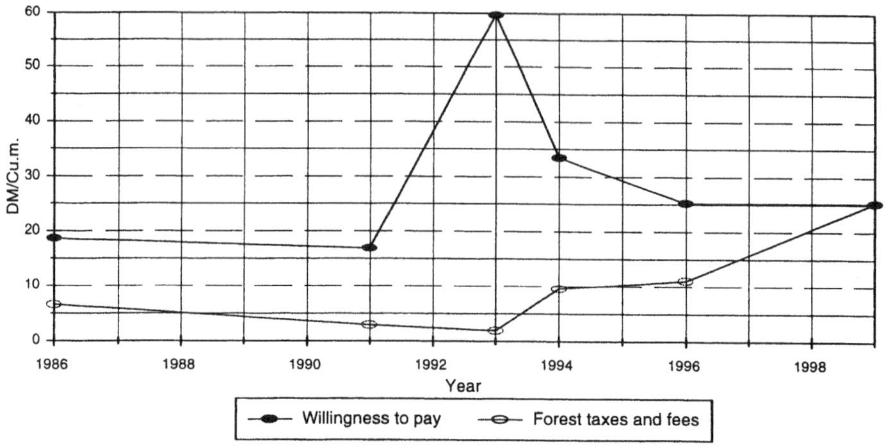

Figure 10.2 **Observed willingness to pay for standing timber compared with official forest fees and taxes for Red Stars (averages)**

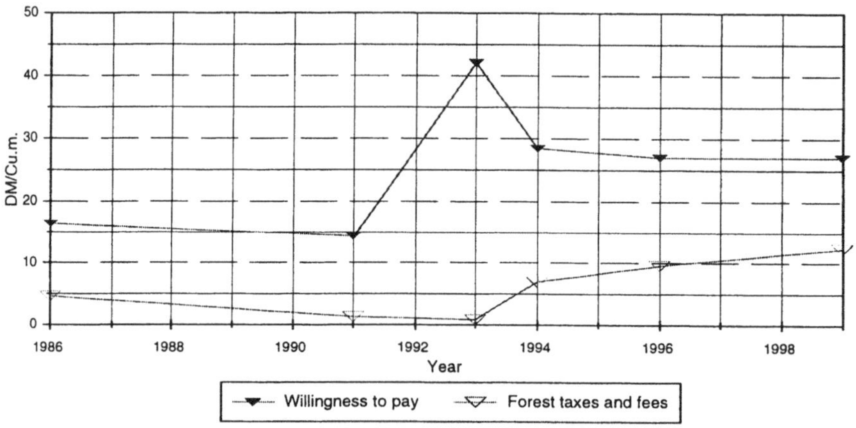

Figure 10.3 **Observed willingness to pay for standing timber compared with official forest fees and taxes for Pink Stars (averages)**

Sources to figures 10.1, 10.2 and 10.3:
World Bank (1986, Working Paper V; 26) [1986, willingness to pay]
Temporary felling permit of Feb. 11, 1991 to MIM Timber [1991, willingness to pay]
GTMO (1994) [domestic log prices, 1993]
Treue and Armstrong-Mensah (1995) [domestic log prices, 1994]
Log sales contracts from 14 companies [domestic log prices, 1994]
Forest Fees (Amendment) Regulations (1986)
Forest Fees (Amendment) Regulations (1989)
Forest Fees (Amendment) Regulations (1993)
Timber Resources Management Regulations (1998)
Trees and Timber Decree (1974)
Trees and Timber (Amendment) Law (1983)
Trees and Timber (Amendment) Act (1994)

10.7.1 Value of the Recorded Extraction

Data on observed willingness to pay for standing timber have only been available for the years 1986, 1991, 1993, 1994 and 1996. Yet, this is enough to demonstrate the substantial difference in value of the recorded timber extraction when calculated in terms of willingness to pay and in terms of official forest fees and taxes. These values are summarised in table 10.8.

Table 10.8 **"Value" in current DM of the total recorded timber extraction in 1986, 1991, 1993, 1994 and 1996**

	Year				
	1986	**1991**	**1993**	**1994**	**1996**
SCARLET STARS					
Value as observed willingness to pay[1]	14,810,629	15,134,698	59,717,260	35,681,984	27,428,021
Value in terms of forest fees and taxes[2]	4,928,334	2,855,675	1,614,995	8,157,899	8,274,483
Difference (forest revenue foregone)	**9,882,295**	**12,279,023**	**58,102,264**	**27,524,085**	**19,153,538**
RED STARS					
Value as observed willingness to pay[1]	1,389,928	1,470,208	31,936,163	13,225,276	5,920,965
Value in terms of forest fees and taxes[2]	932,860	471,945	582,141	4,711,606	3,317,470
Difference (forest revenue foregone)	**457,068**	**998,263**	**31,354,022**	**8,513,670**	**2,603,495**
PINK STARS					
Value as observed willingness to pay[1]	724,780	798,705	7,963,671	751,549	1,579,155
Value in terms of forest fees and taxes[2]	349,216	99,460	176,724	2,154,162	1,012,156
Difference (forest revenue foregone)	**375,564**	**699,246**	**7,786,948**	**-1,402,613**	**566,999**
TOTAL					
Value as observed willingness to pay[1]	16,925,337	17,403,612	99,617,094	49,658,809	34,928,141
Value in terms of forest fees and taxes[2]	6,210,410	3,427,080	2,373,860	15,023,667	12,604,109
Difference (forest revenue foregone)	**10,714,927**	**13,976,532**	**97,243,234**	**34,635,141**	**22,324,031**

Notes:
[1]Only the volume, and hence the value, of species for which the domestic log prices have been available is included.
[2]The volume, and hence forest fees and taxes, of the total recorded extraction is included.

Sources:
FPIB LMC Totals (1986-96)
World Bank (1986, Working Paper V; 26) [1986, willingness to pay]
Temporary felling permit to MIM Timber (1991) [1991, willingness to pay]
GTMO (1994) [domestic log prices, 1993]
Treue and Armstrong-Mensah (1995) [domestic log prices, 1994]
Log sales contracts from 14 companies [domestic log prices, 1994]
Forest Fees (Amendment) Regulations (1986)
Forest Fees (Amendment) Regulations (1989)
Forest Fees (Amendment) Regulations (1993)
Trees and Timber Decree (1974)
Trees and Timber (Amendment) Law (1983)
Trees and Timber (Amendment) Act (1994)

The figures in table 10.8 are quite remarkable. It is seen that in 1996 alone the forest taxation revenue foregone was in the order of DM 22 million. Assuming collection rates of 100%, the official forest fees and taxes generally appear to have captured only 20-40% of the willingness to pay for standing timber. Since table 10.8 counts the forest fees and taxes value for the total annual extractions but only includes the willingness to pay values for species where domestic log prices have been available, the taxation revenue foregone is probably not over-estimated. Accordingly, it is likely that, during the period 1993-96, an annual amount not less than DM 20 million has been available for (i) informal payments, (ii) windfall profits and (iii) keeping inefficient wood industries from going bankrupt.

The huge difference in willingness to pay value, and forest fees and taxes value of the 1993 recorded extraction, is partly caused by the unusually high extraction rate that year (the log export boom) and the low DM value of forest fees and taxes. In addition, domestic log prices for almost all extracted species were available for 1993. Yet, the apparently high, and perhaps more questionable, domestic log prices in 1993 account for most of the difference in value. It is, therefore, doubtful whether the 1993 forest taxation revenue foregone was actually in the order of DM 97 million. On the other hand, it seems reasonable to assume that the revenue foregone was especially high this year.

10.8 Commercial Value of Estimated Annual Allowable Cuts

For obvious reasons it is of interest to estimate the commercial value of on-reserve AACs, since this would indicate the size of the long-term available revenue from on-reserve timber extraction. To this end it is necessary to operate at least with the following scenarios:

- Value of the AAC under large-grained protection only;
- value of the AAC under large-grained protection only, or of the total on- and off-reserve recorded extraction for species which, due to the market situation, are not exploited to the limit of their AACs;
- value of the AAC under large- and fine-grained protection;
- value of the AAC under large- and fine-grained protection, or of the total on- and off-reserve recorded extraction for species which, due to the market situation, are not exploited to the limit of their AACs.

These scenarios are summarised in tables 10.9, 10.10, 10.11 and 10.12. It should be noted that the scenarios operate with average 1994 and 1996 values for willingness to pay for standing timber, which makes the number of species for which an observed willingness to pay is available higher than if only 1996 values were used. Separate calculations are also made that apply minimum star group values for species where no 1994 or 1996 data on willingness to pay for standing timber have been available. To illustrate the effect of the new stumpage fees, both the 1996 and the 1999 value of AACs in terms of official forest fees and taxes are presented in the tables.

Table 10.9 Values in current DM of the AAC under large-grained protection only

SPECIES GROUP	Annual allowable cut (m³)	Value in terms of 1996 royalties and other forest fees (DM)	Value in terms of 1999 stumpage and other forest fees (DM)	Value in terms of willingness to pay for standing timber[1] (DM)	Value in terms of willingness to pay for standing timber[2] (DM)
Scarlet Stars	132,437	1,453,048	3,539,550	4,864,971	5,019,665
Red Stars	224,195	2,298,855	4,267,531	4,182,127	5,342,718
Pink Stars	534,877	5,190,997	6,391,416	6,123,598	12,017,178
TOTAL	**891,509**	**8,942,901**	**14,198,497**	**15,170,697**	**22,379,561**

Notes:
[1] Average of available 1994 and 1996 figures in current DM for willingness to pay for standing timber.
[2] Average of available 1994 and 1996 figures in current DM of willingness to pay for standing timber plus minimum star group values for species where no direct value observations have been available.

Table 10.10 **Values in current DM of the AAC under large-grained protection only, or of the total 1996 extraction of species which have not been exploited to the limit of their AACs**

SPECIES GROUP	Large-grained AAC or total recorded 1996 extraction when this is below the AAC (m³)	Value in terms of 1996 royalties and other forest fees (DM)	Value in terms of 1999 stumpage and other forest fees (DM)	Value in terms of willingness to pay for standing timber[1] (DM)	Value in terms of willingness to pay for standing timber[2] (DM)
Scarlet Stars	132,129	1,448,147	3,519,933	4,849,942	5,004,636
Red Stars	195,520	2,005,471	3,774,239	3,864,431	4,716,934
Pink Stars	102,348	1,012,156	1,387,004	1,815,119	2,634,834
TOTAL	**429,996**	**4,465,775**	**8,681,176**	**10,529,491**	**12,356,404**

Notes:
[1] Average of available 1994 and 1996 figures in current DM for willingness to pay for standing timber.
[2] Average of available 1994 and 1996 figures in current DM of willingness to pay for standing timber plus minimum star group values for species where no direct value observations have been available.

Table 10.11 **Values in current DM of the AAC under large- and fine-grained protection**

SPECIES GROUP	Annual allowable cut (m³)	Value in terms of 1996 royalties and other forest fees (DM)	Value in terms of 1999 stumpage and other forest fees (DM)	Value in terms of willingness to pay for standing timber[1] (DM)	Value in terms of willingness to pay for standing timber[2] (DM)
Scarlet Stars	120,817	1,324,444	3,209,753	4,465,937	4,587,710
Red Stars	201,353	2,062,917	3,810,184	3,736,464	4,770,239
Pink Stars	479,518	4,658,142	5,754,615	5,611,029	10,863,732
TOTAL	**801,688**	**8,045,503**	**12,774,552**	**13,813,430**	**20,221,681**

Notes:
[1]Average of available 1994 and 1996 figures in current DM for willingness to pay for standing timber.
[2]Average of available 1994 and 1996 figures in current DM of willingness to pay for standing timber plus minimum star group values for species where no direct value observations have been available.

Table 10.12 **Values in current DM of the AAC under large- and fine-grained protection, or of the total 1996 extraction of species which have not been exploited to the limit of their AACs**

SPECIES GROUP	Large- and fine-grained AAC or the total 1996 extraction when this is below the AAC (m³)	Value in terms of 1996 royalties and other forest fees (DM)	Value in terms of 1999 stumpage and other forest fees (DM)	Value in terms of willingness to pay for standing timber[1] (DM)	Value in terms of willingness to pay for standing timber[2] (DM)
Scarlet Stars	120,800	1,324,173	3,208,670	4,465,108	4,586,881
Red Stars	180,233	1,845,822	3,430,615	3,515,777	4,309,383
Pink Stars	102,348	1,012,156	1,387,004	1,815,119	2,716,115
TOTAL	**403,381**	**4,182,151**	**8,026,289**	**9,796,003**	**11,612,378**

Notes:
[1]Average of available 1994 and 1996 figures in current DM for willingness to pay for standing timber.
[2]Average of available 1994 and 1996 figures in current DM of willingness to pay for standing timber plus minimum star group values for species where no direct value observations have been available.

Sources to tables 10.9, 10.10, 10.11 and 10.12:
FIMP (1995b) [AAC estimates].
Treue and Armstrong-Mensah (1995) [1994 domestic log prices].
Log sales contracts from 14 companies [1996 domestic log prices].
Forest Fee (Amendment) Regulations (1993).
Trees and Timber (Amendment) Act (1994)
Timber Resources Management Regulations (1998)

The scenarios in tables 10.9 and 10.11 are only theoretical in the sense that the total recorded 1996 extraction of especially Pink Star species was less than 20% the AAC. Accordingly, the value calculations in these two scenarios, where the entire AAC is assumed utilised, should be treated with caution. Yet, if the on-reserve harvest of Scarlet and Red Star species could be reduced to their AACs, the demand for, and hence, the revenue from Pink Star species would most probably increase. In 1996 prices, the long-term sustained forest tax revenue from forest reserves might, therefore, be between DM 14 and 22 million.

The scenarios in tables 10.10 and 10.12 may, on the other hand, be regarded as realistic since they reflect the current market situation. The immediately available annual tax revenue from sustained yield timber harvest on-reserves should, therefore, be between DM 10 and 12 million in 1996 prices. Measured by official 1999 forest fees and taxes, the annual revenue would be around DM 8 million.

Going back to table 10.8 it is seen that official forest fees and taxes due from the, especially for Scarlet Star species, unsustainable 1996 extraction was approximately DM 12.6 million. The scenarios in tables 10.10 and 10.12 suggest that a shift to extraction levels within the species specific AACs *and* a taxation of the timber extraction according to the industry's willingness to pay for standing timber would generate a revenue of approximately the same size from forest reserves only. The 1999 forest fees and taxes could initially generate around DM 8 million from forest reserves. It should, however, be possible to gradually increase the on-reserve revenue through increasing the stumpage fees. In addition, an increased extraction of Pink Star species is expectable.

10.9 Revenue from Timber Off-Reserves

Sustainable or sustained yield management of timber resources off-reserves can only be achieved if it is in the financial interest of the land owners (this issue is further discussed in chapter 12). Such incentives might be possible to establish, but even if this is done it is likely that a sustained off-reserve timber yield will be much lower than the present extraction. Accordingly, the long-term sustainable revenue from off-reserve timber resources is practically impossible to estimate.

The off-reserve lifetime scenarios for star groups presented in section 7.6 may, however, be used to indicate the size of the potential revenue from off-reserve timber resources. If the 1996 off-reserve extraction level continues, it was estimated that (i) Scarlet Star species will be logged out within the coming 15-39 years, (ii) Red Star species will be logged out within the coming 125-126 years and (iii) Pink Star species will be logged out within the coming 315-436 years.

By combining the 1996 off-reserve extraction with the 1996 DM values of willingness to pay and forest taxes and fees, rough estimates of the off-reserve revenues may be calculated. This is presented in table 10.13.

Since the off-reserve inventory data does not provide species specific data, the value figures in table 10.13 should only be regarded as rough estimates. Nevertheless, it is obvious that the off-reserve timber resource represents a substantial revenue potential for many years to come. In terms of the 1999 official forest taxes and fees, the annual revenue could be sustained at more than DM 19 million for at least another 15 years, whereas the actual revenue potential in this period is likely to be between DM 25 and 26 million annually.

Table 10.13 **Estimated annual revenue potentials from off-reserves if the timber extraction continues at the 1996 level**

SPECIES GROUP	1996 recorded off-reserve timber extraction (m³)	Value in terms of 1999 stumpage and other forest fees (DM)	Value in terms of willingness to pay for standing timber[1] (DM)	Value in terms of willingness to pay for standing timber[2] (DM)	Expected resource lifetime if the 1996 extraction level continues (Years)
Scarlet Stars	515,349	13,594,410	19,515,136	19,648,465	15-39
Red Stars	240,633	5,064,730	5,167,362	5,168,475	125-160
Pink Stars	65,068	866,063	1,162,178	1,735,836	315-436
TOTAL	**821,050**	**19,525,203**	**25,844,676**	**26,552,777**	

Notes:
[1]Average of available 1994 and 1996 figures in current DM for willingness to pay for standing timber.
[2]Average of available 1994 and 1996 figures in current DM of willingness to pay for standing timber plus minimum star group values for species where no direct value observations have been available.
Sources:
FIMP (1997b) [estimates of standing timber resources off-reserves]. FPIB LMC Totals (1996) [1996 extraction of timber off-reserves]. Log sales contracts from 14 companies [1996 domestic log prices]
Forest Fee (Amendment) Regulations (1993). Timber Resources Management Regulations (1998)

This is big money, but the off-reserve revenue potential is likely to disappear very soon unless the annual off-reserve timber exploitation is reduced. If the political will is there, it should not be impossible to combine the substantial revenue potential with incentives for land owning communities to grow and protect off-reserve timber resources. In theory, it just a matter of adjusting the income distribution associated with timber harvest. This would help to secure the resource base, which is in the long-term interest for the timber industry (and the national economy) and it would also improve the odds for reducing the on-reserve harvest to the AAC of currently over-exploited species. However, the prognosis for a sustained commercial timber production off-reserves is not particularly good. This is further discussed in section 12.3.7.

10.10 Conclusion on the Economics of Timber Exploitation

Clearly, three major politically determined issues have created problematic economic incentives in the forest sector. These are:

- Log export bans;
- official forest fees and taxes that, until 1999, were far below and probably not even half of the industry's actual willingness to pay for standing timber;
- allocation of rights to standing timber.

The incentives created have kept domestic log prices artificially low and opened the opportunity for log processing companies to:

- Earn windfall profits;

- cover high informal costs;
- operate profitably at very low recovery rates.

The low domestic log prices obviously produce a false signal of all timber species being abundant. Had standing timber been allocated according market forces only, the actual scarcity of especially Scarlet and Red Star species would have led to higher domestic log prices, irrespective of log export bans. Accordingly, the financial incentives have, until 1999, strongly favoured the documented over-exploitation of Scarlet and Red Star species which the wood export based foreign exchange earnings depend so heavily on. Furthermore, it appears that the low forest fees and taxes have, at least during the period 1993-96, resulted in an annual forest revenue foregone of some DM 20 million. Before 1993, the revenue foregone might have been in the order of DM 10 million.

Since practically all Scarlet and most Red Star species are de facto over-exploited on-reserves, harvest control mechanisms other than financial incentives (yield allocation and harvest control) have not been working properly. This implies that the Forestry Department field staff has either been de-motivated to control logging operations or actively motivated to *refrain* from controlling logging operations. The considerable financial room for informal payments, that some companies must have had, suggests the latter possibility cannot be ruled out.

The new legislation on forest revenues (especially stumpage fees) and punishments for offences represents a bold attempt to address the former unfortunate combination of low motivation to control logging operations and funds being available for informal payments (Timber Resources Management Act, 1997 and Timber Resources Management Regulations, 1998). This is a major improvement of the incentives for promoting sustainable timber utilisation which leaves hope for the future.

Off-reserve timber resources contribute significantly to the annual harvest, and hence, to the revenue earned as well as foregone. In 1996 alone, data indicate that approximately DM 17 million of potential forest revenue was foregone. The current level of exploitation off-reserves can, however, not continue indefinitely, and contrary to forest reserves there is no *de jure* prescriptions for a sustained yield management of timber resources off-reserves. Yet, the substantial value of this resource, if measured in terms of the industry's willingness to pay for standing timber or even in terms of 1999 forest fees and taxes, suggests that appropriate political interventions could create sustainable financial incentives for protecting and managing off-reserve timber trees. It should, in other words, be possible to conserve the off-reserve timber resource by getting the distribution of wealth from timber exploitation right, now that the controlled "price" on standing timber has been adjusted to almost equal the market value.

The distribution of costs and benefits from timber exploitation is, in addition to the distorted domestic log market, another key issue affecting the management of timber resources. Particularly off-reserves this is very important because, in practice, financial incentives alone determine the management regime. On-reserves the silvicultural management prescriptions and the control function of the Forestry Department has some influence on the timber harvest, although the pre-1999 financial incentives and the distribution of costs and benefits associated with timber exploitation appears to have de-stabilised rather than supported this function. Accordingly, it is necessary to consider the distributional effects and incentive structure created by the past and newly established system of rights and responsibilities in relation to commercial timber exploitation.

11 Instruments to Regulate the Timber Harvest

The timber harvest may be regulated through demand-side and supply-side measures. Both are politically determined, but they work in different ways and have different strengths and weaknesses. In practice, none of the two can stand alone in a politically regulated timber market as the case is for Ghana. How well demand-side measures can function depends to a large degree on which supply-side measures are in place and vice versa.

Demand-side measures like forest taxes/fees, export levies and export bans have three main objectives:

- To regulate the market demand for standing timber;
- to generate (government) revenue from the utilisation of standing timber;
- to develop the domestic wood industry.

The strength of demand-side measures is that they are fairly cheap to implement and, in most cases, not difficult and costly to monitor. Yet, from a harvest control point of view, their great weakness is that they work indirectly, ie. through the market which is even distorted by the very same measures. Another crucial aspect of demand-side measures is their distributional effects. Politically, it is obviously very important how the costs and benefits created by the package of demand-side measures are distributed among stakeholders. At least five different stakeholder groups should be considered in this context. These are:

- The land-owning communities;
- the average Ghanaian citizen/tax payer;
- the log processing industries;
- the logging companies which are not part of an integrated industry;
- people in or associated with the official forestry/timber administration.

Supply-side measures have one major objective. To keep the timber harvest within socially, environmentally and silviculturally acceptable limits. That is, to ensure that narrow financial interests in utilising standing timber do not lead to destruction or degradation of the forest resource and the values it represents to society at large. The strength of supply-side measures is that, in theory, they can be implemented directly and irrespective of the market forces. The weakness is, of course, that the unofficial side of the market can, in practice, completely undermine the effectiveness of supply-side measures.

11.1 Assessment of Employed Demand-Side Measures

The criteria which employed demand-side measures should be assessed against are their ability to:

- Generate sufficient revenue to manage the forest resource;
- regulate the timber harvest in accordance with the relative scarcity of species;
- promote further and more efficient processing in the wood industry;
- promote an equitable distribution of costs and benefits associated with the timber exploitation.

The previous chapters have already indicated that, in general, the employed package of demand-side measures have until 1999 failed on almost all accounts. General Woods (1992; WD no.3; 26) documents that, during the period 1985-91, timber royalties collected covered between 41% and 95% (on average 69%) of the expenditures from the Forest Improvement Fund which is used to finance the recurrent costs of the Forestry Department and the Forestry School in Sunyani. For the period 1993-96, table 11.1 summarises the available data on overall costs of and revenue generated from forest management and timber extraction. In addition, the table indicates the potential revenues in terms of observed willingness to pay and the new 1999 forest fees and taxes.

Table 11.1 **Available data on cost of and revenue from forest management and timber extraction 1993-96 (current DM)**

	Year			
	1993	**1994**	**1995**	**1996**
Actual Revenues				
Royalties on-reserves	3,070,332	3,111,506	3,789,653	3,368,382
Royalties off-reserves	248,289	868,288	6,076,846	4,266,810
Sub-total royalties	**3,318,621**	**3,979,793**	**9,866,499**	**7,635,192**
Concession rent	67,335	508,190	3,765,678	2,680,279
Property mark renewal	55,768	36,881	99,100	105,380
Silvicultural fees	12,958	25,444	44,627	40,933
Sub-total timber related taxes and fees	**3,454,681**	**4,550,309**	**13,775,904**	**10,461,785**
Other non-timber related fees	1,207,584	754,214	3,253,604	2,489,820
TOTAL	**4,662,265**	**5,304,524**	**17,029,508**	**12,951,605**
On- and off-reserve royalties due	2,338,705	14,995,724	8,658,942	6,110,387
Actual Expenditures				
Payroll	7,128,617	NA	4,197,436	4,694,872
General Expenditure	NA	NA	210,256	234,615
Repair and maintenance of vehicles	NA	NA	948,718	948,718
Depreciation/replacement of equipment	NA	NA	NA	1,288,462
Transitional costs	NA	NA	NA	335,897
Sub-total other than payroll expenditures	**96,862**		**1,158,974**	**2,807,692**
TOTAL	**7,225,479**	NA	**5,356,410**	**7,502,564**
Potential Revenues				
Observed willingness to pay:				
Recorded extraction	99,617,094	49,658,809	NA	**34,928,141**
AAC under large-grained protection only				**22,379,561**
AAC under large-gr. protection only or 1996 extraction				**12,356,404**
1999 Stumpage and other forest fees:				
AAC under large-grained protection only				**14,198,497**
AAC under large-gr. protection only or 1996 extraction				**8,681,176**

Sources: Grut (1996; 5) [revenue collected 1993-96]. FD (1994b) [1993 expenditures]. Touche Ross (1995; Vol II; 35-39) [estimated expenditures 1995-96]. FPIB LMC Totals (1993-96). GTMO (1994) [domestic log prices 1993]. Treue and Armstrong-Mensah (1995) [domestic log prices 1994]. Log sales contracts from 14 companies [domestic log prices 1996]. Forest Fees (Amendment) Regulations (1993). Timber Resources Management Regulations (1998)

It should be noted that the 1995-96 expenditure figures are rough estimates for a proposed new semi-autonomous Forestry Service (to replace the current Forestry Department) (Touche Ross, 1995; Vol II; 35). However, these estimates appear to be taken almost directly out of the Forestry Department's annual accounts allowing for a 30% salary increase in real terms for professional staffs and a total staff number of 4,394 which is slightly below the 1995 staff number of 4,648. Accordingly, the 1995-96 expenditure estimates are not likely to deviate much from the actual expenditures.

From the table it is seen that:

- The revenue collected in 1993 did only cover approximately 64% of actual expenditures;
- the revenue collected in 1995-96 is likely to have exceeded actual expenditures by DM 4-11 million;
- in 1993, approximately DM 1.0 million of old royalty debts was collected;
- in 1994, a royalty debt of approximately DM 11.0 million was built up;
- in 1995, approximately DM 1.2 million of old royalty debts was collected;
- in 1996, approximately DM 1.5 million of old royalty debts was collected;
- if the cost of implementing sustainable/sustained yield forest management is around DM 7.5 million per year, this may be financed from forest reserves alone under the 1999 taxation;
- if fees and taxes were raised to the industry's willingness to pay, the cost of forest management could be covered from forest reserves alone leaving a surplus of approximately DM 4.5 million per year.

It was an old trick of the timber industry to reduce royalty payments in real terms by delaying payments, because royalty debts were not adjusted for inflation under the pre-1999 legislation (Personal Communication, 1995; Francis Bilson-Ogoe and George Nü Mbroh Ashis, Concessions Unit, FD). Judging from the data in table 11.1 this seems to have been going on during the log export boom in 1994 (in 1994 and 1996 the Cedi/DM exchange rate was 615 and 1,131 respectively). Clearly, the Forestry Department/Ministry of Lands and Forestry has foregone substantial tax revenues by allowing this practice. This loophole has now been closed by the Timber Resources Management Regulations (1998; 25) which stipulates bank rates to be charged on royalty debts older than 30 days.

Summing up, it appears that the exploitation of timber has resulted in a net loss for the average Ghanaian taxpayer/citizen during the period 1986-93. Possibly this was also the case for 1994, but since 1995 forest taxes and fees collected have probably exceeded the costs of forest management by several million DM. Whether this has resulted in a net gain for the average tax payer is, however, not certain. It has not been possible to verify how much of the collected revenue has, in fact, been used to finance the Forestry Department and how much has been plowed back to the local communities. If the rules have actually been followed, then the average taxpayer has suffered a considerable net loss (see sections 12.1 and 12.2.2).

Substantial revenue has been foregone by allowing inflation to erode old royalty debts and by the fact that forest fees and taxes were not charged according to the industry's willingness to pay for standing timber. In 1996 the potential revenue from the recorded (non-sustainable) extraction could have been approximately DM 35 million or almost 2.7 times the revenue actually collected. Harvest according to the AAC under large-grained protection only, or according to the 1996 extraction level when this is below the AAC would, from forest reserves alone, have generated almost the same revenue as the actual collected amount from both on- and off-reserves *if* fees and taxes had equalled the actual willingness to pay.

The most important conclusion to be drawn from the above analysis is, however, that forest fees and taxes even from forest reserves alone could easily finance the costs of sustainable or sustained yield management of forest reserves. This conclusion rests on fairly rough cost and, to a lesser degree, potential revenue estimates, but the gap of more that DM 4 million between the two sends a clear message. *It can be done.*

The new stumpage fee system would initially only generate around DM 8.6 million, leaving a surplus of approximately DM 1.1 million from forest reserves if the harvest is held within the AAC. While the potential revenue appears to be somewhat higher, the (long awaited) abolishment of the low timber royalties *does* mark an important shift in policy towards securing that the costs of wise forest management may actually be covered by forest revenues. An outstanding issue in this respect is, however, whether it would be politically acceptable if almost 90% of the revenue from forest reserves should go to fund the Forestry Department (or a future Forestry Service), leaving only 10% to local communities (see section 12.1.2 for further analyses).

11.1.1 Timber Royalties and Stumpage Fees

Among the forest taxes and fees, timber royalties played and stumpage fees play the most important role in relation to harvest control because they differ among species. In terms of generating revenue, timber royalties and especially the new stumpage fees also stand out as a main source.

In section 9.1.1 it was documented that:

- Between revisions of the rates, the real value of timber royalties has been heavily eroded by inflation;
- measured in current DM and as per cent of FOB log and lumber prices, the royalty rates have fluctuated at a very low level;
- the difference in royalty rates between Scarlet, Red and Pink Star species have been rather small;
- the new stumpage fees are significantly higher than the old royalties and they differ substantially among Scarlet, Red and Pink Star species.

From the analysis in chapter 10 it is clear that the timber royalties have, at least since 1986 and probably also before that, stayed far below the industry's actual willingness to pay for standing timber. In section 7.4.1 it was documented that the extraction of Scarlet and Red Star species has remained at a high and fairly constant level during the period 1986-96. It may, therefore, be concluded that:

- Timber royalties have not created incentives for the industry to reduce the harvest of over-exploited species and/or to increase the harvest of "under-utilised" species;
- considered in isolation, timber royalties have not generated the forest revenue they could.

Timber royalties and other pre-1999 forest taxes and fees have clearly not compensated much for the suppressing effect that log export bans have had on domestic log prices. In combination, the log export bans and low timber royalties have probably promoted the establishment of new log processing domestic companies, but they have definitely not created strong incentives for making these very efficient. Compared with a hypothetical log export, Ghana's export of processed wood represented an export revenue foregone of some DM 40 million in 1996 alone. Furthermore, it seems very likely that the combination of low royalties and the policy of banning log exports of an increasing number of, and eventually all, species has contributed to create the apparent over-capacity of the processing industry.

The distributional effects of timber royalties have not been particularly positive either. Since the timber industry could have paid more than it did (at least officially) for standing timber, the pre-1999 level of forest taxes have worked as a net transfer of funds from society to the timber industry and to an unknown number of corrupt officials with whom the industry is likely to have shared some of its potential windfall profits.

The communities who actually own the land from which timber has been extracted also lost out to the industry. This special issue is analysed further in section 12.1. At this stage it may be mentioned that the land owners have in many cases not even received the share of the timber royalties which they were legally entitled to. Even if this had been the case the fact that royalties have been way below the industry's actual willingness to pay for standing timber represents an allocation of available funds from the land owners to the industry.

The new stumpage fees seem to address a lot of the shortcomings that characterised the previous royalty rates. The positive issues are that:

- The taxation level almost equals the industry's willingness to pay for standing timber;
- the substantial difference in stumpage fee rates between species should serve as an incentive to divert harvest away from over-exploited species towards species which are harvested below their AACs.
- the industry gets clear incentives to maximise profits through more efficient processing rather than through informal payments, since the room for such payments is significantly reduced.

Local communities' share of the timber revenue may also be increased through the stumpage fee system as more revenue is collected from the industry. Yet, as mentioned above the actual share of revenue from forest reserves to local communities will first and foremost depend on how much the Forestry Department is allowed to retain to cover its running costs.

11.1.2 Other Forest Taxes and Fees Directly Related to the Timber Harvest

In the group of other forest taxes and fees some are essentially payments for services that are directly related to the harvest and extraction of timber. Others may be considered actual taxes because they either do not relate to a particular service or because they are higher than the costs of providing the service they relate to (see also chapter 9).

Fees that are payments for direct services include:

- Conveyance certificates;
- concession preparation fee;
- "sale" of Log Measurement Certificates.

It may be debatable whether conveyance certificates represent a service to the timber industry. Yet, they are considered a necessary component to control illegal felling and transport of logs. Furthermore, the size of the fee (less than DM 0.65 per m^3 in 1999 prices) is hardly higher than the costs of administration. Concession preparation fees (DM 322 per concession in 1999 prices) and "sale" of Log Measurement Certificates (¢ 200 or DM 0.13 per log in 1999 prices) must be considered service charges that at best cover the costs of their administration.

Fees that may be considered as taxation include:

- Silvicultural fee;
- concession/contract area rent;
- fee for renewal of property marks.

In theory, silvicultural fees should have been a service charge but, in reality, no silvicultural services have been provided by the Forestry Department on concessions where logging has taken place. The size of the fee has also been pathetically small (less than DM 0.05 per ha in 1997 prices), which hardly covered the costs of collection, and in 1998 it was finally abolished (Timber Resources Management Regulations, 1998; 42). Concession rent (DM 0.90 per ha per year in 1997 prices) which was substituted by contract area rents in December 1998 (DM 0.77 per ha per year on-reserves and DM 0.65 per ha per year off-reserves in 1999 prices) as well as the fee for renewal of property marks (DM 258 per year in 1999 prices) probably cover(ed) more than the costs of collection.

Although there are relatively many forest taxes and fees additional to timber royalties their combined impact on logging profitability is and have not been substantial. In chapter 10 it was estimated that, in 1996 prices, the combined value of (i) concession rent, (ii) silvicultural fees, (iii) renewal of property marks and (iv) conveyance certificates was approximately DM 5.58 per m^3. For comparison the average 1996 royalty values for Scarlet, Red and Pink Star species were DM 7.34, 5.51 and 4.00 per m^3, respectively. In 1999 prices the combined value of contract area rents and property mark renewals is approximately DM 2.95 per m^3 and the average stumpage fees are approximately DM 40 per m^3 for Scarlet Star species, DM 22 per m^3 for Red Star species and DM 9 per m^3 for Pink Star species (see appendix 12).

Prior to 1999, the group of forest fees and taxes other than timber royalties have, in other words, not to any significant degree mopped up the substantial potential forest fee revenue represented by the difference between royalties and the industry's willingness to pay for standing timber. Accordingly, a critique similar to that on timber royalties applies to the pre-1999 package of other forest fees and taxes that are directly related to the timber harvest. In the post-1999 forest fee package, forest taxes and fees other than stumpage fees do not represent a high value per m^3, but this has been more than offset by the stumpage fees which are much higher than the previous royalties.

11.1.3 Log Export Bans

The log export bans have been subject to much debate. This is not surprising since the bans are obvious ly a great benefit to saw mills and other log processing industries, while they have successively barred logging companies from the attractive log export market (cf. table 10.3). However, control of the timber exploitation rather than protection of the domestic wood industry has often been used officially to justify the bans (FRMP, 1988; 18).

The World Bank did not readily accept that log export bans was an appropriate policy instrument to promote further processing and protect the timber resource in Ghana. It was, therefore, made a condition for the Forest Resource Management Project's second trance release of loan money that a study of this issue should form the basis for an agreed action plan between the Bank and the Government of Ghana on log export bans and possible alternative measures (FRMP, 1988; 56). This study was made by General Woods (1992), which came out against the bans. The analysis was that, while reducing the harvest of banned species, the effect of the bans in terms of inducing further domestic processing was negligible (General Woods, 1992; WD no. 3; 43-52).

IIED (1993; 110-120) re-analysed the effects of log export bans and concluded that in terms of resource conservation:

- Log export bans should be accompanied with supply-side measures;
- in the short term, log export bans have reduced the harvest of banned species;
- in the longer term the bans have been less effective because they reduce the domestic log price of the banned species, leading to increased domestic consumption and, in combination with the low forest taxes and fees, wasteful processing for export.

From a theoretical point of view, these conclusions are not surprising. Naturally, it will take some time for the timber industry to switch over from a lucrative log export (most log processing industries also exported unprocessed logs) to mainly export processed wood, particularly lumber.

Species specific data on the extraction of timber has only been available since 1986, which precludes a complete analysis of the effect of the 1979 log export ban. However, figure 11.1 clearly shows the short- and longer-term effect of the 1988 and 1993 bans.

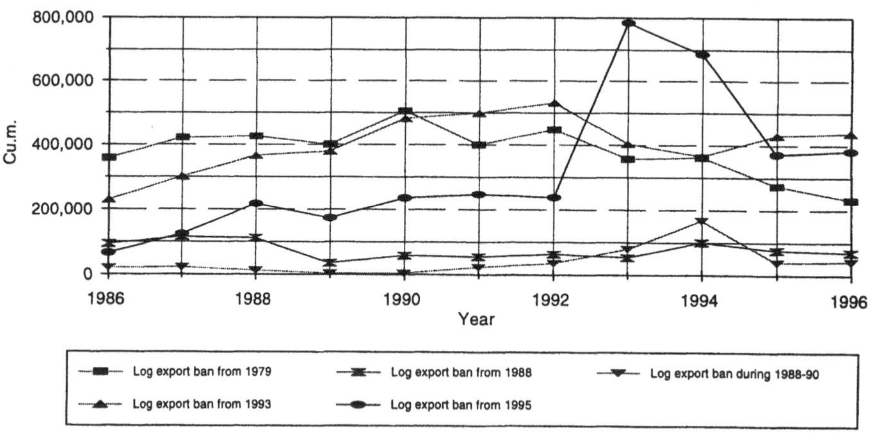

Figure 11.1 Total recorded extraction of species subject to log export bans 1986-96
Source: FPIB LMC Totals (1986-96)

The following observations may be made from figure 11.1:

- The 1988 log export ban appears to have reduced the annual extraction of the affected species by almost 70% in the first year, but already in 1994 the annual extraction was back at the pre-ban level;
- the lifting of the 1988 log export ban for Ofram in 1990 resulted in a rapidly increasing extraction of this species, which suggests that the continued ban on other 1988 banned species has avoided an even bigger increase in their level of extraction than the case has been;
- the 1993, ban on Wawa log exports had some reducing effect on the extraction level in the first year (the was effectuated during 1993), but already in 1995 the extraction level was on the rise again;
- the complete log export ban in 1995 dramatically reduced the extraction level of hitherto un-banned species, but not to the level before the 1993 log export boom, and from 1995 the extraction of these species shows a slight increase;
- the extraction of 1979 banned species culminated in 1990 after which there has been a steady decline in the extraction level for this group.

Accordingly, the graphs in figure 11.1 support the conclusion that log export bans have only had a temporary resource conserving effect. The falling trend, since 1990, in extraction of 1979 banned species is hardly a long-term resource conserving effect of the ban. On the contrary, it probably shows the failure of the ban as a long-term harvest control measure. Most of the 1979 banned species belong to the group of Scarlet Star species and the falling extraction of these species since 1990 is likely to reflect stricter on-reserve harvest control combined with a decreasing availability of these species off-reserves (cf. section 7.4.1). Furthermore, figure 8.3 indicates that, since 1992, the export proportion of extracted Scarlet Star species has gone up, which suggests that species scarcity is a factor contributing to the decreasing extraction of Scarlet Star and therefore 1979 banned species.

In conclusion, the log export bans have failed miserably as long- and even medium-term resource conserving measures. In combination with the pre-1999 low forest taxes and fees, the log export bans have suppressed domestic log prices which has probably induced establishment of new log processing firms, but not produced incentives for efficient processing. In other words, the log export bans and low forest taxes and fees have encouraged inefficiency and created an over-capacity in the log processing industry.

Apart from subsidising domestic processing at the expense of log exporters the distributive effects af log export bans are similar to those created by low forest fees and taxes. People who are employed in the wood industry benefit from this situation, but only temporarily, since the over-exploitation of many species is likely to be a long-term cost for everybody, not least for the wood industry.

Based on the above analyses and the agreement with the World Bank it might appear strange that the Government of Ghana got away with banning log exports of Wawa in 1993 and subsequently all log exports in 1995. Yet, the turn of events during 1993-94 happened so fast that, for political reasons, it must have been impossible for the Bank to insist that log export bans should be avoided.

During 1993, log exports boomed tremendously and it soon became clear that concerns about the resource would eventually force the government to reduce the timber exploitation somehow. For example, during the presentation of the IIED (1993) report, where senior staff from the Forestry Department and the Ministry of Lands and Forestry participated, the then Senior Forester in ODA, John Hudson said the massively increasing timber harvest made "people in London" wonder whether Ghana had completely lost control with its logging sector (Personal observation, 1993).

A complete log export ban must, therefore, have been easy to justify. Essentially, it was also the only measure that could be implemented effectively *and* quickly. In addition, the strong domestic lobby in favour of log export bans would be pleased.

Ironically, the log processing companies significantly contributed to create concerns over the timber harvest going berserk because many mills, during this period, threw all their resources into harvest and export of logs (Personal information, numerous sawmill owners). Saw millers simply found themselves in a win-win-win situation where they could (i) earn quick profits, (ii) clear old debts and (iii) provoke a complete log export ban which would be in their long-term interest. An opportunity too good to let go.

In practice, it is unlikely that the log export ban will ever be lifted even though it may, theoretically, be argued that log export levies would be a better overall policy option. This would, in contrast to the ban, generate government revenue and induce more efficient domestic processing since overseas log buyers would be allowed to compete with domestic mills for logs. The interests already vested in the complete log export ban are, however, likely to be politically very strong. They may also point to the employment generated by further domestic processing and not least to the fact that many other tropical countries have also banned log exports entirely.

The new stumpage fees are likely to form strong incentives for more efficient processing which will offset some of the negative impacts of log export ban. Yet, stumpage fees should be gradually increased. Otherwise the development of the industry will slow down again and the processing efficiency will remain below the standard of developed countries' wood industries. The difference between domestic and FOB log prices will be a good indicator for how much stumpage fees could be further increased.

11.1.4 Export Levies

With the Trees and Timber (Amendment) Act (1994) log export levies between 10% and 30% were introduced for 25 species and export levies on air-dried lumber between 10% and 30% were introduced for nine species. However, it is somewhat unclear when this Act was actually implemented. No log export levies have ever been collected before the complete 1995 ban on log exports, and a circular from FPIB addressed to all saw millers and all commercial banks shows that the levies on air-dried lumber took effect from January 1, 1997 (FPIB, 1997).

As this study does not have the export figures from 1997 onwards it is not possible to directly assess the effect of export levies on air-dried lumber. As a measure to generate revenue at the expense on windfall profits in the saw milling sector and to induce further processing (for which kiln-drying is technically essential) it seems appropriate. Compared with other forest fees and taxes, export levies are also fairly easy and cheap to collect because the system of collecting the 2% of FOB value export levy to finance the FPIB is already in place.

The resource conserving effect of export levies is partially positive because all the nine affected species are over-exploited and belong to the group of Scarlet Stars. However, since the levy does not cover all over-exploited species, this effect might be diluted by an increased pressure on other already over-exploited species. Another issue in this respect is that the levy is likely primarily to promote efficient kiln-drying and only to some degree more efficient conversion of logs to lumber. From a resource conservation viewpoint the levy is, therefore, not as desirable as other demand-side measures that directly increase the "price" on standing timber.

The distributional effects of export levies are positive in the sense that the government revenue is increased. Yet, in relation to the land-owning communities the effect is negative because lumber exports cannot, in practice, be traced back to the stump, which would be necessary to calculate the revenue share that different land-owning communities should rightfully receive. In this respect demand-side measures that directly increase the "price" on standing timber are more equitable.

11.1.5 General Conclusion on Demand-Side Measures

Prior to 1999 the applied demand-side measures undermined rather than supported sustainable forest management. Log export bans and low forest fees and taxes made standing timber a cheap resource. This has promoted (i) over-exploitation of the commercially most valuable species, (ii) wasteful processing and (iii) creation of an over-capacity in the wood industry. Furthermore, considerable sums in terms of potential revenue for the Government and local communities were foregone.

With the 1999 package of stumpage fees and contract area rents introduced by the Timber Resources Management Regulations (1998) it seems that the industry's willingness to pay for standing timber will actually be fully taxed, at least for Scarlet and Red Star species. For Pink Star species there still seems to be some additional revenue available, but this may be a justifiable way to promote species which are harvested below their AACs.

On the whole, the new forest fees and taxes, therefore, appear appropriate and mark an important shift in policy to use demand-side measures rationally in the attempt to make sustainable forest management a reality. Yet, stumpage fees should be gradually increased to promote development of the wood industry and to increase revenues. As long as domestic log prices are below FOB log prices there will be room for improvements.

11.2 Supply-Side Measures

On-reserves supply-side measures encompass the following:

- A 100% stock survey and enumeration of FIP class 1 trees above 50 cm DBH in each compartment;
- in principle, a 10% check survey of all stock surveyed compartments;
- calculation and identification of species specific yield per compartment, based on the stock survey results[38];
- felling checks, marking of stumps and extracted logs and filling out log information forms (since Aug. 1995) supervised/performed by a Forest Guard;

---

[38] See section 6.3.2 for further details.

- issuing of a conveyance certificate for each truck load of logs by the District Forest Officer (since Aug. 1995);
- FPIB road check-points where the truck driver must show his conveyance certificate to receive the necessary log measurement certificates for further transport.

Off-reserves the only supply-side measure was until August, 1995 minimum felling diameters. From 1995, when the "Interim Procedures for Felling Trees Off-reserve" came into effect the following steps should be followed (FD, 1995b; 1):

- Pre-felling inspection of trees by the applicant, community representatives and District Forest Office representatives. On this basis the District Forest Officer indicates which trees may be felled;
- a felling permit will, depending on the application, be issued by the District Forest Officer, the Regional Forest Officer or the Chief Conservator of Forests;
- a post-felling inspection by a FD officer, preferably accompanied by community representatives, must verify proper marking of the stumps and a log information form must be filled out;
- a conveyance certificate valid for three days is issued by the District Forest Officer for every truck load;
- FPIB road check-points where the truck driver must show his conveyance certificate to receive the necessary log measurement certificates for further transport.

Presumably, the above mentioned first three points will also apply for future timber utilisation permits (see section 12.3.5). For future off-reserve timber utilisation contracts the pre- and post-felling inspections of each tree is not likely to be continued, as other provisions in the Timber Resources Management Regulations (1998; 1-7) and the Timber Resources Management Act (1997; 4 (2a-d)) protect the rights of farmers and land owners (see sections 12.3.6 and 12.3.7). Conveyance certificates will still have to be issued for all timber (Timber Resources Management Regulations, 1998; 24). In effect, minimum felling diameters are, therefore, also in the future likely to be the only off-reserve supply-side measure once rights to standing timber have been allocated.

11.2.1 On-Reserve Supply-Side Measures

Although sustained yield management of reserved forests rests entirely on an effective implementation of yield regulation, this issue has received surprisingly little attention. Alder (1996; 11) concludes that in spite of the vast inputs of ODA to the Forestry Department, Planning Branch its capacity in forest management planning is very weak, and there is virtually no effective monitoring.

Essentially, two requirements must be fulfilled to properly implement yield regulation on-reserves. First, accurate stock surveys and yield selection must be carried out. Second, felling checks must ensure that only trees within the yield are felled and that other rules for environmentally sound logging are followed. As mentioned above there is no effective system in place to monitor how well these steps function in practice. Yet, the few studies available on the issue indicate that this is an area where improvements are urgently needed.

Based on a check-survey and careful yield calculation of 35 compartments in 17 districts Kakrada and Maginnis (1992; 12) concluded that of all enumerated stems 42% were over-estimated into a higher diameter-class, 10% were under-estimated into a lower diameter-class and 12% were under-sized (below their minimum DBH) yet promoted into a diameter-class above the felling limit. On average, the result was that 20% of all stems designated as suitable for exploitation (above the minimum DBH) were, in fact, under-sized. For the most valuable species this figure could be as high as 50%. In addition, stock survey data had rarely been verified through a 10% check survey and no felling checks were done.

In a study undertaken by the Forest Research Institute of Ghana (FORIG) on exploitation of under-sized trees, a total of 557 trees marked as originating from forest reserves were re-measured at different FPIB road check-points FORIG (1994). The result of this exercise is summarised in table 11.2.

***Table 11.2* Summary results of re-measuring logs from forest reserves**

Species Group	Observed trees (no.)	Under-sized trees from forest reserves (no.)	(%)
Scarlet Stars	164	97	59.15
Red Stars	254	90	35.43
Pink Stars	139	20	14.39
TOTAL	**557**	**207**	**37.16**

Source: FORIG (1994; 7)

A small post-logging stump check was in 1994 carried out for the purpose of this study. In total 99 stumps in nine compartments of six forest reserves were checked. Summary results of this exercise are presented in table 11.3.

***Table 11.3* Summary results of post-logging stump check in six forest reserves**

Species Group	Stumps checked	Actual diameter compared with stock survey diameter						Actual diameter compared with min. felling diameter			
		correct		above		below		above		below	
	(no.)	(no.)	(%)	(no.)	(%)	(no.)	(%)	(no.)	(%)	(no.)	(%)
Scarlet Stars	56	8	14.29	4	7.14	44	78.57	21	37.50	35	62.50
Red Stars	30	13	43.33	0	0.00	17	56.67	24	80.00	6	20.00
Pink Stars	13	1	7.69	0	0.00	12	92.31	10	76.92	3	23.08
TOTAL	**99**	**22**	**22.22**	**4**	**4.04**	**73**	**73.74**	**55**	**55.56**	**44**	**44.44**

Sources:
Stock tables, yield tables and compartment maps supplied by FD Planning Branch and District Forest Offices.
Table of minimum felling DBHs supplied by FD Planning Branch.
Observations in the field by T.Treue, 1994.

In addition to the data presented in table 11.3 it should be mentioned that only three stumps (3%) represented trees felled outside the approved yield. All other stumps were, through stock survey numbers carved in the bark, positively identified as belonging to trees in the approved yields, even though many of these (44%) were, in fact, under-sized.

The relatively small number of stumps observed in table 11.3 is obviously a statistical weakness. However, the results are quite similar to those of Kakrada and Maginnis (1992), which must build on a large sample (they do not state the total number of measured). Accordingly, it seems safe to conclude that the general tendency in stock surveys to over-estimate the DBH of individual trees is so pronounced that it cannot be explained (away) by normal sampling errors and uncertainties associated with measuring the diameters.

The results presented in tables 11.3 and 11.2 are also quite similar. Approximately 60% of the extracted Scarlet Star species were, in fact, under-sized. For Red Star species the fraction of extracted under-sized trees appears to have been between 20% and 35% and 14% to 23% of Pink Star species were under-sized. The data in table 11.2 is obviously more statistically reliable than the data in table 11.3. Yet, both tables strongly suggest that particularly Scarlet Star species have been felled below their minimum DBH.

In conclusion, the analysis by Alder (1996) combined with the above mentioned technical results indicate that the supply-side measures to control timber exploitation on-reserves have not functioned satisfactorily and clear indications of improvements are still awaited. Yield selection is obscured by over-estimated diameters, and checks at the stump and on the road are not efficient. As a result, under-sized trees and a higher proportion than intended by the interim yield allocation formula of trees above their felling limit have been extracted. This is probably still going on. Overall, the efficiency of on-reserve supply-side measures is indicated by the on-reserve LMC totals which, for 1996, clearly show that most Scarlet and some Red Star species are being exploited above their AACs (cf. section 7.5).

FORIG (1994; 16) points to inadequate field staff and high demand for species with limited stocking as possible reasons for the documented harvest of under-sized trees. Furthermore, it is noted that lack of institutional co-operation between the Forestry Department and FPIB has a tendency to let cases on under-sized logs detected at FPIB check-points end up in the files.

The reasons behind the apparently poor quality of stock surveys are according to Kakrada and Maginnis (1992), that the stock survey teams:

- Rarely receive all their travel and transport allowances
- often lack adequate diameter measurement equipment, especially for buttressed trees
- have poor camping, cooking and transportation equipment
- lack proper supervision in the field.

The above points of Kakrada and Maginnis (1992) are clearly valid but fail to explain the pronounced tendency of tree diameters to be over- rather than under-estimated during stock surveys. FORIG (1994) undoubtedly hits the spot in assuming that the market demand is responsible for exploitation of under-sized trees (what else could it be?), but does not look into how this market affects the yield selection and harvest control.

Interviews with a number of people employed by logging companies and owners of logging companies have added the missing link. After stock surveys and yield allocations are correctly made, the District Forest Officer negotiates with the concessionaire over the cost of enhancing (manipulating) the yield by promoting individual trees to a higher diameter class.

This is a simple exercise which is supposed to be widespread. It explains why tree diameters are being over- rather than under-estimated and why many under-sized trees are actually allocated in the yield. In addition, such cheating is clearly very attractive for loggers since it provides some security against post-logging stump and road checks. Even if a tree is found under-sized it may be documented that it was in the approved yield. In the absence of independent check surveys the District Forest Officer is also fairly well secured from exposure since compromising results of post-logging checks can always be blamed on stock survey teams.

Interestingly, many loggers who told about or admitted to this activity also expressed concerns over its long-term effect. One said that the Forestry Department staff is easy to bribe because their salaries are so low, and in the end he feared the forest would exist on paper only. In fact, many loggers regretted the unfortunate state of affairs and referred to the "good old days" when the Forestry Department was totally free of corruption, and malpractices of any kind by any logger were swiftly and effectively stopped. However, the deteriorating control system has promoted selfish and irresponsible behaviour among loggers, because no one can trust that an under-sized tree left today would actually survive till the next felling cycle.

Post-logging checks may, as here demonstrated, reveal that under-sized trees are extracted from forest reserves. If done at the stump they may also verify whether the under-sized trees were included in the approved yield. Yet, post-logging checks will not be an efficient means to prevent over-exploitation if they are only carried out after a compartment has been logged and not during the operation. In theory, a concessionaire should not be allowed to open a new compartment for logging before a post-logging check has verified that previous compartments have been logged according to the rules.

However, a compartment may be logged during a period of three years before it has to be closed. Commercially, this is reasonable as it may take time to find markets for the available species, but checking three year old stumps is almost impossible. It is our experience that even after two years the bark of stumps is falling off and paint is washed away making identification impossible. After three years it would also be difficult to establish whether an offence has been committed by the District Forest Officer or the Stock Survey Team, should stumps from under-sized trees be identified as allocated in the yield. Furthermore, vines and other pioneer species can after two to three years make a compartment totally inaccessible for post-logging check teams.

Improved harvest control on-reserves must rely on correct selection of the yield based on honest and thorough stock surveys. To this end Kakrada and Maginnis (1992; 26) recommend that stock surveys in general are given higher priority by:

- Making all supervisory staff accountable for the standard of work produced
- paying stock survey teams their entitlements
- providing better supervision, tools and camping equipment to stock survey teams
- ensuring that check surveys are always carried out
- requesting FPIB to immediately furnish the Forestry Department with details on detected under-sized logs
- examining the scope for felling checks.

All of these recommendations, not least the first point, are relevant. FORIG (1994; 24) goes a bit further and recommends that FPIB should be given legal provision to impound logs suspected to stem from under-sized trees pending further investigation. Sometimes the bottom logs of under-sized trees are marked as number two logs of larger trees that never existed, and it will take some time to verify whether this is the case.

However, no system can be better than the people implementing it. If there are no rewards for good performance (at least a decent salary) and/or punishments for bad performance (risk of fines or getting sacked), control and check systems of any kind are not likely to function very well, if at all. The development within Forestry Department over the past 20 years or so is an obvious example.

The rather severe punishments stipulated in the Timber Resources Management Act (1997; 17) and the Timber Resources Management Regulations (1998; 41) for illegal felling of timber may deter officials from involving themselves in such activities, as the rules do now also apply for officials who fail to perform his/her duty of controlling that timber is felled according to the rules (see section 9.1.5). Positive incentives for officials to keep high standards are not yet established.

11.2.2 Off-Reserve Supply-Side Measures

Until August, 1995 the only off-reserve supply-side measure was the species specific minimum felling diameters. Inspections of felling sites or stump checks were rarely carried out by the Lands Commission, which until 1994 was responsible for collection of royalties off-reserves (cf. section 9.1.1).

In the study of FORIG (1994) 3,463 trees marked as originating from areas outside forest reserves were re-measured. The summary result is shown in table 11.4.

Table 11.4 **Summary results of re-measuring logs from off-reserve areas**

Species Group	Observed trees (no.)	Under-sized trees from off-reserve areas (no.)	(%)
Scarlet Stars	1,637	1,121	68.48
Red Stars	1,367	422	30.87
Pink Stars	459	36	7.84
TOTAL	**3,463**	**1,579**	**45.60**

Source:
FORIG (1994)

The table shows that a very high percentage of especially Scarlet Star species were under-sized. The high number of trees checked supports that this result gives a reliable picture of the general situation.

Loggers gave two main reasons for felling under-sized trees. One was that big trees had more technical defects. The other was that if under-sized trees were left off-reserves, farmers would destroy them.

The second objective of the FORIG study was to examine if defects are more abundant in large than in small diameter trees. On this part of the study the results showed that no significant relationship between diameter and the incidence of defects could be proved (FORIG, 1994; 20). In other words, the perception of loggers that defects were more abundant in trees of large diameter cannot be supported and is probably false.

That farmers would destroy off-reserve timber trees left behind by logging companies is, on the other hand, likely to be true in many cases. On several occasions we have seen timber trees being deliberately destroyed by farmers because they saw them as potential risks rather than potential sources of income. The reason being that farmers rarely received any payments for trees being felled on their farms and logging companies rarely payed compensation for damage on crops caused by felling and extraction of timber on farmland (see section 12.2).

The "Interim Procedures for Felling Trees Off-Reserve" attempted to rectify this by specifying that a person who cultivates the land (not the land owner) on which a timber tree grows had the right to stop the tree from being felled, regardless of the person applying for the tree to be felled holds a concession covering the area (FD, 1995b; 3). The idea was to give farmers authority to negotiate a price over and above royalties and other forest taxes for the felling of timber trees on the land they cultivate.

The degree to which this system has benefitted farmers is unclear. Anecdotal information indicate that a few loggers were having difficulties with farmers who resisted felling of trees. Yet, during a field trip in April, 1997 we could observe that farmers were not significantly empowered by the new rules. Informal information from people involved in developing the "Interim Procedures for Felling Trees Off-Reserve" also confirmed that little had actually changed on the ground.

The problem was that farmers did not know of their "new" rights. No information campaign had been launched to alleviate this, and it appeared that the District Forest Officers did not see it as their duty to inform local people about their rights to refuse felling of timber trees in their farms.

The lack of institutional support to the alleged rights of farmers to veto tree felling on land they cultivated may probably be explained by the fact that such rights could be legally challenged in cases where farmers did not cultivate cocoa. The Economic Plants Protection Decree (1979; 4) prevents the felling of trees on farms where "specified" plants are cultivated, and specified plants are defined as cocoa only (Economic Plants Protection Decree, 1979; 7).

The "Interim Procedures for Felling Trees Off-Reserve" were primarily intended to control the chaotic situation of illegal and speculative timber felling off-reserves during the log export boom in 1993-94 (FD, 1994a; 1). They were considered necessary interim measures to be implemented before "the inadequacies and deficiencies in the regulatory system could be systematically tackled" (FD, 1994a; 2). Yet, the complete log export ban, which also came into force in mid 1995, was probably more effective in terms of controlling the off-reserve timber harvest.

Presumably the Timber Resources Management Act (1997) and the Timber Resources Management Regulations (1998) should be considered part of an attempt to address some of the previous regulatory system's shortcomings. Farmers are now legally entitled to veto felling of trees on land they cultivate, irrespective of the agricultural crops (Timber Resources Management Act, 1997; 4 (2d)). This should protect the individual farmer from suffering undue losses, but it is doubtful whether adequate incentives have been created for farmers to actually protect and nurse naturally occurring trees on land they cultivate (see sections 12.2.1 and 12.3.6).

11.3 Summary Assessment of Timber Harvest Regulations

Overall, the exploitation of timber has, during the period 1986-93, resulted in a net loss for the average Ghanaian taxpayer/citizen. Possibly this was also the case in 1994, but since 1995 forest taxes and fees collected have probably exceeded the costs of forest management by several million DM.

Yet, substantial revenues have been foregone by allowing inflation to erode old royalty debts and by the fact that, prior to 1999, forest fees and taxes have not been charged according to the industry's willingness to pay for standing timber. Forest fees and taxes even from forest reserves alone could easily finance the costs of sustainable or sustained yield forest management if timber exploitation was taxed according to the industry's willingness to pay for standing timber. This conclusion rests on fairly rough cost and, to a lesser degree, potential revenue estimates, but the gap of more that DM 4 million between the two sends a clear message. *It can be done*.

The new stumpage fee system would initially only generate around DM 8.6 million, leaving a surplus of approximately DM 1.1 million, from forest reserves if the harvest is held within the AAC. While the potential revenue appears to be somewhat higher, the (long awaited) abolishment of the low timber royalties *does* mark an important shift in policy towards securing that the costs of wise forest management may actually be covered by forest revenues. An outstanding issue in this respect is, however, whether it would be politically acceptable to spend almost 90% of the revenue from forest reserves to finance the Forestry Department (or a future Forestry Service), leaving only 10% to local communities (see section 12.1.2 for further analyses).

11.3.1 Specific Demand-Side Measures

Timber royalties have not created incentives for the industry to reduce the harvest of over-exploited species and/or to increase the harvest of "under-utilised" species. Considered in isolation, timber royalties have not generated the forest revenue they could. In combination with other forest taxes and fees, timber royalties have clearly not compensated much for the suppressing effect that log export bans have had on domestic log prices.

In combination, the log export bans and low timber royalties have probably promoted the establishment of new log processing domestic companies, but they have definitely not created strong incentives for making these very efficient. Compared with a hypothetical log export, Ghana's export of processed wood represented an export revenue foregone of some DM 40 million in 1996 alone. Furthermore, it seems very likely that the combination of low royalties and the policy of banning log exports of an increasing number of, and eventually all, species has contributed to create the apparent over-capacity of the processing industry.

The distributional effects of the low timber royalties and other forest taxes and fees have not been particularly positive either. Since the timber industry could have paid more than it did (at least officially) for standing timber, the pre-1999 level of forest taxes worked as a net transfer of funds from the average Ghanaian taxpayer to the timber industry and to an unknown number of corrupt officials. The communities who actually own the land on which timber is felled and extracted have also lost out to the industry.

The new stumpage fees seem to address a lot of the shortcomings that characterised the previous royalty rates. The positive issues are that:

- The taxation level almost equals the industry's willingness to pay for standing timber
- the substantial difference in stumpage fee rates between species should serve as an incentive to divert harvest away from over-exploited species towards species which are harvested below their AACs
- the industry gets clear incentives to maximise profits through more efficient processing rather than through informal payments, since the room for such payments is significantly reduced.

Local communities' share of the timber revenue may also be increased by the stumpage fee system as more revenue is collected from the industry. Yet, as mentioned above, the actual share of revenue from forest reserves going to local communities will first and foremost depend on how much is retained to finance the Forestry Department.

Log export bans have failed miserably as long- and even medium-term resource conserving measures. The harvest of banned species has only fallen in a few years after the bans, and then the demand has picked up again. Presumably, this reflects the industry's fairly quick ability to shift from export of logs to export of processed wood, especially lumber.

Apart from subsidising domestic processing at the expense of logging companies, which have successively lost access to export markets, the distributive effects af log export bans are similar to those created by the pre-1999 low forest fees and taxes. People who are employed in the wood industry benefit from this situation, but only temporarily, since the over-exploitation of many species is likely to be a long-term cost for everybody, not least for the wood industry.

The export levy on air-dried lumber has a partially positive resource conserving effect because all the nine affected species are over-exploited and belong to the group of Scarlet Stars. However, since the levy does not cover all over-exploited species, this effect might be partly offset by an increased pressure on other already over-exploited species. Furthermore, the levy is likely primarily to promote efficient kiln-drying and only to some degree efficient conversion of logs to lumber. From a resource conservation viewpoint the levy is, therefore, not as desirable as other demand-side measures that directly increase the "price" of standing timber.

The distributional effects of export levies are positive in the sense that government revenue is increased. Yet, in relation to the land-owning communities the effect is negative because lumber exports cannot, in practice, be traced back to the stump, which would be necessary to calculate the revenue share that different land-owning communities should rightfully receive. In this respect demand-side measures which directly increase the "price" of standing timber are more equitable.

With the 1999 package of stumpage fees and contract area rents introduced by the Timber Resources Management Regulations (1998) it seems that the industry's willingness to pay for standing timber will actually be fully taxed at least for Scarlet and Red Star species. For Pink Star species there still seems to be some additional revenue available, but this may be a justifiable way to promote species which are harvested below their AACs.

On the whole, the new forest fees and taxes, therefore, appear appropriate and mark an important shift in policy to use demand-side measures rationally in the attempt to promote sustainable forest management. Yet, stumpage fees should be gradually increased to upgrade the wood industry and to increase revenues. As long as domestic log prices are below FOB log prices there will be room for improvements.

11.3.2 Supply-Side Measures

Supply-side measures to control timber exploitation on-reserves do not appear to function satisfactorily. Collusion between loggers and Forestry Department staff obscures yield selection through deliberate over-estimation of tree diameters. Checks at the stump and on the road are either not efficient or not performed. As a consequence, under-sized trees and a higher proportion, than silviculturally intended, of trees above their felling limit is extracted. The overall result is illustrated by the on-reserve LMC totals for 1996, which clearly show that most Scarlet and some Red Star species are being exploited above their AACs.

Off-reserves control with the timber exploitation has traditionally been very weak and excessive felling of under-sized trees is documented to have happened. The "Interim Procedures for Felling Trees Off-Reserve" (FD, 1995b) and subsequently the rules stipulated in the Timber Resources Management Act (1997) and the Timber Resources Management Regulations (1998) seek to establish control by the Forestry Department of timber exploitation off-reserves. In addition, farmers' rights to oppose felling of trees on land they cultivate have been strengthened. This is likely to slow the off-reserve timber exploitation, but it is questionable whether farmers are given sufficient incentives to nurse naturally occurring trees on farmland (see section 12.3.7).

12 Rights to Timber and Benefits from Timber Exploitation

The distribution of rights to timber and to benefits brought about by timber harvest is one of the most critical forest policy matters in Ghana. Rights over land are closely linked with rights to trees and timber, and the issue of promoting sustainable forest management or sustained yield timber production may be divided into the following main points:

- Rights to land, timber, trees and revenue on-reserves;
- rights to land, timber, trees and revenue off-reserves;
- principles for allocation of rights to standing timber on- and off-reserves.

These subjects have been guided by an array of legislation that has undergone progressive revision since the colonial period. The result has, at least until December 1998, been a successive erosion of customary land-owning communities' ("stools" and "skins") rights. This has alienated local communities from management of forest reserves and hitherto totally undermined economic incentives for growing or tending naturally occurring trees on farms and fallows.

The winners in terms of rights to manage and collect benefits from exploiting the timber resource have been the Lands Commission, the Forestry Department and timber companies. However, as the previous chapters clearly show the outcome has been (i) a notorious over-exploitation of the most valuable species, (ii) wasteful processing and (iii) a tremendous forest taxation revenue foregone. In other words, complete mismanagement.

12.1 Rights to Land, Trees and Revenue On-Reserves

The constitution of a forest reserve does not alter the ownership of the land (Forest Ordinance (Cap. 157), 1927; 18 (1)). However, the management right is given to the President. Forest reserves could either be managed by the land owners under the direction of the Forestry Department or by the Government. In practice, all forest reserves have been managed directly by the Forestry Department, not by the land owners.

Forest reserves were primarily intended to secure environmental services (water supply and protection of agricultural land) and to secure the supply of forest products to the surrounding communities (Forest Ordinance (Cap. 157), 1927; 4 (4)). Accordingly, the declaration of forest reserves in the absence of the land-owners agreement was originally considered exceptional (Cirelli, 1997; 19). Furthermore, the owners of a forest reserve were guaranteed to receive not less than *two thirds* of the *gross revenue* accruing from managing the forest (Forest Ordinance (Cap. 157), 1927; 18 (2)). Except for persons holding admitted rights (usually established during the process of declaring the forest reserve) it is, in forest reserves, prohibited to:

- Fell, uproot, lop, girdle, tap, injure by fire or otherwise to damage any tree or timber;
- establish or cultivate any farm;
- cause any damage by negligence in felling any tree or cutting or removing any timber;
- in any way obstruct the channel of any river, stream, canal or creek;
- hunt, shoot, fish, poison water, or set traps or snares;
- collect, convey or remove any forest produce for any manufacturing process;
- pasture cattle or permit any cattle to trespass.

These prohibitions were originally included in the Forest Ordinance (Cap. 157) (1927; 22 (1-10)) and they have been repeated in subsequent amendments through the Forest Protection Decree (1974) and the Forest Protection (Amendment) Law (1986). Likewise, it has been maintained that the burden of proof that any forest produce has not been taken in contravention with the legislation lies upon the person in whose possession it is found.

In 1962 the President's right to intervene and dispose over revenue was enhanced significantly. Any stool land may be "vested" in the President if it appears to him that this would be in the public interest to do so, and any moneys accruing from such land shall be paid into the "appropriate" account (Administration of Lands Act, 1962; 7 (1-2)). Stool land revenue is collected by the Minister and paid into a Stool Lands Account (Administration of Lands Act (Act 123), 1962; 17 and 18). However, revenue from forest reserves continued to be collected by the Forestry Department and paid into the Forest Improvement Fund from which any funds in excess, after costs of timber exploitation and silvicultural work have been paid (running costs of the Forestry Department), shall be transferred to the Stool Lands Account (Administration of Lands Act, 1962; 17 (3 and 4)). Furthermore, all "timber" and all "trees" on all land including public land, stool land, existing and possible future forest reserves were also vested in the President (Concessions Act, 1962; 16).

The terms "timber" and "tree" are not defined in the Concessions Act (1962). However, the Forest Protection Decree (1974; (10)) defines timber as "including trees when they have fallen or have been felled, and all wood, whether it has been cut up or fashioned or hollowed out for any purpose or not". The Timber Resources Management Act (1997; 20) defines timber as "standing, fallen and felled trees, wood for industrial purposes, logs and sawn wood" (no reference is, however, made to the Forest Protection Decree (1974) which is still in force). Trees are defined as "including palms, bamboos, stumps, brushwood and canes" (Forest Protection Decree, 1974 (10)). That a resource is "vested" in the President means that rights over the resource are split into rights of management (legal rights) and rights to benefit from resource ownership (benefit rights) (Cirelli, 1997; 19 and FD, 1995a).

In effect, the 1962 legislation extinguished the rights of land-owning communities over timber and even other "trees" which are used either for subsistence or commercially, eg. cane for furniture production. It has been increasingly difficult for people to prove that they hold an admitted right to collect forest produce in a forest reserve that was constituted more than 50 years ago, and such rights have become lost or fuzzy (FD, 1995a; 19). In addition, the 1962 legislation made land owners' right to revenue from forest reserves subordinate to the costs of running the Forestry Department.

During the 1970s, the Forestry Department recognised that, in many instances, there was no revenue left for land owners in the Forests Improvement Fund. It was, therefore, administratively decided that 30% of all royalties should go to the land owners before costs of management were deducted (FD, 1995a; 17). From being guaranteed minimum 67% of the *gross revenue* accruing from forest reserves, the 1962 legislation gave land owners right to, effectively, first nothing and then, through an administrative decision, 30% of the *royalties* collected on-reserves.

This is quite a decline, but things were actually much worse, at least for many land-owning communities in the Brong Ahafo Region. As of 18 April, 1997 the Forestry Department paid into the Administrator of Stool Lands' account Cedis 136,167,898.88 (at that time approximately DM 120,000) for on-reserve royalties collected in Brong Ahafo Region during the period *January 1979 to December 1995*

(FD, 1997). Adjustment for inflation was not made, so the real value of the collected revenue had been seriously eroded. Numerous letters to the Forestry Department from disgruntled traditional and elected leaders of local communities from all over the high forest zone indicate that the problem has prevailed a long time and that it has not been confined to Brong Ahafo Region only. Interviews with District Forest Officers have confirmed this.

12.1.1 Future Status of Current and New Forest Reserves

The Constitution (1992) has changed the legal framework on a number of key forestry related issues. In particular, the rights of the President to constitute forest reserve and land-owning communities' rights to revenue are affected.

On stool and skin lands the Constitution (1992; 267 (1)) states "that stool lands vest in the appropriate stools on behalf of the subjects of the stool". Expropriation of stool land by the Government in the public interest is only allowed upon prompt payment of fair and adequate compensation (Constitution, 1992; 20 (2,a)). The provisions of the President under the Administration of Lands Act (1962; 7 (1-2)) to constitute a forest reserve with or without the consent of the land owners does, therefore, not have any basis in the 1992 Constitution (Cirelli, 1997; 32).

Concerning existing forest reserves, FD (1995a; 18) states that the land should not be regarded as stool land any more since the Office of the Administrator of Lands Act (1994; 18) defines stool land as:

> "Any land or interest in or right over any land controlled by a stool or skin, the head of a particular community or the captain of a company, for the benefit of the subjects of that stool or the members of that community or company".

The argument being that it is the Forestry Department, not the stools or skins, which controls the use and management of forest reserves. Accordingly, forest reserves should be considered "public land" as defined in the Constitution (1992; 257 (2)). This would give the Forestry Department full control over revenue, timber and other forest produce. Accordingly, it would attain a high degree of freedom to negotiate terms and conditions for local people to utilise forest produce and to receive a share of timber royalties. It is, however, very likely that this (mis)interpretation of the legislation would be deemed unconstitutional (Cirelli 1997; 23).

New forest reserves can, therefore, only be established with the consent of the land owners unless the Government outright expropriates the land. This would, however, require prompt payment of fair and adequate compensation from the Government, which it probably cannot afford. Furthermore, existing forest reserves must still be considered situated on stool land.

12.1.2 Rights to Revenue from Forest Reserves

The administratively decided share of revenue between the Forestry Department and the land owners probably became unconstitutional in 1992. Since the ownership of land was not changed when forest reserves were originally declared, they must still be considered located on stool or skin land. Regarding such land the Constitution (1992; 267 (6)) stipulates that:

"Ten per cent of the revenue accruing from stool lands shall be paid into the office of the Administrator of Stool Lands to cover administrative expenses; and the remaining revenue shall be disbursed in the following proportions:

a) twenty-five per cent to the stool through the traditional authority for the maintenance of the stool in keeping with its status
b) twenty per cent to the traditional authority[39]
c) fifty-five per cent to the District Assembly, within the area of authority of which the stool lands are situated."

Nothing is mentioned about the administrative expenses of the Forestry Department to manage forest reserves. The above distribution of stool land revenue is repeated in the Office of the Administrator of Lands Act (1994; 8). Furthermore, this Act states that the provisions in the Administrator of Lands Act (1962) and the Concessions Act (1962) "shall have effect with such modifications as may be necessary to give effect to this Act" (Office of the Administrator of Lands Act, 1994; 17). Accordingly, the legal provision of the Forestry Department to retain all revenues from forest reserves necessary to cover its running costs fell away with the 1992 Constitution.

This was a serious problem. Clearly, the Forestry Department could not manage forest reserves if reasonable costs of doing so could not be covered. Since the 1992 Constitution was adopted it, therefore, appears that the Forestry Department has, *de jure,* been financed illegally. How this could be allowed and how the Office of the Administrator of Lands Act (1994) could be passed without specifically considering this rather crucial issue is somewhat mysterious.

Conserving forest resources in Ghana is, however, difficult to imagine without a kind of Forestry Department, and such a body must be financed. A political decision on the matter was, therefore, urgently needed (FD, 1995a; 65). Finally in December, 1998 legal basis for funding the Forestry Department for timber management services out of revenue arising from timber revenue was provided in the Timber Resources Management Regulations (1998; 21 and 26). The amount shall be decided by the Minister for Lands and Forestry in consultation with the Forestry Commission, the Forestry Department and the Administrator of Stool Lands.

It is not specified whether such payments should be deducted from stool land revenue at large or from forest reserve revenue only. Furthermore, it is notable that the new legislation does not set a maximum limit for how much may be deducted from stumpage fee revenue to finance the Forestry Department. The law says that "stumpage fees represent royalties to the land owner and charges for the cost of felled timber which provide a basic return to the landowner and contributes to the cost of forest management and timber regulation" (Timber Resources Management Regulations, 1998; 21 (2)). On one hand, this may allow for financing the Forestry Department through timber related revenue from both on- and off-reserves if the words "timber regulation" are stretched to cover activities of the Forestry Department off-reserves. On the other hand, the word "contribute" should imply that not all revenue may be retained to fund the Forestry Department.

---

39 House of Chiefs or any council or body established or recognised as such under customary law (Office of the Administrator of Stool Lands Act, 1994; 18).

Unless (i) the revenue from timber exploitation is increased significantly, (ii) the cost of running the Forestry Department is reduced significantly or (iii) off-reserve revenue may be used to fund the Forestry Department, there will not be much revenue left for local communities who own land on which forest reserves are located. As the stumpage fee and contract area rent system has recently been introduced, it is not likely that the revenue from forest reserves will be much higher than DM 8.6 million annually in the near future (assuming the harvest is held within the AAC). If the annual costs of running the Forestry Department is approximately DM 7.5 million, this would leave less than 10% of the revenue for the land owners unless other sources may contribute to finance the costs of forest management (see section 11.1). Whether this is reasonable is a political decision, but it is quite far from the original minimum of 67% defined in the Forest Ordinance (Cap 157) (1927, 18 (2)).

Revenue from forest reserves may in the longer term be higher than DM 8.6 million per year since the stumpage fees could be gradually increased, as could the harvest of currently "under-utilised" species. Yet, it appears crucial that the Forestry Department will receive funding through other sources than on-reserve timber revenue to maintain any credibility in relation to local communities who actually own the forest reserves. After all the Forest Ordinance (Cap. 157) (1927; 18 (2b)) says that when the Government is the manager of forest reserves it should be "for the benefit of the owner or owners".

Naturally, it is not only the land owners' rights, but also the conservation of Ghana's forest resources which are at stake. It would seem bizarre if on-reserve timber exploitation could not be controlled because the national forest authority cannot be adequately financed. Yet, it is also reasonable that a national forest authority is not over-staffed or over-financed at the expense of the land-owning communities. In addition, it must be ensured that the forest authority is held accountable for the standard of on-reserve forest management.

12.2 Rights to Land, Trees and Revenue Off-Reserves

The Concessions Act (1962; 16) vests all timber trees, even off-reserves, in the President. All rights to timber trees, whether they stand on cultivated land or not were, therefore, transferred to the company which obtained a timber concession in a given area. With reference to the Economic Plants Protection Decree (1979; 2 (2)) it was required that the concessionaire must pay compensation for damage of crops resulting from felling and extraction of timber.

In other words, the previous legislation intended to indemnify farmers in connection with exploitation of timber, no more. On uncultivated areas farmers are generally in favour of timber exploitation because this clears land for farming. Yet, exploitation of timber on especially cocoa farms normally leads to a net loss for the farmer, because compensation for crop damages was, in general, either inadequate or not paid at all (IIED, 1993; 61 and personal observations).

Logging companies generally want to take all available commercial trees in a concession whether they grow on farms or not and, in reality, the few rights of farmers were not protected by the Economic Plants Protection Decree (1979). No Government or other sufficiently influential authority seriously attempted to enforce the farmers' rights in relation to off-reserve timber exploitation, and an ordinary farmer could not, in practice, drag a logging company to court. Instead the idea spread that timber companies, if necessary assisted by Government authorities, had every right to exploit timber trees on farms (Personal observations).

Accordingly, farmers did not have any positive incentives for growing or tending naturally occurring trees on the land they cultivated, on the contrary. To farmers trees on farmland represented a risk of economic losses rather than a potential source of income.

As at March 1998, the rights of farmers were *de jure* greatly improved. The Timber Resources Management Act (1997; 4 (2d))[40] clearly states that "no timber rights shall be granted in respect of lands with farms without the authorisation in writing of the individual, group or owner concerned". In this respect farmers are "individuals concerned" as they cultivate the land (Cirelli, 1997; 65). It is, however, notable that the Concessions Act (1962; 1 and 16) still apply "with such modifications as may be necessary to give effect to the provisions of the Timber Resources Management Act (1997; 22)". This means that trees on stool land are still vested in the President (see sections 12.3, 12.3.6 and 12.3.7).

12.2.1 Current Rights to Revenue from Timber Harvest Off-Reserves

Farmers rarely own the land they cultivate. Land is generally owned by the stool (Cirelli, 1997; 25). Furthermore, rights to natural assets on stool land that are not the result of differentiated identifiable human labour are held by chiefs on behalf of their people (FD, 1995a; 23).

Revenue from timber exploitation must, according to the Constitution (1992; 267 (6)), be shared between the stool, the traditional authority and the District Assembly (see section 12.1.2). Thus, the individual farmer has no direct right to a share of timber royalties, even though trees are extracted from the land he/she cultivates.

In principle, the Timber Resources Management Act (1997; 4 (2d)) gives a farmer full right to refuse the felling of a timber tree which stands on land he/she cultivates. A farmer should, therefore, be in a position to negotiate a price with a logging company, on top of compensation of crop damage and forest taxes, for the right to fell the tree. Whether this is actually going to happen will fundamentally determine farmers' incentives to tend (not destroy) naturally occurring trees on land they cultivate, because the stool or the traditional authority are not legally required to and do not pay individual farmers for timber exploited on land they cultivate.

12.2.2 Rights of the Forestry Department to Benefit from Off-Reserve Revenue

After a long battle with the Lands Commission the Forestry Department in 1994 gained the right to collect off-reserve timber royalties (cf. section 9.1.1). However, the Timber Resources Management Regulations (1998; 25 (3) and 27) states that stumpage fees and rents from timber utilisation contract areas on stool lands[41] must be collected by the Administrator of Stool Lands.

As mentioned in section 12.1.2 it is not clear how big a share the Forestry Department can claim of on-reserve stumpage fees. On areas off-reserves, the Forestry Department is intended to provide little management service other than a timber inventory, identification of areas suitable for award of timber rights and some control over logging operations (see 12.3.1). On this background it seems difficult to

---

40 The date of assent for this Act was 17 March, 1998.

41 Presumably excluding forest reserves where the District Forest Officer is supposed to collect stumpage fees.

justify that the Forestry Department should receive a significant share of the revenue accruing form off-reserve timber exploitation.

Yet, an efficient control and monitoring of timber exploitation provided by the Forestry Department may ensure that stumpage fees are actually paid. This would greatly benefit local communities and District Assemblies, of which the latter depend heavily on timber revenue (FD, 1995a; 69). The Lands Commission's efficiency in collecting royalties has probably not been very high, since it has never been able to produce any accounts for timber royalties collected up to 1994.

12.3 Principles for Allocation of Rights to Standing Timber

Until 1998 the legal basis for granting rights to standing timber was the Concessions Act (1962; 16). The procedures for granting timber concessions were specified in the Concession Regulations (1962), the Timber Leases and Licences Regulations (1962) and subsequent amendments of these.

A key issue in the allocation of rights to standing timber through timber concessions (long-term rights to standing timber) or leases (short-term rights to standing timber) was that, save the costs of processing applications, such rights were free of other charges than the annual concession rent and other forest fees and taxes (cf. chapter 9). As documented in chapter 10 this left substantial room for unofficial payments and/or windfall profits.

The World Bank did not agree with this policy, and made it conditional to the second trance release of funds for the Forest Resource Management Project that the allocation of concessions were made subject to a tendering process where 50% and subsequently 60% weight should be given to the highest bid (FRMP, 1988; 55). Even though this has never been implemented, the Ghanaian Government acknowledged it had a problem with the allocation of timber concessions. At least this has been the case since the NIC investigations in 1991 brought up the issue of timber concessions being sub-let without the written consent of the Ministry (cf. section 10.6).

In 1991, the then Ministry of Lands and Natural Resources (now the Ministry of Lands and Forestry) froze the granting of new timber leases and thereby, administratively, took over the responsibility of granting rights to standing timber from the Lands Commission, on grounds that the Lands Commission had a poor record on considerations of conservation issues. Yet, the Ministry did not manage to streamline the procedures. On the contrary, it seems, since indications of the Ministry being susceptible to lobbying and pressure began to crop up (Office of the President, 1993;1).

In 1992, by order of the PNDC (the then military Government) the granting of timber leases was, temporarily, made the responsibility of the Chief of Staff, PNDC. He established, within the Ministry and including representatives from relevant agencies[42], a timber concession committee to deal with the allocation of timber concessions (Office of the President, 1993;2).

---

42 Two from the Ministry, the Forestry Commission, one from the Lands Commission, five from the Forestry Department, one the Forest Products Inspection Bureau, one from the Timber Export Development Board and one from the Timber Industry (AGTI) (IIED, 1993; 81).

Apart from allocating concessions, the timber concession committee also worked on designing a new framework for the granting of rights to standing timber on- and off-reserves. The authority of the timber concession committee, established through ad hoc directives and administrative measures, to grant timber concessions could, in principle, be challenged by concessionaires, the AGTI and not least the Lands Commission which, according to the Constitution (1992; 258 (1a) and 267 (3)), is the only body authorised to grant leases (Office of the President, 1993; 3).

The need for a comprehensive legislation on rights to standing timber was, therefore, urgent. The Office of the President wanted the new legislation to avoid using the terms "lease" and/or "concession", as this would definitively prevent the Lands Commission from claiming authority over the granting of rights to standing timber. Furthermore, it was argued that timber leases and concessions are in any case different from other forms of leases and concessions because they cannot be used as security or collateral (Office of the President, 1993; 3).

Much effort has since 1993 gone into developing what finally became the Timber Resources Management Act (1997) which legally defines the concepts of "timber utilisation contracts" and "timber utilisation permits". Following a number of drafts the Timber Resources Management Act (1997) came into force on 17 March, 1998. This Act was followed by the Timber Resources Management Regulations (1998) which came into force on 16 December, 1998. This new legislation on timber rights has introduced the following main concepts and principles:

- A new formula for calculating timber royalties (stumpage fees)
- new rates for contract area rents on- and off-reserves
- a timber utilisation contract which is a long-term right to standing timber on a large area
- the Forestry Department's duty to conduct surveys and timber inventories on- and off-reserves in order to identify suitable areas for timber utilisation contracts
- a coordinated involvement of land owners, and traditional and official authorities during the identification of timber utilisation contract areas off-reserves
- the right of land owners to refuse a timber utilisation contract to cover their land off-reserves
- the obligation of the Forestry Department to obtain written consent of land owners before an off-reserve timber utilisation contract can be granted
- pre-qualification criteria for applicants to timber utilisation contracts
- a timber rights evaluation committee to process applications according to the pre-qualification criteria and to advise the Minister on the granting of timber utilisation contract
- a minimal element of bidding for timber utilisation contracts, but not for timber utilisation permits
- a termination of all previous timber concessions and leases after a period of maximum six months from the date where the Timber Resource Management Act (1997) comes into force
- the right of the Minister, in consultation with the Forestry Commission, to direct such holders of timber utilisation contracts to supply the domestic market with timber products of specified volume, dimensions and species
- a timber utilisation permit which is a short-term right to a specified number standing timber trees in an area not covered by a timber utilisation contract
- the right of the Chief Conservator of Forests to grant timber utilisation permits
- protection of farmers' rights to veto the felling of any tree on land they cultivate
- protection of rights to trees grown by private individuals and groups

- the right of the Minister, through legislative instrument, to make regulations for and prescribe conditions under which timber harvesting may take place inside and outside forest reserves
- the right of the Minister, through legislative instrument, to protect any tree or tree species that may be considered depleted, threatened or endangered.

A key issue in the new legislation is that *all trees on all land are still vested in the President*. Sections 1 and 16 in the Concessions Act (1992) still apply, with such modifications as may be necessary under the new Act (Timber Resources Management Act, 1997; 22 (2)). This secures the right of the President to grant rights to timber on areas off-reserves, and the right of the Ministry of Lands and Forestry to allocate such timber rights, a right that might otherwise easily have fallen to the District, and Assemblies (FD, 1995a; 23).

The principle of acquired rights under a timber utilisations contract not being transferable without the written consent of the Ministry of Lands and Forestry has also been maintained (Timber Resources Management Act, 1997; 16). Rights acquired through a timber utilisation permit are not transferable under any circumstances.

12.3.1 New Timber Royalties and Contract Area Rents

The new formula to calculate stumpage fees which have replaced timber royalties is described in section 9.1.1 and the contract area rent has been considered in section 9.1.2. Here it may be repeated that particularly the stumpage fee system represents a major increase in timber revenue and the package of fees and taxes appears to create appropriate incentives to support sustainable forest management and to develop the wood industry.

12.3.2 Timber Utilisation Contracts

It is envisaged that a timber utilisation contract should run for a period of 40 years inside forest reserves (coinciding with the felling cycle) and 5 to 20 years outside forest reserves (FD, 1995c). The Forestry Department will at certain intervals review timber operations, and logging companies must submit harvest plans every five years (Timber Resources Management Regulations, 1998; 14 (1g-h)). It is paramount that a company's right to timber is acquired through a contract relation with the Ministry of Lands and Forests and that the conditions of the contract are specified by the Ministry. This underscores that the relationship is not one of equals and gives the Ministry authority to cancel the timber utilisation contract if a logging company does not live up to the specified conditions (Timber Resources Management Act, 1997; 8 and 15 & Timber Resources Management Regulations, 1998; 14).

The Forestry Department shall identify areas suitable for grant of timber utilisation contracts. On such areas, which may be inside or outside forest reserves, the Forestry Department will carry out surveys and timber inventories to verify the stocking (Timber Resources Management Regulations, 1998; 1). When suitable areas are identified, a field inspection must take place including two members of the concerned District Assembly, a representative of the Traditional Council, the District Forest Officer and two representatives of the land owners, of whom at least one must be a farmer. This team must through the Regional Forestry Officer submit to the Chief Conservator of Forests a report detailing on a topographic map the area inspected, topographic and man-made peculiarities of the area and the estimated quantity, quality and value of timber (Timber Resources Management Regulations, 1998; 2).

When the Chief Conservator of Forests is satisfied that the inspected area is suitable for a timber utilisation contract he will, for lands other than public land and forest reserves, direct the District Forest Officer to seek the written consent and agreement of the owners of the land. This must be done subject to notices, posted at the concerned District Assembly, Traditional Council and Unit Area as well as at the District Forest Office, which identifies the proposed area. Any person claiming ownership over or interests in the land shall within 21 days inform the District Forest Officer on this. Should there be no objections to the proposed timber utilisation contract area, then the District Forest Officer must obtain the written consent of the land owners (Timber Resources Management Regulations, 1998; 4).

In case of objections from any person claiming ownership over or other interests in the land, the District Forest Officer must refer the matter to a committee consisting of one representative of the Administrator of Stool Lands, two representatives of the concerned District Assembly and a representative of the Traditional Council who will be the committee chairman. This committee shall hear and consider all statements and within 14 days submit a report to the District Forest Officer, stating whether it considers the objections reasonable, and the committee shall also recommend on how the conflicting claims may be resolved. The District Forest Officer will add his comments to the report and send it to the Regional Forestry Officer. After the Regional Forest Officer has satisfied himself about the accuracy of the report and attached maps and plans he will add his comments and submit the case to the Chief Conservator of Forests. The Chief Conservator of Forests will, based on the report, seek to settle the conflict(s). However, no timber utilisation contract on private or stool land may be given out without the written consent of the owners (Timber Resources Management Regulations, 1998; 5-7). When there is a conflict of interests on public land identified as suitable for a timber utilisation contract, the issue must be resolved between the Minister of Lands and Forestry and any other Minister concerned (Timber Resources Management Regulations, 1998; 8).

When timber utilisation contract areas have been identified and eventual conflicts resolved, the Chief Conservator of Forests will advertise such areas indicating, among other things, the estimated yield and call for applications. Three different sizes of timber utilisation contract areas will be made available. Small scale areas not exceeding 2,000 ha and only applicable off-reserves. Medium scale areas between 2,000 ha and 12,500 ha applicable on- and off-reserves. Large scale areas of more than 12,500 ha applicable on-reserves only (Timber Resources Management Regulations, 1998; 9 (2b)).

Applicants for timber utilisation contracts will be evaluated by a Timber Rights Evaluation Committee. This committee will consist of the Chief Conservator of Forests or his representative (chairman), the Forestry Department Planning Officer, the Chief Administrator of the Forestry Commission, the Administrator of Stool Lands, the representative of the Lands Commission on the Forestry Commission, a representative of the Ghana Institute of Professional Foresters and one other person appointed by the Minister (Timber Resources Management Act, 1997; 5). Accordingly, the timber industry will not be represented unless the Minister decides this.

In addition to a number of formalities, applications for timber utilisation contracts will be evaluated according to the following criteria (Timber Resources Management Regulations, 1998; 11):

- Intentions to further process logs extracted from the area;
- number, qualifications, and competence of logging and other technical staff;
- the timber production and operation record (if any) of the applicant;

- previous timber rights (if any) held by the applicant;
- the proposed harvest plan;
- technical and financial capacity of the applicant to operate the area;
- environmental effects of the proposed logging and the applicant's ability to redress such effects;
- a fire protection plan;
- projections of timber production, income and expenditure from operating the area;
- road, logging and (where applicable) log processing facilities;
- proposed measures to secure the area against illegal operations and encroachment;
- plans for disposal of logging and milling residues.

Points will be awarded for each of the above aspects according to a system to be decided by the Timber Rights Evaluation Committee, and applicants scoring above a certain limit (also to be decided) will be invited to submit proposals on (Timber Resources Management Regulations, 1998; 13):

- A reforestation or afforestation plan for the establishment and management of forest plantations of at least 10 ha per 100 ha contract area;
- a social responsibility agreement to assist inhabitants within the contract area with such amenities as shall be specified in the agreement, at a cost of not more than 5% of the annual royalty accruing from the operations under the timber utilisation contract.

On this basis the Timber Rights Evaluation Committee will recommend to the Minister the highest scoring applicant to be awarded the contract (Timber Resources Management Regulations, 1998; 13 (2)). In this way the bidding element of awarding timber utilisation contracts has been reduced to a minimum.

While the criteria for assessing applicants' technical and financial ability to carry out a sound and profitable logging operation appear sensible enough, social aspects such as workers' safety and medical insurance are rather neglected. Furthermore, the system seems vulnerable to abuse as it will be difficult to develop an objective scale of points for many of the assessment criteria. Apart from mopping up revenue that may be available in excess of official forest fees and taxes, presumably the World Bank has been in favour of a competitive bidding element in the allocation of timber rights because the evaluation criteria are objective. The contract would go to the highest bid.

Originally this was also the intention. In order to ensure that applicants for a timber utilisation contract had the capacity (technically, financially and socially) to operate on a large area, they were supposed to pass some pre-qualification criteria much like the ones listed above. The top five companies were then supposed to submit sealed bids specifying in US$ their annual willingness to pay per ha per year (in addition to royalties and all other and fees) for acquiring a given contract . The company placing the highest bid should be recommended to the Minister for obtaining the contract (FD, 1995c; 6, Werbrouck, 1993a, 1993b and Smith, 1993).

The Timber Rights Bill (1996; 5 (2)) still spoke of a "winning bid" among pre-qualified applicants for timber utilisation contracts, but without specifying what kind of bidding process this referred to. In conclusion, it seems that, despite the strong feelings of the World Bank on the issue, bidding in terms of financial offers to be awarded timber rights was never politically acceptable to Ghana.

12.3.3 Termination of Existing Timber Concessions and Leases

As of 18 September, 1998 all existing timber concessions and leases have, in principle, become void, but holders of such rights were granted the option to apply for a timber utilisation contract on such areas (Timber Resources Management Act, 1997; 19). This must have been a hard blow for active concession holders, unless they could be reasonably sure to retain the right to timber on their old concessions and leases. In fact, Cirelli (1997; 65) warned against such a measure as it could prove rather unpopular and recommended to let existing concessions and leases continue, but subject to the new regulations and management procedures.

Assuming that this cancellation of old timber concessions and leases has not been abused by the official system to put undue pressure on active firms, the measure has a commendable and undoubtedly deliberate motive. To get rid of speculative concession holdings which were illegally being sub-led to active companies. Naturally, holders of such concession stand few chances of being rewarded their original areas because they do not have the technical ability to operate them.

Considered in this light the new legislation seems a bold attempt to effectively redress the system of rights to standing timber once and for all. It also suggests that speculative concession holders have lost political influence. Furthermore, a complete shift to the new system of timber utilisation contracts closes all possibilities for concession and lease holders to claim that they are entitled to operate these areas according to the original conditions and not to the conditions stipulated in the Timber Resources Management Act (1997) and the Timber Resources Management Regulations (1998).

The currently moderate difference between the industry's willingness to pay for standing timber and the stumpage fees plus contract area rent should preclude rent seeking behaviour among officials involved in allocation of timber utilisation contracts. Rent seeking behaviour will, however, be a risk if stumpage fees and other taxes do not follow the industry's willingness to pay for standing timber. It is an improvement that stumpage fees are pegged to the export value of lumber, but in the longer term this will not be sufficient. Along with the expected improved efficiency of the processing industries, their willingness to pay for standing timber will increase. Unless this is captured by forest taxes and fees, the room for informal payments will grow and become a temptation that could undermine the credibility of the timber rights allocation system.

12.3.4 Compulsory Supply of Timber to the Domestic Market

It was mentioned in chapter 8 that the supply of timber to the domestic market could become difficult in the not so distant future unless the production of timber was increased or the export volume was reduced. This is obviously an issue of political concern, but there is also a strong political will to maintain the export revenue from timber exports. An attempt to address this dilemma is apparently being made by giving the Minister powers to direct holders of timber utilisation contracts to supply the domestic market with timber products of specified volume, dimensions and species (Timber Resources Management Regulations, 1998; 36).

The social motive of securing Ghanaian timber for the Ghanaians is commendable, but will it work? As timber from Ghana is likely to fetch higher prices overseas than at the domestic market, processing companies would obviously try to avoid compulsory supply for domestic consumption. In addition, the law does not guarantee an equitable distribution among holders of timber utilisation contracts. It is the

Minister who, in consultation with the Forestry Commission, selects the companies that must compulsorily supply the domestic market.

In general, the market always finds a way (officially or unofficially) around political attempts to maintain a price differentiation system for similar products. A more viable way to secure timber for domestic consumption would probably be to give land owners and particularly farmers more rights over trees on their land and land they cultivate, so they could price the resource according to its actual commercial value. Unfortunately, the new legislation does not favour this solution, as off-reserve timber is still vested in the President. This is likely to undermine incentives for farmers to nurse medium-sized naturally occurring timber on land they cultivate (see sections 12.3.6 and 12.3.7).

12.3.5 Timber Utilisation Permits

The Chief Conservator of Forests may on an application by a District Assembly, town committee or any rural community group issue a timber utilisation permit to the applicant. A timber utilisation permit is exclusively for harvesting a specified number of trees on land that is not covered by a timber utilisation contract. Any timber harvested or converted to lumber under a timber utilisation permit may only be used for social purposes and may not be sold or exchanged (Timber Resources Management Regulations, 1998; 37). It also appears that no royalties are supposed to be paid for timber harvested on timber utilisation permits. The obligation to pay stumpage fee and contract area rent is only mentioned for holders of timber utilisation contracts (Timber Resources Management Regulations, 1998; 25 (1)).

As for the Minister's right to demand compulsory supply of timber to the domestic market, the social intention of timber utilisation permits appears commendable. Yet, the concept suffers from the same problem. The official system's attempt to control the commercial utilisation of a politically under-priced resource.

How the Forestry Department should, in practice, be able to ensure that no timber from timber utilisation permits enters the commercial market is difficult to imagine. The law prohibits lumber produced by conversion of logs with chainsaw (chainsaw lumber) from being subject to any commercial exchange (Timber Resources Management Regulations, 1998; 32). In practice, it is, therefore, made compulsory for holders of timber utilisation permits to use chainsaw lumber only, as they are barred from paying a saw mill for converting logs in kind, ie. in lumber or logs. Also, no conveyance certificate may be issued for chainsaw lumber, and all timber being transported shall be covered by a conveyance certificate (Timber Resources Management Regulations, 1998; 24). However, chainsaw lumber can be re-sawn or moulded, making it impossible to distinguish from other lumber, and the fact that no royalties should be paid on timber from timber utilisation permits produces an obvious incentive for finding commercial markets for it.

Accordingly, timber utilisation permits could become an unofficial and illegal backdoor to cheap standing timber through collusion between saw mills and local communities. Only the Forestry Department's ability to monitor activities on timber utilisation permit areas, and to ensure that false conveyance certificates are not issued for logs or lumber from such areas, may prevent abuses. If the number of timber utilisation permits becomes significant, which can be expected over a few years, then the control functions of the Forestry Department could easily become overwhelming.

12.3.6 Protection of Farmers' Rights

"No timber rights shall be granted in respect of lands with farms without the authorisation in writing of the individual, group or owner concerned" (Timber Resources Management Act 1997; 4 (2d)).

The individual farmer is, in other words, given a *de jure* right to veto felling of trees on land he/she cultivates. In principle, farmers may, therefore, set a price for allowing a certain tree to be felled. Whether this will *de facto* happen only time can tell, but without information campaigns and actual support to farmers who attempt to use their new rights the chances appear rather slim. This is particularly so because previous legislation, rules and practices have created a perception among farmers that they have no rights whatsoever to trees on land they cultivate. Furthermore, individual farmers may come under pressure by the Stool, the Traditional Council and the District Assembly. These bodies will all benefit from timber exploitation through their constitutional right to a share of the timber revenue, but they are not obliged to pay anything to the individual farmer.

Had farmers been granted general ownership to trees on land they cultivate, very clear and unambiguous incentives for growing and tending naturally occurring timber trees on farmland would have been produced. Indigenous land and tree tenure before the colonial period actually supports that cultivators of land should also hold rights to naturally occurring trees on such land (FD 1995a; 23). Yet, a return to pre-colonial rules would constitute a major change in the fortune of chiefs and District Assemblies. In addition, it would probably be in conflict with the constitutional provisions on sharing of stool land revenue (Cirelli, 1997; 25).

To sum up, the strongest incentive for promoting off-reserve timber production, giving farmers full ownership over trees on farmland, appears impossible to implement for legal and cultural reasons. The probably next best (easiest) way to promote off-reserve timber production, to divest the President of rights in trees off-reserves, is not being pursued for political reasons.

This consolidation of farmers's *lack* of rights to trees on land they cultivate is likely to become the root of many future forest policy problems in Ghana. Instead of creating clear and unambiguous incentives to nurse naturally occurring trees on farmland, the new legislation reflects a political preoccupation with rights to benefit from exploitation of timber off-reserves until the resource is gone. This focus on timber extraction rather than production undermines the potentials of making farmers timber growers. Accordingly, the timber related income, which is not the result of a deliberate and planned promotion of timber production, to Stools, Traditional Councils and District Assemblies will dwindle together with the off-reserve timber resource. As a result the timber industry is likely to be short of raw material.

12.3.7 Protection of Rights to Cultivated Trees

To at least produce some incentive for growing trees off-reserves the new legislation states that "no timber rights shall be granted on lands with timber grown or owned by any individual [or] group without the written authorisation of the individual, group or owner concerned" (Timber Resources Management Act, 1997; 4 (2b)). The term "grown" is not specified, but it must be assumed to cover planted trees and naturally occurring trees which have been nursed from sapling, ie. not mature and medium-sized naturally occurring trees which are or will soon become commercially valuable for the industry.

The intention seems to be that, in time, all trees off-reserves will be planted or nursed by individuals or groups who will then have full ownership to such trees (FD, 1995a; 23). This is a very long but currently not particularly attractive investment perspective, considering that many naturally occurring mature and medium-sized timber tree are still found off-reserves. The industry's access to timber off-reserves will be secured through timber utilisation contracts. Local communities, NGOs and the like may get access to cheap timber through timber utilisation permits. So why grow trees on farmland? This will only become really attractive when other resources are running out. However, by then the market for at least export oriented industrial timber may have collapsed.

Meanwhile, farmers are not likely to invest anything in nursing naturally occurring timber trees which are bigger than seedlings on land they cultivate. A lot of pole size timber trees which would need another 20-50 years before they are suitable for industrial use will probably end up as firewood long before then. Timber utilisation contracts or not. In the end, no logging company or official institution can prevent farmers, who are the actual managers of trees off-reserves, from doing what they find fit. Protection without a guaranteed payment of existing medium-sized timber trees for the future benefit of the timber industry is not likely to rank high on any farmer's list of priorities.

12.3.8 Protection of Forest Plantations

The Timber Resources Management Act (1997; 4 (2a)) protects land with forest plantations from becoming part of a timber utilisation contract. At least this provides a minimum guarantee that investments in timber plantations will benefit the investor and not some other company which acquires rights to timber in a given area that happens to include a timber plantation (actually such guarantees did previously not exist). The question is, however, whether this is enough to initiate investments in timber plantations.

Under the established incentive structure it seems likely that, unless investments in fairly big tree planting programmes off-reserves or inside degraded forest reserves are undertaken, industrial timber will be available from forest reserves only, once the existing naturally occurring off-reserve resource has been logged out. In terms of securing its resource base the domestic industry should, therefore, be interested in timber plantations since it cannot expect farmers to invest in planting or tending naturally occurring trees on farmland.

Timber plantations would require long-term contracts between land owners and the investor plus substantial funds. Long-term contracts with land owners should be possible to organise (oil palm plantations have been established through such arrangements). Yet, funds for establishing timber plantations will only be available if the investment is likely to be profitable.

This is the weak point. Investments in timber plantations are *not* likely to be profitable. Why? Mainly because the expected selling price of plantation logs is too low. Why? Because domestic log prices of competing species from forest reserves are kept artificially low through the log export ban and the package of forest taxes and fees (HTS, 1995, appendix D; 22).

One might expect that timber plantations will be relatively more valuable in the future than today, because domestic log prices are bound to rise when existing off-reserve timber resources are increasingly exhausted. Yet, in a controlled market it is basically up to the politicians to secure that this actually happens. In other words, the profitability of plantation investments is highly dependent on political decisions, which does not guarantee that prices will reflect supply and demand in a meaningful way.

The study of HTS (1995) on plantation forestry in Ghana was a consultancy for ODA. One of the objectives was to assess the potentials for investing in plantation forestry in Ghana (mainly Teak) and consultations on this issue were held between HTS and the Commonwealth Development Corporation (HTS, 1995; 3). It speaks for itself that the Commonwealth Development Corporation has decided not to invest in plantations in Ghana.

In theory, plantation forestry might be profitable for domestic timber industries since they could use the produced logs as raw material instead of having to sell them at the domestic market. In practice, the industry is not particularly interested in plantation wood because its machinery and skills is geared for big diameter logs (Grut, 1996; 16).

Already existing plantations have not received much interest from the industry either. In the early 1970s around 50,000 ha of mainly Teak plantations were established. Today it is estimated that only approximately 15,000 ha of these plantations have survived (FD, 1993; 4). About 50,000 transmission poles are annually taken out of these plantations and for the period 2005-2015 an annual volume of approximately 300,000 m^3 saw logs may be extracted from clear fellings of approximately 35 year old stands (FD, 1993; 5). This will, however, be relatively small dimension logs as it would probably require about 80 years to produce Teak of good quality with a high proportion of the valuable and durable heartwood (Grut, 1996; 15).

Accordingly, existing Teak plantations are not likely to deliver logs that would be of interest to the current wood processing industries for another 60 years. Another option is to clear fell the stands at the age of 35, but for which buyers? Since the resource might run out over a 10 year period, investments in new machinery specifically for small dimension timber appears rather unattractive.

12.3.9 Powers of the Minister

The rights of the Minister to prescribe conditions under which timber harvesting may take place inside and outside forest reserves are fairly elaborate in the Timber Resources Management Regulations (1998). They appear to represent a genuine attempt to avoid any unclarity on who has authority to define (i) rules for logging and (ii) conditions for timber utilisation contracts and permits. The Forestry Department has some bitter experience is this respect as logging companies have previously succeeded to legally outmanoeuvre attempts of the Department to enforce better logging practices.

In 1992 the Forestry Department submitted a Logging Manual (FD, 1992) which all logging operations were supposed to follow. The manual generally stipulated sensible technical and environmental standards for good logging practice. Furthermore, it introduced a "completion certificate" which, subject to inspection by the District Forest Officer, should be issued for each logged-over compartment before the concession holder was allowed to begin logging in a new compartment (FD, 1992; 7). Yet, this never happened because the Logging Manual had no legal backing. A District Forest Officer told that he was "booted out of court" when in one case he tried to stop a contractor who refused to follow the Logging Manual (FD, 1992) (Personal information). In any case, the Forestry Department never tried very hard to enforce it.

The rights of the Minister to protect trees and tree species which are considered endangered refer to various international treaties and conventions that Ghana is signatory to and to concerns over the depletion of different species. It should, for example, be possible for the Minister to ban the felling of species listed by CITES. Such rights are, however, not new. With reference to the Trees and Timber

Decree (1974; 17 (1f)), the harvest of Subaha was in 1993 banned in order to protect banks of rivers and streams where this species is normally found. Special felling permits, which must be issued by the Chief Conservator of Forests, are today required for the following species (Timber Resources Management Regulations, 1998; 40):

- Edinam (Entandophragma angolense).
- Sapele (Entandophragma cylindricum).
- Utile (Entandophragma utile).
- Candollei (Entandophragma candollei).
- Mahogany (Khaya anthotheca/grandifolia/ivorensis).
- Odum (Milicia excelsa/regia).
- Kusia (Nauclea diderrichii).
- Makore (Tieghmella heckelii).
- Hyedua (Guibourtia ehie).
- Afrormosia (Pericopsis elata).

It is seen that all species except Candollei belong to the group of Scarlet Stars. As mentioned in section 7.5.2 Candollei should be moved from the Red to the Scarlet Star group, so the list of species is very sensible.

Whether the special felling permit system will serve to especially protect these species, which are all in very high demand, depends on how it is implemented. A constructive way of using the system would, for each of the restricted species, be to require that a certain minimum number of trees above a certain diameter must be situated on the same compartment as the tree for which a felling permit is requested.

12.4 Summary Assessment of Rights to Timber and Related Benefits

Past legislation has successively eroded the rights of land-owning communities to benefits and revenue from forest reserves. Local communities' rights to extract forest produce have become lost or fuzzy because such rights were established when the forest reserves were declared some 50 years ago. Furthermore, extensive rights were given to the President to constitute forest reserves with or without the consent of the land owners and without paying any compensation to land owners (Administration of Lands Act, 1962; 7 (1-2)).

Today the right to constitute forest reserves without paying compensation to land owners must be considered unconstitutional (Cirelli, 1997; 23 and Constitution, 1992; 20 (2a)). The rights of land-owning communities to extract forest produce is, however, an outstanding issue. It appears that the Forest Department would like to be given authority to control not only the extraction of timber but also all forest produce in order to secure a sustainable utilisation of the resource (FD, 1995a; 18). Another approach would be to grant land owners extensive rights to extract forest produce for domestic consumption, while allowing for some control of the Forestry Department over some produce for which there is a documented risk of over-exploitation. This would avoid a lot of unnecessary control functions, which the Forestry Department may not be able to lift anyway, and bring some substance to the original intention that forest reserves should the managed for the benefit of the land owners (Forest Ordinance (Cap. 157), 1927 (18; 2b).

12.4.1 Revenue from Forest Reserves

The right of land owning communities (stools or skins) to revenue accruing from forest reserves was with the Administration of Lands Act (1962; 17 (3 and 4)) reduced from minimum 67% of the gross revenue to what was left over after the Forestry Department had deducted the costs of its management. In practice, nothing was left for the land owners, so it was administratively decided to pay minimum 30% of the timber royalties from forest reserves to land owners. Yet, the Forestry Department apparently did not consider this decision binding, as it did not pay anything to many stools over a long period.

The Constitution (1992; 267 (6)) clearly stipulates that stool land revenue, after a retention of 10% for administrative costs of the Administrator of Stool Lands Office, should be divided as follows:

a) twenty-five per cent to the stool through the traditional authority for the maintenance of the stool in keeping with its status;
b) twenty per cent to the traditional authority;
c) fifty-five per cent to the District Assembly, within the area of authority of which the stool lands are situated.

The ownership of land did not change when forest reserves were constituted (Forest Ordinance (Cap. 157), 1924; 18 (1)), so practically all forest reserves are situated on stool land. Accordingly, the retention of funds by the Forestry Department to cover the costs of management probably became unconstitutional after 1992. The continued practice of the Forestry Department to retain revenue from forest reserves was, therefore, probably illegal.

In December, 1998 legal basis to pay the Forestry Department for management services out of timber revenue was provided in the Timber Resources Management Regulations (1998; 21 and 26). It is not specified whether such payments should be deducted from stool land revenue at large or from forest reserve revenue only. Neither does the new legislation set a maximum limit for how much may be deducted from stumpage fee revenue to finance the Forestry Department.

Unless (i) the revenue from timber exploitation is increased significantly, (ii) the cost of running the Forestry Department is reduced significantly or (iii) off-reserve revenue may be used to fund the Forestry Department, there will not be much revenue left for local communities who own land on which forest reserves are located. The new system of stumpage fees and contract area rent is not likely to generate much more than DM 8.6 million annually from forest reserves in the near future (assuming the harvest is held within the AAC). If the annual costs of running the Forestry Department is approximately DM 7.5 million, this would leave less than 10% of the revenue for the land owners unless other sources may contribute to finance the costs of forest management (cf. section 11.1). Whether this is reasonable is a political decision, but it is quite far from the original minimum of 67% defined in the Forest Ordinance (Cap 157) (1927, 18 (2)).

Revenue from forest reserves may in the longer term be higher than DM 8.6 million per year, since the stumpage fees could be gradually increased as could the harvest of currently "under-utilised" species. Yet, it appears crucial that the Forestry Department receives funding through other sources than timber revenue from forest reserves if it is to maintain any credibility in relation to local communities who actually own the land carrying forest reserves. After all the Forest Ordinance (Cap. 157) (1927; 18

(2b)) says that when the Government is managing forest reserves it should be done "for the benefit of the owner or owners".

Naturally, it is not only the land owners' rights, but also the conservation of Ghana's forest resources which are at stake. It would seem bizarre if on-reserve timber exploitation could not be controlled because the national forest authority cannot be adequately financed. Yet, it is also reasonable that a national forest authority is not over-staffed or over-financed at the expense of the land-owning communities. In addition, it must be ensured that the forest authority is held accountable for the standard of on-reserve forest management.

12.4.2 Revenue from Off-Reserve Areas

It is not clear how big a share of off-reserve stumpage fees the Forestry Department can claim. In fact, the new legislation seems to be deliberately held in unspecific terms concerning how the Forestry Department may receive funding through stumpage fees. Off-reserves, the Forestry Department is intended to provide little management service other than a timber inventory, identification of areas suitable for award of timber rights and some control over logging operations. On this background it seems difficult to justify that the Forestry Department should receive a significant share of the revenue accruing from off-reserve timber exploitation.

Nevertheless, the new legislation appears to be designed such that the Forestry Department may claim a fairly big share of the timber revenue accruing from areas off-reserves. Otherwise it would, as indicated above, not be in a position to fund its management of forest reserves *and* pay a reasonable share of the on-reserve timber revenue to land-owning communities.

Today timber grown or owned by any individual or group is protected from being included in a timber utilisation contract area and farmers' right to veto felling of trees on land they cultivate has also been given full legal backing (Timber Resources Management Act, 1997; 4 (2)). However, the Timber Resources Management Act (1997; 21) has maintained the general vestiture of trees on stool land in the President. This secures that the authority to grant rights to naturally occurring trees off-reserves remains with the Government. As a result, the authority to regulate and allocate such rights stays with the Forestry Department. The Forestry Department can, therefore, claim that it is entitled to a share of off-reserve stumpage fees in accordance with the Timber Resources Management Regulations (1998; 12 (2)), because it is managing and regulating off-reserve timber resources.

12.4.3 Poor Incentives to Tend Natural Trees on Farmland

By securing the Forestry Department a formal right to off-reserve timber revenue the Timber Resources Management Act (1997) and the Timber Resources Management Regulations (1998) consolidate farmers' *lack* of rights to currently valuable trees on land they cultivate. Instead of creating clear and unambiguous incentives to nurse naturally occurring trees on farmland, the new legislation seems to reflect a political preoccupation with rights to benefit from off-reserve timber exploitation until the resource is gone. This focus on extraction rather than production undermines the potentials of making farmers timber growers. Accordingly, timber related income to Stools, Traditional Councils and District Assemblies will dwindle together with the off-reserve timber resource. As a result, the timber industry is likely to be short of raw material.

It may be argued that the constitutional provisions on sharing of stool land revenue rule out the strongest incentive for promoting off-reserve timber production, giving farmers full ownership over trees on farmland. However, the probably next best (easiest) way to promote off-reserve timber production, to divest the President of rights in trees off-reserves, has not been pursued either. Political concerns over how to finance the Forestry Department appear to be an underlying political reason.

12.4.4 Supply of Timber to the Domestic Market and for Export

The dual objective of being able to meet the future domestic demand for timber and to maintain foreign exchange earnings from wood exports is rightfully a source of concern. The AAC of approximately 0.85 million m^3 from forest reserves alone will obviously not be able to serve both objectives, since the domestic demand alone is already some 0.70 million m^3 (cf. section 8.2).

The Government's approach to this problem is an attempt to control the domestic timber market and the off-reserve timber exploitation rather than one of creating incentives for increasing the off-reserve timber production. On one hand, the Government is subsidising export of value added products through log export bans and forest fees that result in domestic log prices which are significantly lower than the FOB log prices. On the other hand, the Government reserves the right to direct holders of timber utilisation contracts to supply wood products to the domestic market (Timber Resources Management Regulations, 1998; 36).

Whether such a system can actually work when it becomes necessary (when the sustainable supply of raw logs cannot meet the combined export driven and domestic demand) appears doubtful. The obvious alternative, over-exploitation of the remaining resource, will definitely be a risk.

In any case, the very introduction of rights to demand compulsory supply of wood products to the domestic market shows that the Government is aware that the production of timber may not be able to keep up with the demand. The irony is that strong incentives for sustaining a substantial off-reserve timber production have deliberately not been established. Presumably, the background is (at least partly) a wish to secure the Forestry Department a right to off-reserve timber revenue such that it can finance the management of forest reserves (cf. sections 12.4.2 and 12.4.3).

Establishment of plantations to prevent future shortages of timber is not likely to be undertaken and plans of establishing big mainly Teak plantations through foreign investments have recently been scraped. The main reason appears to be that the profitability of plantations is undermined by the artificially low domestic log prices on naturally occurring species, with which wood from plantations would have to compete. The domestic industry is not geared to process small diameter plantation logs, which is a reason for it not to invest in plantations, at least not in the foreseeable future.

Accordingly, the medium long-term timber supply scenario is likely to show:

- Decreasing volumes being extracted from areas off-reserves as the resource dwindles;
- substantial volumes of medium-sized naturally occurring timber trees on farmland being cleared for non-industrial purposes;
- inadequate investments in plantations to keep up with the combined demand for domestic consumption and wood exports.

Assuming that existing medium-sized timber off-reserves will be cleared for other than industrial purposes (which seems likely), the lifetime of Scarlet Star species off-reserves may only be 15 years (cf. section 7.6). By that time plantations would not be yielding any saw logs even if they were established today. The approximately 15,000 ha of reasonably good condition Teak plantations may offer some temporary relief, but the stands are generally less than 25 years old. Even 20 years from today, saw logs from these plantations will still be of small dimension compared with the indigenous timber that the industry is geared to handle. Overall, the industry is, therefore, likely to become increasingly desperate for logs to maintain export earnings. Meanwhile, the domestic demand for timber is likely to have reached some one million m^3. This is approximately 0.15 million m^3 above the on-reserve AAC.

12.4.5 Conditions for Managing Forest Reserves

Following the above analysis, a tremendous pressure on forest reserves for timber may be expected within the coming 10-15 years. Whether the Forestry Department will be able to control the on-reserve timber harvest is an open question. In terms of harvest control the recent record of the Forestry Department is not exactly impressing, so it is paramount that the supply-side regulative function of the Forestry Department is re-habilitated.

The Timber Resources Management Act (1997) and the Timber Resources Management Regulations (1998) appear to provide a reasonably good framework for managing on-reserve timber harvest. The right to define standards for logging operations is clearly with the official authorities and punishments for breaking the rules and regulations are quite severe for both loggers and officials.

Positive incentives, such as reasonable salaries, for officials to efficiently perform logging control on the ground are yet to be established. This is pursued through lifting the existing Forestry Department out of the civil service system and making it a semi-autonomous body, to be named the Ghana Forest Service. Among other things, this would allow for paying employees higher wages than under the civil service structure.

However, the combination of (i) an AAC from forest reserves which is expected to be below the domestic demand in about 10 years, (ii) a high demand for raw logs from a wood industry that is politically strongly encouraged to be export oriented and (iii) poor incentives to invest in timber production, presents a serious risk of failure no matter how well structured and well motivated a future Forest Service might be. This is likely to become the cardinal problem for managing the forest reserves.

13 Forest and Timber Resources Management in the Future

During the implementation of the Forest Resource Management Project and other associated forestry sector related projects, a process of revising the 1948 forest policy took place. It was not difficult to see that the 1948 policy on several points had been outdated for some time. Particularly the environmental and economic wisdom of continuing a deliberate deforestation of areas outside forest reserves to make way for agriculture had come into question. The growth of the timber industry, its impact on the resource base as well as its economic importance and potential had also created a need for re-assessing and clarifying the policy on commercial timber utilisation. Furthermore, the philosophy of enhancing public participation in forest and timber resources management had increasingly gained ground in decision-making circles.

Finally in 1994, following numerous workshops and stakeholder consultations, the Ministry of Lands and Forestry submitted a new Forest and Wildlife Policy (1994), which stipulates the principles for development of the forestry sector. It is recognised that, during the past decade, the Governments' attempts to stimulate wood processing and exports have "yielded varied results", which is explained by inadequate institutional capacity (Forest and Wildlife Policy, 1994; 4). The previous chapters in the present study suggest that a more direct way of putting it would be to state that, during the past decade, commercial timber harvest has led to over-exploitation of the resource and yielded questionable economic results for the country.

In any case, the 1994 Forest and Wildlife Policy is very comprehensive and rather detailed on aims and objectives of managing the forest resource, not least in relation to commercial timber production. Reference is also made to the International Tropical Timber Organisation, of which Ghana is a member, and its "Target 2000" aimed at ensuring that all trade in tropical timber will be sourced from sustainably managed forests by the year 2000 (Forest and Wildlife Policy, 1994; 5).

The overall aim of the 1994 Forest and Wildlife Policy is "conservation and sustainable development of the country's forests and wildlife resource for maintenance of the environmental quality and perpetual flow of optimum benefits to all segments of society" (Forest and Wildlife Policy, 1994; 8). With respect to different stakeholder groups, the policy objectives and strategies may be summarised as follows:

The Timber Industry:

- The private sector is acknowledged as the engine of economic growth;
- promotion of responsible resource utilisation through long-term concessions;
- allocation of rights to standing timber on the basis of competitive bidding among properly equipped processors and entrepreneurs;
- encouragement of more rational resource utilisation and in particular production of value added products for export through forest fees and taxes that are regularly reviewed to reflect the value of the resource and to recover optimum revenue for resource management;
- transformation from a high volume low value to a low volume high value sector through promotion of secondary and tertiary processing;
- development of the marketability of lesser used timber species.

Institutions:

- Forest fees and taxes to be regularly reviewed to reflect the value of the resource and to recover optimum revenue for resource management;
- effective capability to be developed at national, regional and district level to ensure sustainable management of forest and wildlife resources;
- unreserved forests to be included under the Forestry Department's management system for regulating timber harvest and collection of forest revenue;
- reorganising the Forestry and Wildlife Departments as autonomous or semi-autonomous agencies with a view to upgrade staff performance and to improve monitoring, coordination and accountability.

Local People:

- Rights to have access to natural resources for maintaining a basic standard of living;
- incorporation of traditional methods of resource management where appropriate;
- rights to a share of the financial benefits arising from resource utilisation;
- rights to be involved in matters concerning their welfare;
- consultative and participatory mechanisms to be developed with a view to enhance land and tree tenure for farmers;
- support with respect to management and legal protection of natural resources which local communities decide to protect and manage for domestic and economic purposes.

Concerning the resource it is the objective of the 1994 Forest and Wildlife Policy to:

- Protect and manage areas already set aside for permanent forest and wildlife conservation/preservation;
- expand the conservation area through a dedication of various land categories with potential for nature protection and production of timber and other products;
- include off-reserve timber resources under Forestry Department's management system for regulating timber harvest and collection of forest revenue.

13.1 Development of the Timber Industry

As described in chapters 10 and 11 the intentions on developing the timber industry have to a large degree been followed up by the Timber Resources Management Act (1997) and the Timber Resources Management Regulations (1998). The system of awarding rights to standing timber has been revamped in a way that should benefit active and dedicated companies, although the bidding element has, in effect, been scraped, which could make the system somewhat vulnerable to abuse.

The forest fee and taxation system has also undergone a highly needed transformation that (i) links stumpage fees to the FOB value of air-dried lumber, (ii) differentiates adequately between species according to their abundance and (iii) has increased the taxation level significantly such that forest taxes and fees almost equal the current willingness of the industry to pay for standing timber. In combination with the log export ban, forest taxes and fees, therefore, produce clear incentives for further and more efficient processing.

The log export ban and the current level of forest fees and taxes secure cheap logs for the domestic industries. Accordingly, a transformation of the timber industry from a high volume low value to a low volume high value industry will to a high degree depend on how quickly the gap between domestic and FOB log prices can be closed.

The log export ban and the forest fees are subsidising domestic processing by keeping domestic log prices below world market prices, but by doing so incentives to grow timber or invest in tending naturally occurring trees are obviously reduced. This anti-production effect is even amplified through the rules and regulations on rights to standing timber (see section 13.3). In fact, it seems that no matter how efficient the industry may become over the next decade, it could easily be short of raw material if incentives to produce timber are not improved.

13.2 A Semi-Autonomous Forest Service

The Forestry Department should be lifted out of the civil service system and restructured under a new name, The Ghana Forest Service. It is not considered possible to create an effective national forestry institution without loosening it from the constraints associated with being a part of the civil service sector. In particular, salary levels are critical in this respect.

The financial base for a new Forest Service has obviously be improved through the stumpage fee system. Yet, as illustrated in table 11.1 the annual costs of running a new Forest Service are likely to be approximately DM 7.5 million, excluding development costs (Touche Ross, 1995; 21). This is almost 90% of the expected revenue from forest reserves, which would leave the new Forest Service with a credibility problem if additional sources of income cannot be found. To this end it appears that the intention is to claim a share of the off-reserve timber revenue with reference to the legally defined regulatory functions of the Forestry Department (cf. section 12.4.2). Reducing the costs of running the Forestry Service may, theoretically, be another option, but for political reasons this is not likely to happen as it would require a sizable reduction of the number of employees.

Financially, the viability of a new Forest Service will, therefore, depend on a sustained flow of funds from off-reserve areas. This is a big challenge as the incentives for investing in a sustained timber yield from off-reserve areas are currently rather poor. Individual farmers are not likely to invest in naturally occurring trees from which the Forestry Service could claim a share of the stumpage fee. Existing plantations may generate some temporary revenue for the Forestry Service, but new plantations, if ever established, are likely to be private. The only viable solution seems to be a massive promotion of dedicated forests (see sections 13.4 and 13.4.2).

13.3 Interests of Local People

Existing rights of local people to extract goods (NTFPs) from forest reserves date back to when forest reserves were established, so they have become increasingly difficult to verify. As the 1994 Forest and Wildlife Policy re-affirms the original rights af local people to draw subsistence benefits from forest reserves, it would seem appropriate to grant communities, who own the land on which forest reserves are situated, extensive rights to extract goods from the reserves.

Granting extensive rights of managing NTFPs to the Forest Service, including the right to collect fees from NTFPs, would not be in the spirit of the 1994 Policy on Forests and Wildlife which requires incorporation of traditional methods of resource management where appropriate. In order to avoid over-exploitation of certain goods, the new Forest Service should be obliged to develop participatory plans for managing NTFPs on-reserves. Land-owning communities' rights to a share of the financial benefits from timber utilisation have been re-affirmed in the Timber Resources Management Regulations (1998; 21 (2)). How big this share is going to be will be decided by the Minister for Lands and Forestry in consultation with the Forestry Commission, the Forestry Department, and the Administrator of Stool Lands. The latter institution is presumably supposed to represent the interests of the land owners. Yet, the arrangement appears undemocratic and more concerned with securing a legal basis for funding the Forestry Department (or a future Forest Service) than securing a minimum share of the financial benefits from timber harvest to the land owners. What if the land-owning communities are not satisfied with the share of stumpage fees they are awarded? According to which legislation can they file a case?

The Timber Resources Management Act (1997) and the Timber Resources Management Regulations (1998) have institutionalised consultative and participatory mechanisms for identification of timber utilisation contract areas. Enhancement of farmers' rights to land and trees is limited to veto felling of trees on land they cultivate, but it is even doubtful if this right can, in practice, be exercised. In fact, it appears that the 1994 Forest and Wildlife Policy is promising more than it can keep in terms of granting farmers more rights to land and trees, since the Constitution (1992; 267) clearly vests all stool lands in the stools and stipulates stool land revenue to be shared between stool, the traditional authority and the District Assembly. In addition, the Timber Resources Management Act (1997; 22) has maintained the vestiture of naturally occurring trees off-reserves in the President.

Support with respect to management and legal protection of natural resources, which local communities decide to protect and manage for domestic and economic purposes, is still to be institutionalised and legally backed. Areas which local communities wish to set aside for protection of flora and fauna or cultural/religious reasons should be relatively straightforward to legally back if this does not involve large areas which would otherwise be deemed suitable for an off-reserve timber utilisation contract.

Areas which local communities wish to set aside for their own economic purposes are likely to become a more thorny issue, at least as far as timber production is concerned. Nevertheless, it appears crucial that solutions which are attractive for the land owners are found. Otherwise, the supply of timber could soon be too low to meet both the domestic and export driven demand. This would undermine foreign exchange earnings from wood exports and make it difficult to keep the on-reserve harvest within the AAC. In addition, the source of revenue that off-reserve timber harvest has represented to stools, traditional authorities, District Assemblies and other official bodies like the Forestry Department will dry out.

13.4 Management of Forest and Timber Resources

It will be difficult for the Government to establish more forest reserves, as this would require substantial compensation payments to the land owners (cf. section 12.1.1). Enhancing nature and timber conservation is rather going to be pursued through dedicated forests, a new concept introduced by the Forest and Wildlife Policy (1994; 5 (3)). Dedicated forests and other nature areas are to be set aside by the land-owning communities for protection of natural resources and cultural heritages like sacred groves (FD, 1995a; 9). The establishment of such areas appears a flexible and cost effective way of enhancing the conservation of forests of cultural and religious importance off-reserves.

13.4.1 Production of Timber in Dedicated Forests

Dedicated forests are envisaged to be set aside by the land owners (not the District Assemblies) and managed by them in accordance with a plan which the Forestry Department must endorse (FD, 1995a; 20). The Forestry Department could also, subject to a reasonable fee, deliver technical assistance to prepare management plans for dedicated forests. Yet, as recommended by FD (1995a; 28), the right to naturally occurring trees on stool land is still vested in the President (Timber Resources Management Act, 1997; 22). The financial incentives of creating dedicated forests for timber production are, therefore, not clear. At least not yet.

Guarantees against dedicated forests being included in timber utilisation contract areas must naturally be provided. In addition, it is obvious that land owners will require additional rights to trees in dedicated forest where long-term timber production is one of the management objectives. Otherwise it would be rather pointless for them to consider timber production. A full transfer to the land owners of ownership to all trees in dedicated forests would produce the best incentives for timber production, but this is likely to run against the constitutional provisions on how revenue from stool land must be shared between the land-owning stool, the traditional authority and the District Assembly (Constitution, 1992; 267 (6)).

Accordingly, the only way to promote a sustained timber yield from dedicated forests is through a flexible interpretation of the term "stool land revenue". In order to stimulate off-reserve timber production it is paramount that provisions are made for deducting a fee to the land owners for forest management from the gross revenue *before* the rest is distributed in accordance with the constitutional provisions.

13.4.2 Off-Reserve Timber; The Key to Sustainable Forest Management

It should be a top priority of the Ministry of Lands and Forests to ensure that adequate financial incentives are established for local communities to manage off-reserve forest and timber resources. In the medium-long term this is the only feasible way to secure a sustained flow of timber for export and domestic consumption. Four fundamental facts support this conclusion:

- The AAC from forest reserves is currently slightly higher than the domestic demand, but this is likely to change in a few years;
- existing plantations are not big enough to substitute the current off-reserve harvest, and new plantations, if ever established, need minimum 40 years before they will yield saw logs;
- farmers have currently no incentives to tend medium-sized naturally occurring trees on land they cultivate;
- The stock of Scarlet Star species off-reserves is likely to be logged out within the next 15 years *if incentives to produce timber off-reserves are not significantly improved.*

These are the core problems of managing timber resources in the high forest zone and they should be carefully considered in relation to the likely demand for timber in the medium-long term. There are undoubtedly strong interests invested in maintaining the revenue from wood exports.

Realistically this renders a drastic reduction of the current export harvest impossible[43]. At the same time domestic timber demands will undoubtedly be met officially or unofficially.

Accordingly, the feasibility of implementing sustainable management of forest reserves relies heavily on the degree to which a sustained timber yield from off-reserve areas may be realised. Dedicated forests are the only realistic option.

In the process towards sustainable forest and timber resources management, decision-makers may have acknowledged that there will be no timber without trees. Now they are well advised to look beyond this obvious fact and realise that there will be no trees for timber if there are no incentives to grow them.

---

43 In 1996 the export harvest was almost 1,000,000 m^3, of which a minimum of 570,000 m^3 were Scarlet Star species.

References

Administration of Lands Act 1962: *Administration of Lands Act (Act 123) 1962*. Government Printing Department, Accra, Ghana, p. 15

Ahadome, E.C. 1992: *Forestry Taxation System in Ghana.* Forestry Commission, Accra Ghana, p.10. Unpublished.

Alder, D. 1996: *Technical Assistance to the Forest Sector Development Project, Ghana; Consultancy Report No. 4; Report on a Consultancy to the Planning Branch, Ghana Forestry Department.* ODA Contract No. CNTR 955528A. Fountain Renewable Resources Limited, UK, p. 15. Unpublished.

Alder, D. 1992: *Simple Methods for Calculating Minimum Diameter and Sustainable Yield in Mixed Tropical Forest.* In Miller, F.R. and Adam, K.L. (eds.) Wise management of tropical forest. Proceedings of the Oxford Conference on Tropical Forests. OFI, Oxford. pp.189-200.

Alder, D. 1990: *GHAFOSIM: A Projection System for Natural Forest Growth and Yield in Ghana.* Manas Systems Ltd, Systems Consultants, p. 58 + Annexes I-IV. Unpublished.

Alder, D. 1989: *Natural Forest Increment, Growth and Yield.* In FIP (1989a; 47-52).

Bach, C.F., Gram, S. and Treue, T. 1995: *Incentives for Sustainable Production of Tropical Timber.* Department of Economics and Natural Resources, Royal Veterinary and Agricultural University, Copenhagen, p. 211. Unpublished report to Danida.

Benneh, G. and Agyepong, G.T. 1989: *Land Degradation in Ghana.* London Commonwealth Secretariat, London, p.138.

Blackett, H.L. 1989: *Forest Inventory Methods.* In FIP (1989a; 14-22).

Cirelli, M.T., 1997: *Forestry Legislation; Interim report to the Government of Ghana.* ODA Forest Sector Development Project, p. 68. Unpublished.

Concessions Act 1962: *Concessions Act, 1962 (Act 124).* The Government Printing Department, Accra, Ghana, p. 10.

Concession Regulations 1962: *Concession Regulations 1962 (L.I. 228).* The Government Printing Department, Accra, Ghana, p. 12.

Constitution 1992: Constitution of the Republic of Ghana. Tema Press of Ghana Publishing Corporation, p. 210.

Economic Plants Protection Decree 1979: *Economic Plants Protection Decree 1979 (A.F.R.C.D. 47).* Ghana Publishing Corporation, Printing Division, Accra-Tema, Ghana, p. 2.

Enevoldsen, K., Kronslev, P., and Treue, T. 1991: *Regnskovsdrift i Ghana; Analyse af Bæredygtighed [Rainforest Management in Ghana; Assessment of Sustainability].* Department of Economics and Natural Resources, Royal Veterinary and Agricultural University, Copenhagen, p. 292. Unpublished.

Falconer, J. 1993: *Non-timber Forest Products in Southern Ghana; Main Report.* Natural Resources Institute, Chatham Maritime, UK, p. 244.

Falconer, J. 1992: *Non-timber Forest Products in Southern Ghana; A Summary Report.* Natural Resources Institute, Chatham Maritime, UK, p. 23.

FAO 1992: *Forest Products and Prices 1971-1990.* FAO Forestry Paper 104, Food and Agricultural Organisation of the United Nations, Rome, p. 254.

FAO 1983: *Forest Revenue Systems in Developing Countries.* FAO Forestry Paper 43, Food and Agricultural Organisation of the United Nations, Rome, p. 261.

FC 1989: *Revision of Ghana's Forest Policy.* Forestry Commission Symposia Series No.3: Proceedings of a National Conference 2nd-8th April, 1989, Accra Ghana. J.G.K. Owusu (ed.), Forestry Commission, Ghana p. 238.

FD undated: *Notes for all Regional & District Officers on the Calculation of Yield from Stock Survey.* Forestry Department, Ghana. Unpublished.

FD 1997: *Letter of 18 April, 1997 titled "Payment of Royalties Inside Reserves".* The letter refers to Bank of Ghana Cheque No. GG^{I}_{M} 0400462 for Cedis 136,167,989.88 (of which the Lands Commission has kindly provided a copy). Unpublished.

FD 1995a: *The proposed Contents of a New Forest Act: Discussion Paper.* Committee for the Review of Laws in the Forestry Sector for the Ministry of Lands and Forestry. Forestry Department, Accra, p. 79. Unpublished.

FD 1995b: *Interim Procedures for Felling Trees Off-reserve.* Forestry Department, Headquarters, Accra, p. 23. Unpublished.

FD 1995c: *Draft Timber Concession Allocation Procedures.* Forestry Department, Accra, p. 6. Unpublished.

FD 1994a: *Interim Measures to Control Illegal Timber Harvesting outside Forest Reserves; Recommendations of the Forestry Department [Working Group] on Control of Illegal Felling Outside Forest Reserves.* Planning Branch, Forestry Department, Kumasi, Ghana, p. 77 + annexes 1-7. Unpublished.

FD 1994b: *Total and Budgetary Allocations and Loans.* Summary accounts obtained from the Forestry Department Chief Accountant. Forestry Department Headquarters, Accra. Unpublished.

FD 1993: *A Strategy for the Development of Plantations in the High Forest Zone of Ghana.* Forestry Department, Planning Branch, Kumasi, p. 24. Unpublished.

FD 1992: *Logging Manual for Forest Management in Ghana.* Forestry Department, p. 13. Unpublished.

FIMP 1997a: *Manual of Procedures for Off Reserve Inventory.* Planning Branch, Forestry Department, Ghana, p. 8. Unpublished.

FIMP 1997b: *Off-reserve Inventory Results.* Planning Branch, Forestry Department, Ghana, p. 7. Unpublished.

FIMP 1995a: *Mean volume m^3/km^2 by cm diameter classes for Ghana.* Volume table by species and diameter class produced by FIMP, Forestry Department Planning Branch, Kumasi. Unpublished.

FIMP 1995b: *Timber yields from the Forest Reserves of Ghana: An analysis of the implications of sustainable management.* FIMP, Kumasi, p. 31. Unpublished.

FIMP 1994a: *Study of the implications of the Forest Protection Proposals, Definition of Scenarios.* FIMP Discussion Paper, Kumasi, p. 15. Unpublished.

FIMP 1994b: *Timber species classification and the assessment of exploitation patterns.* FIMP Discussion Paper 4, Kumasi, p. 8. Unpublished.

FIMP 1994c: *Concession status and rationalisation .* FIMP Discussion Paper 2, Kumasi, p. 9. Unpublished.

FIMP 1988: *Forest Inventory and Management Project, Revised project document.* ODA, UK, p. 38 + Appendices A-H. Unpublished.

Financial Times 1989: *Forests pay as Ghana loses out.* Article of Feb. 8 by William Keeling.

Financial Times 1984: *Ghana builds on timber revival.* Article of June 4 by Peter Blackburn, Abidjan.

FIP 1989a: *Ghana Forest Inventory Project, Seminar Proceedings 29-30 March 1989.* Wong, J.L.G. (ed.). ODA, UK and Forestry Department, Ghana, p. 101.

FIP 1989b: *Forest Inventory Project National Results.* Forestry Department, Ghana and ODA, UK, p.14 + Appendices 1-9.

Flint, M.E.S. and Hardcastle, P.D. 1992: *Evaluation of the Forest Inventory Project, Ghana.* Evaluation Report EV 534, ODA, London, p. xiii + 115.

Foggie, A. 1962: *The Role of Forestry in the Agricultural Economy* In Wills (ed.), 1962, pp. 236-251.

Forest and Wildlife Policy 1994: *Republic of Ghana, Forest and Wildlife Policy.* Ministry of Lands and Forestry, Accra, p. 16.

Forest Fees (Amendment) Regulations 1993: *Forest Fees (Amendment) Regulations 1993 (LI 1576).* Ghana Publishing Corporation, Printing Division, A&O, p.2.

Forest Fees (Amendment) Regulations 1989: *Forest Fees (Amendment) Regulations 1989 (LI 1483).* Ghana Publishing Corporation, Printing Division, Accra-Tema, Ghana, p. 2.

Forest Fees (Amendment) Regulations 1986: *Forest Fees (Amendment) Regulations 1986 (LI 1322).* Ghana Publishing Corporation, Printing Division, Accra-Tema, Ghana, p. 2.

Forest Fees (Amendment) Regulations 1983: *Forest Fees (Amendment) Regulations 1983 (LI 1275).* Ghana Publishing Corporation, Printing Division, Accra-Tema, Ghana, p. 2.

Forest Fees Regulations 1976: *Forest Fees Regulations 1976 (LI 1089).* Ghana Publishing Corporation, Printing Division, Accra-Tema, Ghana, p. 3.

Forest Ordinance (Cap. 157) 1927: *An Ordinance for the Protection of Forests and for the Constitution of Forest Reserves.* The Gold Coast, 30th March, 1927, p. 17.

Forest Products Inspection Bureau Law 1985: *Forest Products Inspection Bureau Law, 1985 (P.N.D.C. Law 117).* Ghana Publishing Corporation, Printing Division, Accra-Tema, Ghana, p. 4.

Forest Protection (Amendment) Law 1986: *Forest Protection (Amendment) Law (P.N.D.C. Law 142).* Ghana Publishing Corporation, Printing Division, Accra-Tema, Ghana, p. 1.

Forest Protection Decree 1974: *Forest Protection Decree, 1974 (N.R.C. Decree 243).* Ghana Publishing Corporation, Printing Division, Accra-Tema, Ghana, p. 4.

FORIG 1994: *Exploitation of Under-Sized Trees, Draft Report.* Forestry Research Institute of Ghana and Institute of Renewable Resources, University of Technology and Science, Kumasi, Ghana, p. 56. Unpublished.

FPIB 1997: *Circular No. FPIB/3/97 to all saw millers and all commercial banks on export levies on air-dried lumber.* Forest Products Inspection Bureau, Takoradi, Ghana, p. 1.

FPIB Export Permits 1986-96: *Wood Export Permits, 1986-96 [Total annual wood exports by products and species, 1986-96].* Forest Products Inspection Bureau, Takoradi, Ghana. Unpublished.

FPIB LMC Totals 1986-96: *Log Measurement Certificate Totals, 1986-96 [Total annual recorded timber extraction by species, 1986-96].* Forest Products Inspection Bureau, Takoradi, Ghana. Unpublished.

Francois, J. 1989a: *Overview of Ghana's Forest Policy.* In FC (1989).

Francois, J. 1989b: *Introduction and History of the Project.* In FIP (1989; 6-7).

Friends of the Earth 1992: *Plunder in Ghana's Rainforest for Illegal Profit.* Vol. 2: Research Report, Friends of the Earth, London, p. 121.

Frimpong-Mensah, K. 1989: *Requirement of the Timber Industry.* In FIP (1989; 70-78).

FRMP 1992: *Mid-Term Review: Forest Resource Management Project.* World Bank, Washington, p. 31.Unpublished

FRMP 1988: *Staff Appraisal Report, Ghana: Forest Resource Management Project.* World Bank, Washington, p. 119. Unpublished.

General Woods 1992: *Technical and Financial Audit of The Ghana Timber Industry.* Working Documents no. 1 (*Forest Resource Analysis*, p.33 + App.1-4); no. 2 *(Marketing of Forest Products*, p. 72 + App. 1-10); no. 3 (*Forest Fiscal Study*, p. 67 + App. 1-6); no. 4 (*Survey of Processing Plants and Analysis*, p. 25 + App. 1-4); no. 5 (*Financial Audit*, p. 60 + App. 1-4) and no. 6 (*Forest Institutions*, p. 47 + App. 1-7). General Woods and Veneers Consultants International Ltd. Unpublished.

Ghana Timber Association 1994: *Towards a New Fiscal Policy for The Timber Sector; Submission to Parliament on the Trees and Timber (Amendment) Bill, 1994.* Ghana Timber Association, Kumasi, p.22. Unpublished.

Ghartey, K.K.F. 1990: *Yield regulation within the high forest.* Circular note of May 30, 1990 to all Regional Forestry Offices. Unpublished.

Ghartey, K.K.F. 1989: *Results of the Inventory.* In FIP (1989; 32-46).

Grut, M. 1996: *Technical Assistance to the Forest Sector Development Project, Ghana; Consultancy Report No. 3; Interim Report of the Economics and Finance Consultant: A Discussion Paper.* ODA Contract No. CNTR 955528A. Fountain Renewable Resources Limited, UK, p. 19 + App. 1-4. Unpublished.

GTMO 1994: *List of Guiding Prices for Logs at the Domestic Market.* Ghana Timber Millers Organisation, Kumasi. Unpublished.

Hall, J.B. 1987: *Conservation of Forest In Ghana.* Universitas 8:33-42 (University of Legon, Ghana).

Hall, J.B. and Swaine, M.D. 1981: *Distribution and ecology of vascular plants in a tropical rainforest. Forest vegetation in Ghana.* Geobotany 1. Dr. W. Junk Hague.

Hall, J.B. and Swaine, M.D. 1976: *Classification and Ecology of Closed Canopy Forest in Ghana.* Journal of Ecology Vol 64, pp. 913-951.

Hawthorne, W.D. 1993a: *FROGGIE; Forest Reserves of Ghana Graphical Information Exhibitor, Part 1 (Manual for the programme).* Forest Inventory and Management Project, Forestry Department, Planning Branch, Kumasi, Ghana. p. 117. Unpublished.

Hawthorne, W.D. 1993b: *FROGGIE; Forest Reserves of Ghana Graphical Information Exhibitor, Part 2 (First Draft).* Forest Inventory and Management Project, Forestry Department, Planning Branch, Kumasi, Ghana, p. 62 + Appendices A and B. Unpublished.

Hawthorne, W.D. 1990: *Field guide to the forest trees of Ghana.* Ghana Forestry Series 1. Natural Resources Institute, Ghana

Hawthorne, W.D. 1989: *The Flora and Vegetation of Ghana's Forests.* In FIP (1989a; 8-13).

Hawthorne, W.D. and Musah, A.J. 1993: *Forest Protection In Ghana*, Forest Inventory and Management Project, Planning Branch, Forestry Department, Kumasi, Ghana and ODA, UK, p. 64 plus appendices A-C.

Howard, W. 1989: *The Forest Resource Management Project (World Bank, ODA, Danida).* In FIP (1989; 59-62).

HTS 1995: *Ghana Forest Plantations Preparation Project.* Hunting Technical Services Ltd., Hertfordshire, UK, p. 130 plus appendices A-H. Unpublished.

Huq, M.M. 1989: *The Economy of Ghana, The First 25 Years since Independence.* Macmillan Press, London, p. 355.

IIED 1993: *Study of Incentives for the Sustainable Management of the Tropical High Forest of Ghana.* International Institute for Environment and Development, London, p. 249. Unpublished.

ITTO 1996: *Tropical Forest Update.* Vol. 6 No 1, March 1996/1. International Tropical Timber Organisation, Yokohama, p. 31.

IUCN 1992: *The Conservation Atlas of Tropical Forests: Africa.* World Conservation Union. Macmillan, UK.

Kakrada, A.R. and Maginnis, S. 1992: *Stock Survey and Yield Allocation in the High Forest Zone of Ghana: A review of current procedures.* Forest Resources Management Project, p. 27 + Appendices 1-8. Unpublished.

Land and Concessions Bulletin 1977: *The Forests (Sukusuku Forest Reserve) Order, 1977.* Lands and Concessions Bulletin, 17th June, 1977, pp. 83-88, Ghana Publishing Corporation (Printing Division), Accra.

LC 1991: *Collection of Royalties on Timber.* Letter of July 12, 1991 from the Ashanti Regional Lands Secretary (Mr. J.O. Sarpong) to The Forestry Department with copies to all PNDC District Secretaries in the Ashanti Region, the Executive Secretary, Lands Commission Secretariat, All Revenue Collectors, Inspectors, etc., Lands Commission Secretariat, Ashanti and the Registrator, Ashanti Regional House of Chiefs, Kumasi, p.1. Unpublished.

Mensah-Ntiamoa, A.Y. 1989: *Pre-Feasibility Study on Wildlife Potentials in the Kakum and Asin-Attandoso Forest Reserves - Central Region - Ghana.* Department of Game and Wildlife, Accra. Unpublished.

MLF 1994: *Collection of Royalty Fees Outside Reserves.* Letter of July 25, 1994 from the Minister of Lands and Forestry (Dr. Kwabena Adjet) to the acting Executive Secretary, Lands Commission Secretariat, Accra with copy to the Chief Conservator of Forests, p.1. Unpublished.

MLG 1992: *PNDC Secretary for Local Government's Visit to the Northern Districts of the Western Region.* Three letters of similar wording from the PNDC Secretary of Local Government (Mr. Kwabena Ahwoi) to the Chief Conservator of Forests complaining about non-payment of timber royalties to the districts of Wassa-Amenfi, Aowin-Suaman and Sefwi-Wiawso, p.3. Unpublished.

MLNR, 1992: *Collection and Payment of Royalties Outside Forest Reserves.* Letter of April 7, 1992 from the PNDC Secretary for Lands and Natural Resources (Mr. J.A. Danso) to the Chief Conservator of Forests with copy to the Executive Secretary, Lands Commission Secretariat, p.1. Unpublished.

MLNR, 1991: *Collection of Royalties on Timber Trees.* Letter of April 4, 1991 from the PNDC Secretary for Lands and Natural Resources (Mr. J.A. Danso) to the Chief Conservator of Forests with copy to the Lands Commission Secretariat, the Forestry Commission and the Chairman of the PNDC Oversight Committee on Timber, p.1. Unpublished.

NIC Report, 1991. *Report by the Sub-Committee on Timber.* The report is highly confidential, but we have unofficially been allowed to look through it but not to make any photocopies. It summarises the results and recommendations arising from the National Investigation Committee's investigation into various parts of the timber industry. National Investigation Committee. Unpublished.

Nolan, T.M. 1989: *Increment, Annual Cut and the Growth of the Forest.* In FIP (1989a; 53-58).

ODA 1987: *Ghana: Forest inventory project.* ODA project approval submission. Unpublished.

Office of the Administrator of Lands Act 1994: *Office of the Administrator of Lands Act (Act 481).* GPC, Printing Division A&O, p. 5.

Office of the President 1993: *Confidential Memorandum on Forest Concessions, 08 April, 1993.* Office of the President, p. 4. Unpublished.

Overseas Consultancy Service 1995: *Policy recommendations for the sustainable management of the forest resource in Ghana.* UK Forestry Commission, p. 44 + App. 1-10. Unpublished.

PNDC Law 42 1982: *Provisional National Defence Council (Establishment) Proclamation (Supplementary and Consequential Provisions) Law, 1982.* Ghana Publishing Corporation, Printing Division, Accra-Tema, Ghana, p. 43.

Silviconsult 1985: *The Forest Department Review and the Requirements of the Forest Products Inspection Bureau and the Timber Export Development Board.* Unpublished, p. 186.

Smith, K. 1993: *Letter of 14 April from Smith, the Ministry of Lands and Forestry, Ghana to Werbrouck, the World Bank on timber concession acquisition.* Unpublished, p. 2.

Statistisches Bundesamt Deutschland 1999: *Preisindex für die Lebenshaltung aller privaten Haushalte [General Consumer Price Index].* Internet address: "http://194.95.119.6/basis/d/preis/preistab4.htm".

TEDB 1995: *The Ghana Timber Industry; Basic Information, Facts and Figures.* Timber Export Development Board, Takoradi, Ghana, p. 16.

TEDB 1993: *List of 1993 FOB log prices for from West-African countries including species that are banned from log export in Ghana.* Timber Export Development Board, Takoradi, Ghana, p. 2. Unpublished.

Temporary Felling Permit to MIM Timber 1991: *Letter of 11 February, 1991 from the Forestry Department allowing MIM Timber to temporarily continue its operation on another company's concession, provided the payment for timber felled (Cedis per m^3 specified by species) is paid to the Government instead of to the concession holder.* Forestry Department, Accra, Ghana, p.1. Unpublished.

Timber Export Development Board Law 1985: *Timber Export Development Board Law, 1985 (P.N.D.C. Law 123).* Ghana Publishing Corporation, Printing Division, Accra-Tema, Ghana, p. 7.

Timber Leases and Licences Regulations 1962: *Timber Leases and Licences Regulations, 1962 (L.I. 229).* Government Printing Department, Accra, Ghana, p. 6.

Timber Leases and Licences (Amendment) Regulations 1979: *Timber Leases and Licences (Amendment) Regulations, 1979 (L.I. 1215).* Ghana Publishing Corporation, Printing Division, Accra-Tema, Ghana, p. 3.

Timber Resources Management Act 1997: *Timber Resources Management Act 1997 (Act 547).* Government Printer, Accra, p. 10.

Timber Resources Management Regulations 1998: *Timber Resources Management Regulations 1998 (L.I. 1649).* Government Printer, Assembly Press, Accra, p. 24.

Timber Rights Bill 1996: *Timber Rights Bill 1996; A Bill Entitled the Timber Rights Act 1996.* Ministry of Lands and Forestry, Accra, p. 10.

Touche Ross 1995: *Institutional Reform of The Forestry Department; Final Report Vol I-III.* Touche Ross Management Consultants, London, p. 29 (Vol I), 71 (Vol II) and 48 (Vol III). Unpublished.

Townson, I.M. 1995: *Incomes from Non-timber Forest Products: Patterns of Enterprise Activity in the Forest Zone of Southern Ghana, Summary Report.* Oxford Forestry Institute, UK, p. 40.

Trees and Timber (Amendment) Act 1994: *Trees and Timber (Amendment) Act , 1994 (Act 493).* Ghana Publishing Corporation, Printing Division, Accra-Tema, Ghana, p. 5.

Trees and Timber (Amendment) Law 1983: *Trees and Timber (Amendment) Law , 1983 (P.N.D.C. Law 70).* Ghana Publishing Corporation, Printing Division, Accra-Tema, Ghana, p. 1.

Trees and Timber Decree 1974: *Trees and Timber Decree, 1974 (N.R.C. Decree 273).* Ghana Publishing Corporation, Printing Division, Accra-Tema, Ghana, p. 7.

Treue, T. and Odoom, F.K. 1996: *WVLC Concession Study, Ghana.* Consultancy for EAC Timber Ltd., Copenhagen, p. 73. Unpublished.

Treue, T. and Armstrong-Mensah, K.O. 1995: *Economic Analysis of the Potential for Enhancing Exports of Processed Products under Alternative Government Restrictions in Log Exports.* Consultancy report prepared for the Ministry of Trade and Industry, Ghana through Plan Consult, Accra and Cargill Technical Services Ltd., UK, p. 105. Unpublished.

Vanclay, J.K. 1993: *Review of the Forest Inventory and Management Project.* Annex IV of Kemp, R.H., Flint, M. and Vanclay J.K. Forest Inventory and Management Project, Review and project preparation report. July 1993. Ministry of Lands and Forestry, Ghana and ODA. Unpublished.

Werbrouck, P. 1993a: *Letter of 12 March from Werbrouck, the World Bank to Smith, the Ministry of Lands and Forestry, Ghana on timber concession acquisition.* Unpublished, p. 2.

Werbrouck, P. 1993b: *Fax of 20 April from Werbrouck, the World Bank to Smith, the Ministry of Lands and Forestry, Ghana on timber concession acquisition.* Unpublished, p. 1

Westoby, J. 1987: *The Purpose of Forests.* Blackwell, Oxford, p. 343.

Westoby, J. 1978: *Forest Industries for Socio-Economic Development.* Guest speaker's address, Eighth World Forestry Congress, Jakarta, Sep. 1978. First published in *Commonwealth Forestry Review, 58:2, pp.107-16.*

Westoby, J. 1962: *The Role of Forest Industries in the Attack on Economic Underdevelopment.* In *The State of Food and Agriculture 1962,* Chapter III pp. 88-128, FAO, Rome.

Wills, J.B., ed. 1962: *Agriculture and Land use i Ghana.* Oxford University Press, London.

Wong, J.L.G. 1990: *Volume Sampling Technical Note.* ODA, UK and Forestry Department, Ghana, Forest Inventory Project, p. 36.

Wong, J.L.G 1989: *Data Preparation and Analysis.* In FIP (1989a; pp. 23-31).

World Bank 1986: *Ghana, Forestry Sector Review.* Working Papers I - VII. Unpublished.

Appendix 1

GHANA
TIMBER SPECIES IN ALPHABETICAL ORDER BY BOTANICAL NAME

Trade Name	Botanical Name	Species Classifications: Star Group FIMP (1994b)	Star Group Hawthorne (1993b)	Silviconsult 1985	FIP Class 1989
	Afrosersalisia afzelii				2
Afzelia/Papao	Afzelia africana/bellea	Red	Red	A	1
Albizia	Albizia adianthifolia	Pink	Green	C	2
Albizia	Albizia ferruginea	Scarlet	Scarlet	A	1
	Albizia glaberrima	Green	Green	C	2
Albizia	Albizia zygia	Pink	Green	C	1
Alstonia	Alstonia boonei	Pink	Green	C	1
Yaya	Amphimas pterocarpoides	Pink	Green	B	1
Asanfina	Aningeria altissima/robusta	Scarlet	Pink	A	1
	Aningeria spp.				2
Kokote	Anopyxis klaineana	Pink	Pink	A	1
Kyenkyen	Antiaris toxicaria	Red	Pink	A	1
Aprokuma	Antrocaryon micraster	Red	Red	B	1
	Balanites wilsoniana				2
Berlinia	Berlinia confusa	Pink	Green	C	1
Berlinia	Berlinia spp.	Pink	Green	C	1
	Blighia spp.		Green		2
	Bussea occidentalis		Green		2
	Calpocalyx brevibracteatus				2
Canarium	Canarium schweinfurthii	Red	Pink	A	1
Ceiba	Ceiba pentandra	Red	Green	B	1
	Celtis adolfi-friderici	Green	Green	B	2
Celtis/Essa	Celtis mildbraedii/zenkeri	Pink	Green	B	1
	Celtis wightii		Green		2
	Chidlowia sanguinea				2
	Chrysophyllum perpulchrum		Green		2
	Chrysophyllum pruniforme				2
Akasa	Chrysophyllum spp.	Red	Pink	B	1
	Cleistopholis patens				2
	Cola gigantea		Green		2
Cordia	Cordia spp.	Green	Green	C	1
Pamprana	Coryanthe pachyceras	Pink	Green	C	2
	Coula edulis				2
Denya	Cylicodiscus gabonensis	Pink	Blue	A	1
Ananta	Cynometra antana	Pink	Pink	B	1
	Dacryodes klaineana		Green		2
Daniella	Daniella ogea/thurifera	Scarlet	Pink	B	1
Duabankye	Dialium aubrevillei	Pink	Green	C	1
Bonsamdua/Ayan	Distemonanthus benthamianus	Red	Pink	A	1
	Dubroscia viridiflora				2
Edinam	Entandophragma angolense	Scarlet	Red	A	1
Candollei	Entandophragma candollei	Red	Scarlet	A	1
Sapele	Entandophragma cylindricum	Scarlet	Scarlet	A	1
Utile	Entandophragma utile	Scarlet	Scarlet	A	1
	Erthroxylum manii				2
Potrodum	Erytrophleum sauaveolens	Pink	Pink	B	1
	Figus spp. (non-stranglers)				2
	Gilbertiodendron spp.				2
Guarea	Guarea spp	Red	Pink	A	1
Hyedua	Guibourtia ehie	Scarlet	Red	A	1
Subaha	Hallea spp./Mitragyna spp.	Pink	Red	A	1
	Hannoa klaineana		Green		2
Niangon	Heritiera utilis	Red	Red	A	1
	Hexalobus crispiflorus		Green		2
Nakwa	Holoptelea grandis	Pink	Green	B	2
	Homalium letestui				2
	Homalium stripulaceum/dewev.				2
	Irvingia gabonensis			B	2
Mahogany	Khaya anthotheca/grandifolia	Scarlet	Scarlet/Red	A	1
Mahogany	Khaya ivorensis	Scarlet	Scarlet	A	1
Kroma	Klaindoxa gabonensis	Pink	Green	C	1

Appendix 1

TIMBER SPECIES IN ALPHABETICAL ORDER BY BOTANICAL NAME

Trade Name	Botanical Name	Species Classifications Star Group FIMP (1994b)	Star Group Hawthorne (1993b)	Silviconsult 1985	FIP Class 1989
Kumanini	Lannea welwitschii	Pink	Green	C	2
	Lonchocarpus sericeus				2
Kaku/Ekki	Lophira alata	Red	Red	A	1
Walnut	Lovoa trichilioides	Red	Red	A	1
Bompagya	Mammea africana	Pink	Pink	B	1
Mansonia	Mansonia altissima	Red	Pink	A	1
	Maranthes spp.				2
	Margaritaria discoidea				2
Odum	Milicia excelsa/regia	Scarlet	Scarlet	A	1
	Monodora myristica				2
Wonton	Morus mesozygia	Pink	Green	B	2
Kusia	Nauclea diderrichii	Scarlet	Scarlet	A	1
Danta	Nesogordonia papaverifera	Pink	Pink	A	1
Akwana	Ongokea gore	Pink	Green	C	2
	Pachypodanthium staudtii				2
	Panda oleosa				2
	Pantaclethra macrophylla		Green		2
Afam	Parinari exelsa	Pink	Green	C	2
Asoma	Parkia bicolor	Pink	Green	B	1
	Parkia filicoidea				2
Afrormosia	Pericopsis elata	Scarlet	Scarlet	A	1
Esia	Petersianthus macrocarpus	Pink	Green	B	1
	Phyllocosmus africanus				2
Dahoma	Piptadenisatrum africanum	Red	Pink	A	1
	Protomegabaria stapfiana				2
	Pseudospondias microcarpa				2
	Pteleopsis hylodendron				2
Kyere	Pterygota macrocarpa	Scarlet	Red	C	1
Otie	Pycnanthus angolensis	Pink	Pink	A	1
Bombax	Rhodognaphalon/Bombax brevicuspe	Red	Red	A	1
Bombax	Rhodognaphalon/Bombax buonopozense	Pink	Green	B	1
Wama	Ricinodendron heudelotii	Pink	Green	C	2
	Scottellia klaineana		Pink	B	2
Sterculia	Sterculia oblonga		Green	B	2
Sterculia	Sterculia rhinopetala	Pink	Pink	B	1
Sterculia	Sterculia tragacantha				2
	Stereospermum acuminatissimum				2
Afena	Strombosia glaucescens	Pink	Green	A	1
	Tabernaemontana spp.				2
	Talbotiella gentii				2
Emire	Terminalia ivorensis	Red	Scarlet	A	1
Ofram	Terminalia superba	Pink	Pink	A	1
Makore	Tieghemella heckelii	Scarlet	Scarlet	A	1
	Treculia africana		Green		2
	Trichilia prieuriana				2
Tanuronini	Trichilia tessmannii	Pink	Green	C	2
Okure	Trilepisium madagascariense	Pink	Green	-	2
Wawa	Triplochiton scleroxylon	Scarlet	Scarlet	A	1
Avodire	Turreanthus africanus	Pink	Pink	A	1
	Uapaca guineensis	Green	Green		2
	Xylia evansii		Green		2
	Zanthoxylum spp.				2

Appendix 1

GHANA

TIMBER SPECIES IN ALPHABETICAL ORDER BY TRADE NAME

Trade Name	Botanical Name	Species Classifications			
		Star Group FIMP (1994b)	Star Group Hawthorne (1993b)	Silviconsult 1985	FIP Class 1989
Afam	Parinari exelsa	Pink	Green	C	2
Afena	Strombosia glaucescens	Pink	Green	A	1
Afrormosia	Pericopsis elata	Scarlet	Scarlet	A	1
Afzelia/Papao	Afzelia africana/bellea	Red	Red	A	1
Akasa	Chrysophyllum spp.	Red	Pink	B	1
Akwana	Ongokea gore	Pink	Green	C	2
Albizia	Albizia zygia	Pink	Green	C	1
Albizia	Albizia adianthifolia	Pink	Green	C	2
Albizia	Albizia ferruginea	Scarlet	Scarlet	A	1
Alstonia	Alstonia boonei	Pink	Green	C	1
Ananta	Cynometra antana	Pink	Pink	B	1
Aprokuma	Antrocaryon micraster	Red	Red	B	1
Asanfina	Aningeria altissima/robusta	Scarlet	Pink	A	1
Asoma	Parkia bicolor	Pink	Green	B	1
Avodire	Turreanthus africanus	Pink	Pink	A	1
Berlinia	Berlinia confusa	Pink	Green	C	1
Berlinia	Berlinia spp.	Pink	Green	C	1
Bombax	Rhodognaphalon/Bombax brevicuspe	Red	Red	A	1
Bombax	Rhodognaphalon/Bombax buonopozense	Pink	Green	B	1
Bompagya	Mammea africana	Pink	Pink	B	1
Bonsamdua/Ayan	Distemonanthus benthamianus	Red	Pink	A	1
Canarium	Canarium schweinfurthii	Red	Pink	A	1
Candollei	Entandophragma candollei	Red	Scarlet	A	1
Ceiba	Ceiba pentandra	Red	Green	B	1
Celtis/Essa	Celtis mildbraedii/zenkeri	Pink	Green	B	1
Cordia	Cordia spp.	Green	Green	C	1
Dahoma	Piptadenisatrum africanum	Red	Pink	A	1
Daniella	Daniella ogea/thurifera	Scarlet	Pink	B	1
Danta	Nesogordonia papaverifera	Pink	Pink	A	1
Denya	Cylicodiscus gabonensis	Pink	Blue	A	1
Duabankye	Dialium aubrevillei	Pink	Green	C	1
Edinam	Entandophragma angolense	Scarlet	Red	A	1
Emire	Terminalia ivorensis	Red	Scarlet	A	1
Esia	Petersianthus macrocarpus	Pink	Green	B	1
Guarea	Guarea spp	Red	Pink	A	1
Hyedua	Guibourtia ehie	Scarlet	Red	A	1
Kaku/Ekki	Lophira alata	Red	Red	A	1
Kokote	Anopyxis klaineana	Pink	Pink	A	1
Kroma	Klaindoxa gabonensis	Pink	Green	C	1
Kumanini	Lannea welwitschii	Pink	Green	C	2
Kusia	Nauclea diderrichii	Scarlet	Scarlet	A	1
Kyenkyen	Antiaris toxicaria	Red	Pink	A	1
Kyere	Pterygota macrocarpa	Scarlet	Red	C	1
Mahogany	Khaya ivorensis	Scarlet	Scarlet	A	1
Mahogany	Khaya anthotheca/grandifolia	Scarlet	Scarlet/Red	A	1
Makore	Tieghemella heckelii	Scarlet	Scarlet	A	1
Mansonia	Mansonia altissima	Red	Pink	A	1
Nakwa	Holoptelea grandis	Pink	Green	B	2
Niangon	Heritiera utilis	Red	Red	A	1
Odum	Milicia excelsa/regia	Scarlet	Scarlet	A	1
Ofram	Terminalia superba	Pink	Pink	A	1
Okure	Trilepisium madagascariense	Pink	Green	-	2
Otie	Pycnanthus angolensis	Pink	Pink	A	1
Pamprana	Coryanthe pachyceras	Pink	Green	C	2
Potrodum	Erytrophleum sauaveolens	Pink	Pink	B	1
Sapele	Entandophragma cylindricum	Scarlet	Scarlet	A	1
Sterculia	Sterculia oblonga		Green	B	2
Sterculia	Sterculia rhinopetala	Pink	Pink	B	1
Sterculia	Sterculia tragacantha				2
Subaha	Hallea spp./Mitragyna spp.	Pink	Red	A	1

Appendix 1

GHANA

TIMBER SPECIES IN ALPHABETICAL ORDER BY TRADE NAME

Trade Name	Botanical Name	Species Classifications			
		Star Group FIMP (1994b)	Star Group Hawthorne (1993b)	Silviconsult 1985	FIP Class 1989
Tanuronini	Trichilia tessmannii	Pink	Green	C	2
Utile	Entandophragma utile	Scarlet	Scarlet	A	1
Walnut	Lovoa trichilioides	Red	Red	A	1
Wama	Ricinodendron heudelotii	Pink	Green	C	2
Wawa	Triplochiton scleroxylon	Scarlet	Scarlet	A	1
Wonton	Morus mesozygia	Pink	Green	B	2
Yaya	Amphimas pterocarpoides	Pink	Green	B	1
	Afrosersalisia afzelii				2
	Albizia glaberrima	Green	Green	C	2
	Aningeria spp.				2
	Balanites wilsoniana				2
	Blighia spp.		Green		2
	Bussea occidentalis		Green		2
	Calpocalyx brevibracteatus				2
	Celtis adolfi-friderici	Green	Green	B	2
	Celtis wightii		Green		2
	Chidlowia sanguinea				2
	Chrysophyllum perpulchrum		Green		2
	Chrysophyllum pruniforme				2
	Cleistopholis patens				2
	Cola gigantea		Green		2
	Coula edulis				2
	Dacryodes klaineana		Green		2
	Dubroscia viridiflora				2
	Erthroxylum manii				2
	Figus spp. (non-stranglers)				2
	Gilbertiodendron spp.				2
	Hannoa klaineana		Green		2
	Hexalobus crispiflorus		Green		2
	Homalium letestui				2
	Homalium stripulaceum/dewev.				2
	Irvingia gabonensis			B	2
	Lonchocarpus sericeus				2
	Maranthes spp.				2
	Margaritaria discoidea				2
	Monodora myristica				2
	Pachypodanthium staudtii				2
	Panda oleosa				2
	Pantaclethra macrophylla		Green		2
	Parkia filicoidea				2
	Phyllocosmus africanus				2
	Protomegabaria stapfiana				2
	Pseudospondias microcarpa				2
	Pteleopsis hylodendron				2
	Scottellia klaineana		Pink	B	2
	Stereospermum acuminatissimum				2
	Tabernaemontana spp.				2
	Talbotiella gentii				2
	Treculia africana		Green		2
	Trichilia prieuriana				2
	Uapaca guineensis	Green	Green		2
	Xylia evansii		Green		2
	Zanthoxylum spp.				2

Appendix 2

GHANA

NATIONAL RESULTS; MEAN VOLUMES BY SPECIES AND DIAMETER CLASSES (Cu.m.)

Survey done 1985-88

Trade name	Botanical name	Diameter Classes 30-49	50-69	70-89	90-109	110-129	>130
SCARLET STARS							
Albizia	Albizia ferruginea	243.390	394.060	405.650	243.390	197.030	139.080
Asanfina 2/	Aningeria altissima/robusta	498.370	533.140	591.090	463.600	347.700	162.260
Daniella	Daniella ogea/thurifera	185.440	254.980	220.210	197.030	150.670	139.080
Edinam 1/	Entandophragma angolense	405.650	382.470	544.730	336.110	208.620	278.160
Sapele 1/	Entandophragma cylindricum	162.260	243.390	382.470	440.420	336.110	336.110
Utile 1/	Entandophragma utile	81.130	115.900	150.670	162.260	150.670	231.800
Hyedua 1/	Guibourtia ehie	197.030	324.520	173.850	0	0	0
Mahogany 1/	Khaya anthotheca/grandifolia/ivorensis	208.620	463.600	764.940	614.270	602.680	301.340
Odum 1/	Milicia excelsa/regia	243.390	452.010	614.270	695.400	637.450	846.070
Kusia	Nauclea diderrichii	57.950	127.490	231.800	254.980	104.310	23.180
Afrormosia 1/	Pericopsis elata	11.590	11.590	57.950	81.130	0	23.180
Kyere	Pterygota macrocarpa	544.730	1.159.000	1.159.000	591.090	324.520	92.720
Makore 1/	Tieghemella heckelii	57.950	46.360	104.310	69.540	173.850	567.910
Wawa 3/	Triplochiton scleroxylon	1.483.520	3.500.180	5.748.640	5.968.850	3.986.960	4.207.170
SUB TOTAL		**4.381.020**	**8.008.690**	**11.149.580**	**10.118.070**	**7.220.570**	**7.348.060**
RED STARS		0	0	0	0	0	0
Afzelia/Papao	Afzelia africana/bellea	220.210	243.390	254.980	127.490	46.360	34.770
Kyenkyen	Antiaris toxicaria	567.910	741.760	1.228.540	1.240.130	880.840	625.860
Aprokuma	Antrocaryon micraster	92.720	162.260	92.720	127.490	69.540	46.360
Canarium	Canarium schweinfurthii	46.360	69.540	81.130	46.360	92.720	115.900
Ceiba	Ceiba pentandra	440.420	567.910	857.660	1.251.720	1.761.680	6.988.770
Akasa	Chrysophyllum spp.	1.019.920	1.008.330	776.530	278.160	11.590	0
Bonsamdua/Ayan	Distemonanthus benthamianus	150.670	243.390	197.030	104.310	81.130	34.770
Candollei 1/	Entandophragma candollei	46.360	34.770	69.540	69.540	0	92.720
Guarea	Guarea spp	394.060	428.830	289.750	139.080	81.130	0
Niangon 1/	Heritiera utilis	150.670	185.440	139.080	34.770	57.950	0
Kaku/Ekki	Lophira alata	11.590	34.770	69.540	115.900	81.130	104.310
Walnut 1/	Lovoa trichilioides	69.540	139.080	162.260	231.800	23.180	0
Mansonia 1/	Mansonia altissima	591.090	695.400	301.340	81.130	11.590	0
Dahoma	Piptadenisatrum africanum	788.120	1.483.520	1.981.890	1.865.990	1.274.900	904.020
Bombax	Rhodognaphalon/Bombax brevicuspe	197.030	243.390	336.110	173.850	0	0
Emire 2/	Terminalia ivorensis	57.950	57.950	127.490	208.620	266.570	162.260
SUB TOTAL		**4.844.620**	**6.339.730**	**6.965.590**	**6.096.340**	**4.740.310**	**9.109.740**
PINK STARS		0	0	0	0	0	0
Alstonia	Alstonia boonei	347.700	475.190	579.500	463.600	312.930	278.160
Yaya	Amphimas pterocarpoides	150.670	197.030	266.570	208.620	127.490	57.950
Kokote	Anopyxis klaineana	23.180	57.950	150.670	104.310	23.180	23.180
Berlinia	Berlinia confusa	0	0	0	0	0	0
Berlinia	Berlinia spp.	625.860	452.010	370.880	162.260	69.540	0
Celtis/Essa	Celtis mildbraedii/zenkeri	6.629.480	5.945.670	4.427.380	1.842.810	440.420	127.490
Bubinga	Copaifera salikounda	23.180	11.590	11.590	11.590	0	34.770
Cordia	Cordia spp.	46.360	57.950	81.130	23.180	34.770	46.360
Denya	Cylicodiscus gabonensis	104.310	197.030	336.110	683.810	857.660	1.541.470
Ananta	Cynometra antana	266.570	533.140	452.010	104.310	23.180	0
Duabankye	Dialium aubrevillei	509.960	533.140	254.980	69.540	23.180	0
	Diospyros sanza-minika	197.030	11.590	0	0	0	0
Potrodum	Erytrophleum sauaveolens	34.770	57.950	104.310	81.130	57.950	23.180
Subaha 2/, 4/	Hallea spp./Mitragyna spp.	46.360	46.360	23.180	23.180	34.770	0
Kroma	Klaindoxa gabonensis	139.080	231.800	312.930	185.440	185.440	104.310
Bompagya	Mammea africana	92.720	104.310	104.310	46.360	23.180	23.180
Danta 2/	Nesogordonia papaverifera	1.900.760	1.634.190	637.450	173.850	57.950	81.130
Asoma	Parkia bicolor	185.440	347.700	521.550	359.290	289.750	127.490
Esia	Petersianthus macrocarpus	846.070	1.147.410	1.019.920	799.710	185.440	139.080
Otie	Pycnanthus angolensis	834.480	1.622.600	1.332.850	486.780	46.360	11.590
Bombax	Rhodognaphalon/Bombax buonopozense	92.720	127.490	243.390	127.490	46.360	46.360
Sterculia	Sterculia spp.	1.645.780	1.622.600	417.240	127.490	11.590	0
Afena	Strombosia glaucescens	1.576.240	382.470	57.950	0	0	0
Ofram 2a/	Terminalia superba	243.390	1.077.870	2.097.790	1.564.650	730.170	463.600
Avodire 1/	Turreanthus africanus	1.031.510	915.610	336.110	57.950	11.590	0
SUB TOTAL		**22.693.220**	**24.431.720**	**21.568.990**	**14.186.160**	**8.599.780**	**12.401.300**
GRAND TOTAL FOR FIP-CLASS 1		**26.819.260**	**32.139.070**	**32.254.970**	**23.921.760**	**15.553.780**	**19.587.100**
		0	**0**	**0**	**0**	**0**	**0**
GRAND TOTAL FOR FIP-CLASS 2		**14.487.500**	**11.068.450**	**6.548.350**	**3.036.580**	**1.367.620**	**637.450**
		0	**0**	**0**	**0**	**0**	**0**
GRAND TOTAL FOR FIP-CLASS 3		**7.440.780**	**2.341.180**	**614.270**	**254.980**	**11.590**	**57.950**
		0	**0**	**0**	**0**	**0**	**0**
GRAND TOTAL FOR ALL CLASSES		**48.735.950**	**45.571.880**	**39.429.180**	**27.248.090**	**16.921.400**	**20.305.680**

Notes:
1/ Log export banned in 1979
2/ Log export banned in 1988
2a/ Log export banned in 1988 but lifted in 1990
3/ Log export banned in 1993
4/ Harvest banned in 1993
Log exports was entirely banned in 1995

Source: FIP National Results, November 1989
Figures are estimated from a 0.25% sample. In this case 1332 plots on 1 ha were sampled in 43 forest reserves.
Volumes include only trees with FIP quality codes and form codes of 1 or 2.
Volumes are calculated from FIP volume equations,
which relate DBH to bole volume, and assuming a production forest area of 1,159,000 ha.

Continued

Appendix 2

GHANA

NATIONAL RESULTS; MEAN VOLUMES BY SPECIES AND DIAMETER CLASSES (Cu.m.)

Survey done 1985-88

Trade name	Botanical name	Diameter Classes 30-49	Total	TOT>70	TOT>110	TOT>F-LIMIT
SCARLET STARS						
Albizia	Albizia ferruginea	243.390	1.622.600	985.150	336.110	985.150
Asanfina 2/	Aningeria altissima/robusta	498.370	2.596.160	1.564.650	509.960	509.960
Daniella	Daniella ogea/thurifera	185.440	1.147.410	706.990	289.750	486.780
Edinam 1/	Entandophragma angolense	405.650	2.155.740	1.367.620	486.780	486.780
Sapele 1/	Entandophragma cylindricum	162.260	1.900.760	1.495.110	672.220	672.220
Utile 1/	Entandophragma utile	81.130	892.430	695.400	382.470	382.470
Hyedua 1/	Guibourtia ehie	197.030	695.400	173.850	0	173.850
Mahogany 1/	Khaya anthotheca/grandifolia/ivorensis	208.620	2.955.450	2.283.230	904.020	904.020
Odum 1/	Milicia excelsa/regia	243.390	3.488.590	2.793.190	1.483.520	1.483.520
Kusia	Nauclea diderrichii	57.950	799.710	614.270	127.490	127.490
Afrormosia 1/	Pericopsis elata	11.590	185.440	162.260	23.180	0
Kyere	Pterygota macrocarpa	544.730	3.871.060	2.167.330	417.240	417.240
Makore 1/	Tieghemella heckelii	57.950	1.019.920	915.610	741.760	0
Wawa 3/	Triplochiton scleroxylon	1.483.520	24.895.320	19.911.620	8.194.130	8.194.130
SUB TOTAL		**4.381.020**	**48.225.990**	**35.836.280**	**14.568.630**	**14.823.610**
RED STARS		0	0	0	0	0
Afzelia/Papao	Afzelia africana/bellea	220.210	927.200	463.600	81.130	463.600
Kyenkyen	Antiaris toxicaria	567.910	5.285.040	3.975.370	1.506.700	3.975.370
Aprokuma	Antrocaryon micraster	92.720	591.090	336.110	115.900	498.370
Canarium	Canarium schweinfurthii	46.360	452.010	336.110	208.620	336.110
Ceiba	Ceiba pentandra	440.420	11.868.160	10.859.830	8.750.450	8.750.450
Akasa	Chrysophyllum spp.	1.019.920	3.094.530	1.066.280	11.590	289.750
Bonsamdua/Ayan	Distemonanthus benthamianus	150.670	811.300	417.240	115.900	417.240
Candollei 1/	Entandophragma candollei	46.360	312.930	231.800	92.720	92.720
Guarea	Guarea spp	394.060	1.332.850	509.960	81.130	509.960
Niangon 1/	Heritiera utilis	150.670	567.910	231.800	57.950	231.800
Kaku/Ekki	Lophira alata	11.590	417.240	370.880	185.440	370.880
Walnut 1/	Lovoa trichilioides	69.540	625.860	417.240	23.180	417.240
Mansonia 1/	Mansonia altissima	591.090	1.680.550	394.060	11.590	394.060
Dahoma	Piptadenisatrum africanum	788.120	8.298.440	6.026.800	2.178.920	2.178.920
Bombax	Rhodognaphalon/Bombax brevicuspe	197.030	950.380	509.960	0	509.960
Emire 2/	Terminalia ivorensis	57.950	880.840	764.940	428.830	764.940
SUB TOTAL		**4.844.620**	**38.096.330**	**26.911.980**	**13.850.050**	**20.201.370**
PINK STARS		0	0	0	0	0
Alstonia	Alstonia boonei	347.700	2.457.080	1.634.190	591.090	591.090
Yaya	Amphimas pterocarpoides	150.670	1.008.330	660.630	185.440	394.060
Kokote	Anopyxis klaineana	23.180	382.470	301.340	46.360	301.340
Berlinia	Berlinia confusa	0	0	0	0	0
Berlinia	Berlinia spp.	625.860	1.680.550	602.680	69.540	602.680
Celtis/Essa	Celtis mildbraedii/zenkeri	6.629.480	19.413.250	5.945.670	4.427.380	1.842.810
Bubinga	Copaifera salikounda	23.180	92.720	57.950	34.770	69.540
Cordia	Cordia spp.	46.360	289.750	185.440	81.130	185.440
Denya	Cylicodiscus gabonensis	104.310	3.720.390	3.419.050	2.399.130	3.419.050
Ananta	Cynometra antana	266.570	1.379.210	579.500	23.180	579.500
Duabankye	Dialium aubrevillei	509.960	1.390.800	347.700	23.180	347.700
	Diospyros sanza-minika	197.030	208.620	0	0	0
Potrodum	Erytrophleum sauaveolens	34.770	359.290	266.570	81.130	266.570
Subaha 2/, 4/	Hallea spp./Mitragyna spp.	46.360	173.850	81.130	34.770	0
Kroma	Klaindoxa gabonensis	139.080	1.159.000	788.120	289.750	475.190
Bompagya	Mammea africana	92.720	394.060	197.030	46.360	197.030
Danta 2/	Nesogordonia papaverifera	1.900.760	4.485.330	950.380	139.080	950.380
Asoma	Parkia bicolor	185.440	1.831.220	1.298.080	417.240	417.240
Esia	Petersianthus macrocarpus	846.070	4.137.630	2.144.150	324.520	2.144.150
Otie	Pycnanthus angolensis	834.480	4.334.660	1.877.580	57.950	1.877.580
Bombax	Rhodognaphalon/Bombax buonopozense	92.720	683.810	463.600	92.720	463.600
Sterculia	Sterculia spp.	1.645.780	3.824.700	556.320	11.590	556.320
Afena	Strombosia glaucescens	1.576.240	2.016.660	57.950	0	440.420
Ofram 2a/	Terminalia superba	243.390	6.177.470	4.856.210	1.193.770	4.856.210
Avodire 1/	Turreanthus africanus	1.031.510	2.352.770	405.650	11.590	405.650
SUB TOTAL		**22.693.220**	**103.881.170**	**56.756.230**	**21.001.080**	**42.859.820**
GRAND TOTAL FOR FIP-CLASS 1		**26.819.260**	**150.275.940**	**90.425.180**	**39.000.350**	**56.408.530**
		0	**0**	**0**	**0**	**0**
GRAND TOTAL FOR FIP-CLASS 2		**14.487.500**	**37.145.950**	**11.590.000**	**2.005.070**	**0**
		0	**0**	**0**	**0**	**0**
GRAND TOTAL FOR FIP-CLASS 3		**7.440.780**	**10.720.750**	**938.790**	**69.540**	**0**
		0	**0**	**0**	**0**	**0**
GRAND TOTAL FOR ALL CLASSES		**48.735.950**	**198.223.770**	**103.904.350**	**37.227.080**	**0**

Notes:
1/ Log export banned in 1979
2/ Log export banned in 1988
2a/ Log export banned in 1988 but lifted in 1990
3/ Log export banned in 1993
4/ Harvest banned in 1993
Log exports was entirely banned in 1995

Source: FIP National Results, November 1989
Figures are estimated from a 0.25% sample.
In this case 1332 plots on 1 ha were sampled in 43 forest reserves.
Volumes include only trees with FIP quality codes and form codes of 1 or 2.
Volumes are calculated from FIP volume equations,
which relate DBH to bole volume, and assuming a production forest area of 1,159,000 ha.

Appendix 2

GHANA
ESTIMATED ANNUAL ALLOWABLE CUT (AAC) IN ALL FOREST RESERVES

Based on:
- Forty (40) year felling cycle.
- Estimated total production area: 11.590 km2
- Estimated annual production area: 290 km2/yr
- Normal yield formula Z=0.5Y+0.2X (for all species) 5/

Trade name	Botanical name	Minimum felling DBH (cm)	Estimated AAC for each species (Cu.m.)	AAC as % of volume above felling limit on the annual production area (%)
SCARLET STARS				
Albizia	Albizia ferruginea	70	14.285	58
Asanfina 2/	Aningeria altissima/robusta	110	8.693	68
Daniella	Daniella ogea/thurifera	90	7.186	59
Edinam 1/	Entandophragma angolense	110	7.765	64
Sapele 1/	Entandophragma cylindricum	110	10.605	63
Utile 1/	Entandophragma utile	110	5.592	58
Hyedua 1/	Guibourtia ehie	70	3.796	87
Mahogany 1/	Khaya anthotheca/grandifolia/ivorensis	110	14.372	64
Odum 1/	Milicia excelsa/regia	110	22.021	59
Kusia	Nauclea diderrichii	110	2.869	90
Afrormosia 1/	Pericopsis elata	permit required		
Kyere	Pterygota macrocarpa	110	8.171	78
Makore 1/	Tieghemella heckelii	permit required		
Wawa 3/	Triplochiton scleroxylon	110	132.271	65
SUB TOTAL			**237.624**	**64**
RED STARS				
Afzelia/Papao	Afzelia africana/bellea	70	7.012	61
Kyenkyen	Antiaris toxicaria	70	53.401	54
Aprokuma	Antrocaryon micraster	50	6.693	54
Canarium	Canarium schweinfurthii	70	4.549	54
Ceiba	Ceiba pentandra	110	115.639	53
Akasa	Chrysophyllum spp.	90	7.505	104
Bonsamdua/Ayan	Distemonanthus benthamianus	70	6.432	62
Candollei 1/	Entandophragma candollei	110	1.507	65
Guarea	Guarea spp	70	8.519	67
Niangon 1/	Heritiera utilis	70	3.825	66
Kaku/Ekki	Lophira alata	70	4.810	52
Walnut 1/	Lovoa trichilioides	70	5.911	57
Mansonia 1/	Mansonia altissima	70	8.403	85
Dahoma	Piptadenisatrum africanum	110	36.566	67
Bombax	Rhodognaphalon/Bombax brevicuspe	70	7.591	60
Emire 2/	Terminalia ivorensis	70	9.852	52
SUB TOTAL			**288.214**	**57**
PINK STARS				
Alstonia	Alstonia boonei	110	9.707	66
Yaya	Amphimas pterocarpoides	90	6.259	64
Kokote	Anopyxis klaineana	70	4.057	54
Berlinia	Berlinia confusa	70		
Berlinia	Berlinia spp.	70	9.794	65
Celtis/Essa	Celtis mildbraedii/zenkeri	70	104.049	226
Bubinga	Copaifera salikounda	50	985	57
Cordia	Cordia spp.	70	2.608	56
Denya	Cylicodiscus gabonensis	70	43.723	51
Ananta	Cynometra antana	70	9.909	68
Duabankye	Dialium aubrevillei	70	7.012	81
Potrodum	Erytrophleum sauaveolens	70	3.622	54
Kroma	Klaindoxa gabonensis	90	7.505	63
Bompagya	Mammea africana	70	2.984	61
Danta 2/	Nesogordonia papaverifera	70	20.051	84
Asoma	Parkia bicolor	110	7.012	67
Esia	Petersianthus macrocarpus	70	32.539	61
Otie	Pycnanthus angolensis	70	31.583	67
Bombax	Rhodognaphalon/Bombax buonopozense	70	6.432	56
Sterculia	Sterculia spp.	70	15.067	108
Afena	Strombosia glaucescens	50	13.386	122
Ofram 2a/	Terminalia superba	70	66.092	54
Avodire 1/	Turreanthus africanus	70	9.649	95
Subaha 2/, 4/	Hallea spp./Mitragyna spp.	-		
SUB TOTAL			**414.024**	**77**
GRAND TOTAL (FIP class 1 species)			**939.862**	**67**
Estimated AAC within the production area (Cu.m./ha/year):			**0,81**	

Notes:
1/ Log export banned in 1979
2/ Log export banned in 1988
2a/ Log export banned in 1988 but lifted in 1990
3/ Log export banned in 1993
4/ Harvest banned in 1993
Log exports was entirely banned in 1995
5/ The yield formula is applied on volumes of species in diameter classes. Z is the volume which may be felled, Y is the standing volume above felling limit and X is the standing volume in the diameter class below the felling limit.

Source: FIP National Results, November 1989

Appendix 3

GHANA

NATIONAL RESULTS; MEAN VOLUMES BY SPECIES AND DIAMETER CLASSES (Cu.m.)

Survey done 1985-94

Trade name	Botanical name	Diameter Classes (cm) 30-49	50-69	70-89	90-109	110-129	>130
SCARLET STARS							
Albizia	Albizia ferruginea	243.390	382.470	382.470	289.750	162.260	127.490
Asanfina 2/	Aningeria altissima/robusta	463.600	486.780	498.370	359.290	312.930	139.080
Daniella	Daniella ogea/thurifera	243.390	324.520	347.700	278.160	197.030	104.310
Edinam 1/	Entandophragma angolense	475.190	463.600	649.040	336.110	185.440	208.620
Sapele 1/	Entandophragma cylindricum	150.670	243.390	417.240	382.470	336.110	266.570
Utile 1/	Entandophragma utile	57.950	81.130	127.490	127.490	139.080	220.210
Hyedua 1/	Guibourtia ehie	162.260	266.570	173.850	0	0	0
Mahogany 1/	Khaya anthotheca/grandifolia/ivorensis	208.620	452.010	672.220	556.320	405.650	301.340
Odum 1/	Milicia excelsa/regia	185.440	312.930	428.830	452.010	440.420	567.910
Kusia	Nauclea diderrichii	57.950	115.900	231.800	220.210	104.310	23.180
Afrormosia 1/	Pericopsis elata	11.590	11.590	46.360	46.360	0	11.590
Kyere	Pterygota macrocarpa	509.960	996.740	938.790	382.470	185.440	81.130
Makore 1/	Tieghemella heckelii	69.540	81.130	139.080	104.310	127.490	417.240
Wawa 3/	Triplochiton scleroxylon	1.193.770	2.955.450	4.925.750	4.334.660	3.372.690	2.978.630
SUB TOTAL		**4.033.320**	**7.174.210**	**9.978.990**	**7.869.610**	**5.968.850**	**5.447.300**
RED STARS							
Afzelia/Papao	Afzelia africana/bellea	150.670	185.440	185.440	81.130	34.770	11.590
Kyenkyen	Antiaris toxicaria	475.190	625.860	1.008.330	950.380	764.940	486.780
Aprokuma	Antrocaryon micraster	92.720	150.670	127.490	115.900	81.130	46.360
Canarium	Canarium schweinfurthii	57.950	92.720	92.720	92.720	81.130	57.950
Ceiba	Ceiba pentandra	301.340	394.060	672.220	996.740	1.668.960	5.551.610
Akasa	Chrysophyllum spp.	730.170	706.990	463.600	150.670	11.590	0
Bonsamdua/Ayan	Distemonanthus benthamianus	139.080	220.210	173.850	104.310	57.950	34.770
Candollei 1/	Entandophragma candollei	34.770	34.770	57.950	57.950	0	46.360
Guarea	Guarea spp	347.700	347.700	278.160	127.490	46.360	0
Niangon 1/	Heritiera utilis	312.930	405.650	347.700	104.310	34.770	23.180
Kaku/Ekki	Lophira alata	23.180	81.130	162.260	208.620	150.670	173.850
Walnut 1/	Lovoa trichilioides	57.950	115.900	115.900	150.670	11.590	11.590
Mansonia 1/	Mansonia altissima	417.240	475.190	208.620	46.360	11.590	0
Dahoma	Piptadenisatrum africanum	904.020	1.460.340	1.981.890	1.611.010	1.216.950	741.760
Bombax	Rhodognaphalon/Bombax brevicuspe	208.620	266.570	266.570	139.080	11.590	23.180
Emire 2/	Terminalia ivorensis	81.130	104.310	162.260	197.030	185.440	92.720
SUB TOTAL		**4.334.660**	**5.667.510**	**6.304.960**	**5.134.370**	**4.369.430**	**7.301.700**
PINK STARS							
Alstonia	Alstonia boonei	278.160	405.650	556.320	394.060	312.930	173.850
Yaya	Amphimas pterocarpoides	150.670	254.980	324.520	173.850	104.310	34.770
Kokote	Anopyxis klaineana	34.770	81.130	173.850	150.670	92.720	23.180
Berlinia	Berlinia confusa	162.260	173.850	139.080	46.360	23.180	0
Berlinia	Berlinia spp.	753.350	579.500	440.420	185.440	57.950	11.590
Celtis/Essa	Celtis mildbraedii/zenkeri	6.073.160	5.760.230	3.813.110	1.356.030	359.290	139.080
Bubinga	Copaifera salikounda	11.590	23.180	23.180	34.770	23.180	69.540
Cordia	Cordia spp.	46.360	46.360	57.950	34.770	23.180	46.360
Denya	Cylicodiscus gabonensis	104.310	185.440	324.520	533.140	799.710	1.298.080
Ananta	Cynometra antana	324.520	637.450	533.140	139.080	46.360	0
Duabankye	Dialium aubrevillei	672.220	649.040	289.750	81.130	23.180	11.590
	Diospyros sanza-minika	324.520	34.770	0	0	0	0
Potrodum	Erytrophleum sauaveolens	34.770	57.950	104.310	104.310	46.360	11.590
Kroma	Klaindoxa gabonensis	197.030	243.390	324.520	220.210	162.260	104.310
Bompagya	Mammea africana	115.900	139.080	139.080	81.130	57.950	34.770
Danta 2/	Nesogordonia papaverifera	1.877.580	1.448.750	579.500	127.490	46.360	34.770
Asoma	Parkia bicolor	231.800	452.010	637.450	370.880	289.750	127.490
Esia	Petersianthus macrocarpus	927.200	1.298.080	1.298.080	788.120	301.340	139.080
Otie	Pycnanthus angolensis	880.840	1.773.270	1.448.750	405.650	69.540	23.180
Bombax	Rhodognaphalon/Bombax buonopozense	81.130	150.670	208.620	150.670	57.950	46.360
Sterculia	Sterculia spp.	1.240.130	1.182.180	289.750	104.310	0	0
Afena	Strombosia glaucescens	1.865.990	452.010	57.950	0	0	0
Ofram 2a/	Terminalia superba	220.210	938.790	1.773.270	1.216.950	556.320	243.390
Avodire 1/	Turreanthus africanus	1.182.180	1.135.820	591.090	127.490	11.590	11.590
Subaha 2/, 4/	Hallea spp./Mitragyna spp.	69.540	127.490	173.850	81.130	34.770	0
SUB TOTAL		**17.860.190**	**18.231.070**	**14.302.060**	**6.907.640**	**3.500.180**	**2.584.570**
GRAND TOTAL FOR FIP-CLASS 1		**26.228.170**	**31.072.790**	**30.586.010**	**19.911.620**	**13.838.460**	**15.333.570**
GRAND TOTAL FOR FIP-CLASS 2		**12.633.100**	**9.584.930**	**5.667.510**	**2.399.130**	**985.150**	**382.470**
GRAND TOTAL FOR FIP-CLASS 3		**9.167.690**	**3.500.180**	**1.182.180**	**417.240**	**115.900**	**127.490**
GRAND TOTAL FOR ALL CLASSES 5/		**49.199.550**	**44.992.380**	**38.026.790**	**22.925.020**	**15.043.820**	**15.855.120**

Notes:

1/ Log export banned in 1979
2/ Log export banned in 1988
2a/ Log export banned in 1988 but lifted in 1990
3/ Log export banned in 1993
4/ Harvest banned in 1993
Log exports was entirely banned in 1995
5/ Volumes for species within diameter classes are rounded down to the nearest cubic metre. Therefore, sums do not always stem.

Source:
FIMP National Results
Figures are estimated from a 0.25% sample.
In this case 3101 plots on 1 ha were sampled in 128 forest reserves.
Volumes include only trees with FIP quality codes and form codes of 1 or 2.
Volumes are calcutaled from FIP volume equations which relate DBH to bole volume, and assuming a production forest area of 1,159,000 ha (see Appendix 2).

Appendix 3

GHANA

NATIONAL RESULTS; MEAN VOLUMES BY SPECIES AND DIAMETER CLASSES (Cu.m.)

Survey done 1985-94

Trade name	Botanical name	Diameter Classes (cm) 30-49	Total	TOT>70	TOT>110	TOT>F-LIMIT
SCARLET STARS						
Albizia	Albizia ferruginea	243.390	1.587.830	961.970	289.750	961.970
Asanfina 2/	Aningeria altissima/robusta	463.600	2.260.050	1.309.670	452.010	452.010
Daniella	Daniella ogea/thurifera	243.390	1.495.110	927.200	301.340	579.500
Edinam 1/	Entandophragma angolense	475.190	2.318.000	1.379.210	394.060	394.060
Sapele 1/	Entandophragma cylindricum	150.670	1.796.450	1.402.390	602.680	602.680
Utile 1/	Entandophragma utile	57.950	753.350	614.270	359.290	359.290
Hyedua 1/	Guibourtia ehie	162.260	602.680	173.850	0	173.850
Mahogany 1/	Khaya anthotheca/grandifolia/ivorensis	208.620	2.596.160	1.935.530	706.990	706.990
Odum 1/	Milicia excelsa/regia	185.440	2.387.540	1.889.170	1.008.330	1.008.330
Kusia	Nauclea diderrichii	57.950	753.350	579.500	127.490	127.490
Afrormosia 1/	Pericopsis elata	11.590	127.490	104.310	11.590	0
Kyere	Pterygota macrocarpa	509.960	3.094.530	1.587.830	266.570	266.570
Makore 1/	Tieghemella heckelii	69.540	938.790	788.120	544.730	0
Wawa 3/	Triplochiton scleroxylon	1.193.770	19.760.950	15.611.730	6.351.320	6.351.320
SUB TOTAL		**4.033.320**	**40.472.280**	**29.264.750**	**11.416.150**	11.984.060
RED STARS						
Afzelia/Papao	Afzelia africana/bellea	150.670	649.040	312.930	46.360	312.930
Kyenkyen	Antiaris toxicaria	475.190	4.311.480	3.210.430	1.251.720	3.210.430
Aprokuma	Antrocaryon micraster	92.720	614.270	370.880	127.490	521.550
Canarium	Canarium schweinfurthii	57.950	475.190	324.520	139.080	324.520
Ceiba	Ceiba pentandra	301.340	9.584.930	8.889.530	7.220.570	7.220.570
Akasa	Chrysophyllum spp.	730.170	2.063.020	625.860	11.590	162.260
Bonsamdua/Ayan	Distemonanthus benthamianus	139.080	730.170	370.880	92.720	370.880
Candollei 1/	Entandophragma candollei	34.770	231.800	162.260	46.360	46.360
Guarea	Guarea spp	347.700	1.147.410	452.010	46.360	452.010
Niangon 1/	Heritiera utilis	312.930	1.228.540	509.960	57.950	509.960
Kaku/Ekki	Lophira alata	23.180	799.710	695.400	324.520	695.400
Walnut 1/	Lovoa trichilioides	57.950	463.600	289.750	23.180	289.750
Mansonia 1/	Mansonia altissima	417.240	1.159.000	266.570	11.590	266.570
Dahoma	Piptadenisatrum africanum	904.020	7.915.970	5.551.610	1.958.710	1.958.710
Bombax	Rhodognaphalon/Bombax brevicuspe	208.620	915.610	440.420	34.770	440.420
Emire 2/	Terminalia ivorensis	81.130	822.890	637.450	278.160	637.450
SUB TOTAL		**4.334.660**	**33.112.630**	**23.110.460**	**11.671.130**	17.419.770
PINK STARS						
Alstonia	Alstonia boonei	278.160	2.120.970	1.437.160	486.780	486.780
Yaya	Amphimas pterocarpoides	150.670	1.043.100	637.450	139.080	312.930
Kokote	Anopyxis klaineana	34.770	556.320	440.420	115.900	440.420
Berlinia	Berlinia confusa	162.260	544.730	208.620	23.180	208.620
Berlinia	Berlinia spp.	753.350	2.028.250	695.400	69.540	695.400
Celtis/Essa	Celtis mildbraedii/zenkeri	6.073.160	17.500.900	5.667.510	498.370	5.667.510
Bubinga	Copaifera salikounda	11.590	185.440	150.670	92.720	173.850
Cordia	Cordia spp.	46.360	254.980	162.260	69.540	162.260
Denya	Cylicodiscus gabonensis	104.310	3.245.200	2.955.450	2.097.790	2.955.450
Ananta	Cynometra antana	324.520	1.680.550	718.580	46.360	718.580
Duabankye	Dialium aubrevillei	672.220	1.726.910	405.650	34.770	405.650
	Diospyros sanza-minika	324.520	359.290	0	0	0
Potrodum	Erytrophleum sauaveolens	34.770	359.290	266.570	57.950	266.570
Kroma	Klaindoxa gabonensis	197.030	1.251.720	811.300	266.570	486.780
Bompagya	Mammea africana	115.900	567.910	312.930	92.720	312.930
Danta 2/	Nesogordonia papaverifera	1.877.580	4.114.450	788.120	81.130	788.120
Asoma	Parkia bicolor	231.800	2.109.380	1.425.570	417.240	417.240
Esia	Petersianthus macrocarpus	927.200	4.751.900	2.526.620	440.420	2.526.620
Otie	Pycnanthus angolensis	880.840	4.601.230	1.947.120	92.720	1.947.120
Bombax	Rhodognaphalon/Bombax buonopozense	81.130	695.400	463.600	104.310	463.600
Sterculia	Sterculia spp.	1.240.130	2.816.370	394.060	0	394.060
Afena	Strombosia glaucescens	1.865.990	2.375.950	57.950	0	509.960
Ofram 2a/	Terminalia superba	220.210	4.948.930	3.789.930	799.710	3.789.930
Avodire 1/	Turreanthus africanus	1.182.180	3.059.760	741.760	23.180	741.760
Subaha 2/, 4/	Hallea spp./Mitragyna spp.	69.540	486.780	289.750	34.770	0
SUB TOTAL		**17.860.190**	**63.385.710**	**27.294.450**	**6.084.750**	**24.872.140**
GRAND TOTAL FOR FIP-CLASS 1		**26.228.170**	**136.970.620**	**79.669.660**	**29.172.030**	**54.275.970**
GRAND TOTAL FOR FIP-CLASS 2		**12.633.100**	**31.652.290**	**9.434.260**	**1.367.620**	
GRAND TOTAL FOR FIP-CLASS 3		**9.167.690**	**14.510.680**	**1.842.810**	**243.390**	
GRAND TOTAL FOR ALL CLASSES 5/		**49.199.550**	**191.258.180**	**91.839.160**	**30.898.940**	

Notes:
1/ Log export banned in 1979
2/ Log export banned in 1988
2a/ Log export banned in 1988 but lifted in 1990
3/ Log export banned in 1993
4/ Harvest banned in 1993
Log exports was entirely banned in 1995
5/ Volumes for species within diameter classes are rounded down to the nearest cubic metre. Therefore, sums do not always stem.

Source:
FIMP National Results
Figures are estimated from a 0.25% sample.
In this case 3101 plots on 1 ha were sampled in 128 forest reserves.
Volumes include only trees with FIP quality codes and form codes of 1 or 2.
Volumes are calculaled from FIP volume equations which relate DBH to bole volume, and assuming a production forest area of 1,159,000 ha (see Appendix 2).

Appendix 4

ESTIMATED ON-RESERVE AACs COMPARED WITH THE 1996 RECORDED EXTRACTION ON-RESERVES (Cu.m. ROUNDWOOD)

		Large-grained protection only (production forest = 762,400 ha)			Large- and fine-grained protection (production forest = 645,080 ha)				
SPECIES Trade name	Botanical name	AAC total volume (Cu.m.)	1996 on-reserve extraction (Cu.m.)	1996 extraction as % of AAC	AAC total volume (Cu.m.)	1996 on-reserve extraction (Cu.m.)	1996 extraction as % of AAC	Volume of over-extracted species (Cu.m.)	Over-extracted specis as % of extraction
SCARLET STARS									
Albizia	Albizia ferruginea	4.628	1.089	24	4.264	1.089	26		
Asanfina 2/	Aningeria altissima/robusta	3.450	9.789	284	3.098	9.789	316	9.789	
Daniella	Daniella ogea/thurifera	7.167	3.182	44	5.642	3.182	56		
Edinam 1/	Entandophragma angolense	3.564	9.668	271	3.269	9.668	296	9.668	
Sapele 1/	Entandophragma cylindricum	5.175	12.723	246	4.433	12.723	287	12.723	
Utile 1/	Entandophragma utile	2.397	4.280	179	2.321	4.280	184	4.280	
Hyedua 1/	Guibourtia ehie	2.847	1.460	51	2.556	1.460	57		
Mahogany 1/	Khaya anthotheca/grandifolia	558	1.207	216	522	1.207	231	1.207	
Mahogany 1/	Khaya ivorensis	6.138	18.427	300	5.382	18.427	342	18.427	
Odum 1/	Milicia excelsa/regia	6.820	20.598	302	6.659	20.598	309	20.598	
Kusia	Nauclea diderrichii	2.342	1.000	43	2.074	1.000	48		
Afrormosia 1/	Pericopsis elata	616	391	64	583	391	67		
Kyere	Pterygota macrocarpa	4.888	6.391	131	4.525	6.391	141	6.391	
Makore 1/	Tieghemella heckelii	3.293	5.962	181	2.764	5.962	216	5.962	
Wawa 3/	Triplochiton scleroxylon	78.555	87.702	112	72.727	87.702	121	87.702	
SUB TOTAL	**Scarlet stars**	**132.437**	**183.869**	**139**	**120.817**	**183.869**	**152**	**176.747**	**96**
RED STARS									
Afzelia/Papao	Afzelia africana/bellea	3.175	1.043	33	2.618	1.043	40		
Kyenkyen	Antiaris toxicaria	41.405	9.178	22	38.087	9.178	24		
Aprokuma	Antrocaryon micraster	6.610	832	13	5.962	832	14		
Canarium	Canarium schweinfurthii	4.207	598	14	3.454	598	17		
Ceiba	Ceiba pentandra	80.906	7.723	10	73.292	7.723	11		
Akasa	Chrysophyllum spp.	4.176	1.521	36	3.712	1.521	41		
Bonsamdua/Ayan	Distemonanthus benthamianus	5.892	2.013	34	5.171	2.013	39		
Candollei 1/	Entandophragma candollei	638	1.577	247	638	1.577	247	1.577	
Guarea	Guarea spp	6.440	5.290	82	5.605	5.290	94		
Niangon 1/	Heritiera utilis	10.222	7.745	76	8.444	7.745	92		
Kaku/Ekki	Lophira alata	9.815	1.228	13	8.918	1.228	14		
Walnut 1/	Lovoa trichilioides	3.758	2.405	64	3.175	2.405	76		
Mansonia 1/	Mansonia altissima	5.280	1.753	33	5.040	1.753	35		
Dahoma	Piptadenisatrum africanum	27.522	17.967	65	24.680	17.967	73		
Bombax	Rhodognaphalon/Bombax brevicuspe	7.843	2.705	34	6.864	2.705	39		
Emire 2/	Terminalia ivorensis	6.307	2.892	46	5.692	2.892	51		
SUB TOTAL	**Red stars**	**224.195**	**66.470**	**30**	**201.353**	**66.470**	**33**	**1.577**	**2**
PINK STARS									
Albizia	Albizia adianthifolia	3.629			3.437				
Albizia	Albizia zygia	6.653	16	0	6.132	16	0		
Alstonia	Alstonia boonei	15.115			13.166				
Yaya	Amphimas pterocarpoides	5.650	210	4	5.108	210	4		
Kokote	Anopyxis klaineana	5.975	454	8	5.131	454	9		
Berlinia	Berlinia confusa	4.682			4.019				
Berlinia	Berlinia spp.	14.086			11.629				
Celtis/Essa	Celtis mildbraedii/zenkeri	138.200	1.624	1	127.028	1.624	1		
Pamprana	Coryanthe pachyceras	9.520			8.738				
Denya	Cylicodiscus gabonensis	33.627	936	3	29.096	936	3		
Ananta	Cynometra antana	18.131	203	1	15.445	203	1		
Duabankye	Dialium aubrevillei	13.448	91	1	10.742	91	1		
Potrodum	Erytrophleum sauaveolens	7.465	138	2	6.374	138	2		
Nakwa	Holoptelea grandis	3.898			3.756				
Kroma	Klaindoxa gabonensis	8.466	176	2	7.068	176	2		
Kumanini	Lannea welwitschii	5.100			4.575				
Bompagya	Mammea africana	4.893	560	11	3.933	560	14		
Wonton	Morus mesozygia	6.029			5.780				
Danta 2/	Nesogordonia papaverifera	19.465	4.514	23	17.680	4.514	26		
Akwana	Ongokea gore	4.147			3.696				
Afam	Parinari exelsa	7.402			6.541				
Asoma	Parkia bicolor	7.642			6.874				
Esia	Petersianthus macrocarpus	37.391	1.701	5	33.317	1.701	5		
Otie	Pycnanthus angolensis	32.070	7.316	23	28.395	7.316	26		
Bombax	Rhodognaphalon/Bombax buonopozense	5.868	30	1	5.106	30	1		
Wama	Ricinodendron heudelotii	9.065			8.523				
Sterculia	Sterculia spp.	8.880	2.810	32	8.288	2.810	34		
Afena	Strombosia glaucescens	11.560	13	0	10.200	13	0		
Ofram 2a/	Terminalia superba	56.055	9.043	16	51.404	9.043	18		
Tanuronini	Trichilia tessmannii	6.491			5.925				
Okure	Trilepisium madagascariense	4.176			3.816				
Avodire 1/	Turreanthus africanus	20.100	3.585	18	18.600	3.585	19		
SUB TOTAL	**Pink stars**	**534.877**	**33.420**	**6**	**479.518**	**33.420**	**7**		
Subaha 2/, 4/	Hallea spp./Mitragyna spp.								
Other species			1.173			1.173			
GRAND TOTAL		**891.509**	**284.932**	**32**	**801.688**	**284.932**	**36**	**178.324**	**63**

Notes:
1/ Log export banned in 1979
2/ Log export banned in 1988
2a/ Log export banned in 1988 but lifted in 1990
3/ Log export banned in 1993
4/ Harvest banned in 1993
Log exports was entirely banned during 1995.

Sources:
FIMP (1995b)
FPIB LMC Totals (1996)

Appendix 5

GHANA

TOTAL RECORDED TIMBER EXTRACTION BY SPECIES 1986-91 (Cu.m. ROUNDWOOD)

SPECIES		YEAR					
Trade name	Botanical name	1986	1987	1988	1989	1990	1991
SCARLET STARS							
Albizia	Albizia ferruginea			6.092	12.580	8.997	19.209
Asanfina 2/	Aningeria altissima/robusta	22.989	36.367	35.736	8.985	13.740	25.570
Daniella	Daniella ogea/thurifera			632	2.406	3.586	3.054
Edinam 1/	Entandophragma angolense	24.254	50.031	33.167	27.883	42.609	39.403
Sapele 1/	Entandophragma cylindricum	35.245	41.878	41.135	39.805	60.738	46.077
Utile 1/	Entandophragma utile	56.676	53.759	31.891	23.379	29.052	16.758
Hyedua 1/	Guibourtia ehie	6.045	9.268	10.620	7.670	9.747	8.990
Mahogany 1/	Khaya anthotheca/grandifoliola						
Mahogany 1/	Khaya ivorensis	45.008	57.340	66.877	71.291	101.248	85.833
Odum 1/	Milicia excelsa/regia	136.911	155.797	172.983	169.413	188.538	142.152
Kusia	Nauclea diderrichii			7.813	12.508	8.382	12.396
Afrormosia 1/	Pericopsis elata	21.843	11.547	7.190	9.230	9.945	7.063
Kyere	Pterygota macrocarpa	29.045	43.167	49.556	39.785	41.420	42.090
Makore 1/	Tieghemella heckelii	17.355	26.309	39.642	30.558	29.504	22.472
Wawa 3/	Triplochiton scleroxylon	230.618	300.943	366.064	378.938	481.932	499.602
SUB TOTAL	**Scarlet stars**	**625.989**	**786.406**	**869.398**	**834.431**	**1.029.438**	**970.669**
RED STARS							
Afzelia/Papao	Afzelia africana/bellea	2.078	6.609	8.571	3.876	19.176	12.063
Kyenkyen	Antiaris toxicaria	2.103	12.727	14.801	13.193	14.962	12.990
Aprokuma	Antrocaryon micraster			4.114	2.615	9.743	4.517
Canarium	Canarium schweinfurthii			281	221	553	1.768
Ceiba	Ceiba pentandra	5.890	27.848	26.238	9.904	19.191	37.855
Akasa	Chrysophyllum spp.			7.565	1.618	5.634	4.416
Bonsamdua/Ayan	Distemonanthus benthamianus			5.695	2.108	8.652	5.381
Candollei 1/	Entandophragma candollei	4.693	5.302	8.222	7.923	9.977	8.758
Guarea	Guarea spp	6.607	10.369	10.972	6.548	14.245	11.445
Niangon 1/	Heritiera utilis	2.796	3.327	3.716	4.362	8.419	11.897
Kaku/Ekki	Lophira alata			3.004	5.968	1.196	11.814
Walnut 1/	Lovoa trichilioides	2.107	2.053	4.340	4.804	6.625	5.877
Mansonia 1/	Mansonia altissima	4.872	5.341	5.830	3.360	4.923	2.625
Dahoma	Piptadenisatrum africanum	8.994	8.612	14.915	20.491	38.078	29.213
Bombax	Rhodognaphalon/Bombax brevicuspe			11.104	7.073	12.082	9.719
Emire 2/	Terminalia ivorensis	55.921	50.031	42.242	20.095	34.031	20.691
SUB TOTAL	**Red stars**	**96.061**	**132.219**	**171.610**	**114.159**	**207.487**	**191.029**
PINK STARS							
Albizia	Albizia adianthifolia			342	57	69	40
Albizia	Albizia zygia			506	499	397	160
Alstonia	Alstonia boonei						
Yaya	Amphimas pterocarpoides			791	794	1.513	2.019
Kokote	Anopyxis klaineana			157	67	247	17
Berlinia	Berlinia confusa			35	44	82	24
Berlinia	Berlinia spp.						
Celtis/Essa	Celtis mildbraedii/zenkeri			7.705	6.904	4.583	493
Pamprana	Coryanthe pachyceras						
Denya	Cylicodiscus gabonensis			1.594	1.296	327	1.916
Ananta	Cynometra antana			122	988	0	0
Duabankye	Dialium aubrevillei						
Potrodum	Erytrophleum sauaveolens			3.924	1.278	315	458
Nakwa	Holoptelea grandis						4
Kroma	Klaindoxa gabonensis						10
Kumanini	Lannea welwitschii					11	44
Bompagya	Mammea africana			479	346	285	335
Wonton	Morus mesozygia			39	18	60	17
Danta 2/	Nesogordonia papaverifera	15.229	29.858	24.787	4.063	9.997	8.589
Akwana	Ongokea gore			16	12		
Afam	Parinari exelsa						
Asoma	Parkia bicolor			127	42	20	95
Esia	Petersianthus macrocarpus			1.999	1.590	1.782	1.819
Otie	Pycnanthus angolensis	11.214	14.797	28.253	17.626	18.509	20.704
Bombax	Rhodognaphalon/Bombax buonopozense						
Wama	Ricinodendron heudelotii						
Sterculia	Sterculia spp.			271	1.125	2.271	1.885
Afena	Strombosia glaucescens						
Ofram 2a/	Terminalia superba	20.210	23.300	12.984	4.340	4.314	23.286
Tanuronini	Trichilia tessmannii						
Okure	Trilepisium madagascariense						
Avodire 1/	Turreanthus africanus			269	1.290	4.489	1.922
SUB TOTAL	**Pink stars**	**46.653**	**67.955**	**84.400**	**42.379**	**49.271**	**63.835**
Subaha 2/, 4/	Hallea spp./Mitragyna spp.			9.195	2.712	695	0
Other species 5/		35.388	38.020	2.144	2.526	2.732	3.900
GRAND TOTAL		**804.091**	**1.024.600**	**1.136.747**	**996.207**	**1.289.623**	**1.229.433**

Notes:
1/ Log export banned in 1979
2/ Log export banned in 1988
2a/ Log export banned in 1988 but lifted in 1990
3/ Log export banned in 1993

Source:
FPIB LMC Totals (1986-96)

Appendix 5

GHANA

TOTAL RECORDED TIMBER EXTRACTION BY SPECIES 1992-96 (Cu.m. ROUNDWOOD)

SPECIES		YEAR				
Trade name	**Botanical name**	**1992**	**1993**	**1994**	**1995**	**1996**
SCARLET STARS						
Albizia	Albizia ferruginea	7.010	4.102	7.804	8.317	9.924
Asanfina 2/	Aningeria altissima/robusta	29.251	23.264	33.253	42.256	40.295
Daniella	Daniella ogea/thurifera	6.246	38.004	31.301	10.046	9.682
Edinam 1/	Entandophragma angolense	41.823	28.760	33.056	26.817	22.583
Sapele 1/	Entandophragma cylindricum	40.607	28.612	25.064	22.524	21.905
Utile 1/	Entandophragma utile	12.587	13.994	21.089	10.693	7.361
Hyedua 1/	Guibourtia ehie	14.271	170	1.422	1.569	2.539
Mahogany 1/	Khaya anthotheca/grandifoliola			0	7.115	9.979
Mahogany 1/	Khaya ivorensis	98.667	83.964	82.330	52.740	49.671
Odum 1/	Milicia excelsa/regia	168.434	128.933	134.708	89.620	66.770
Kusia	Nauclea diderrichii	11.953	14.987	13.992	14.157	6.837
Afrormosia 1/	Pericopsis elata	8.319	6.492	2.124	1.341	920
Kyere	Pterygota macrocarpa	37.495	39.709	40.907	40.634	42.669
Makore 1/	Tieghemella heckelii	29.225	24.866	21.214	13.007	12.609
Wawa 3/	Triplochiton scleroxylon	531.438	404.410	365.243	430.197	436.790
SUB TOTAL	**Scarlet stars**	**1.037.326**	**840.267**	**813.507**	**771.033**	**740.534**
RED STARS						
Afzelia/Papao	Afzelia africana/bellea	12.872	14.439	21.990	12.991	8.361
Kyenkyen	Antiaris toxicaria	8.668	156.209	227.236	41.104	31.616
Aprokuma	Antrocaryon micraster	5.706	3.641	5.119	4.008	2.299
Canarium	Canarium schweinfurthii	1.458	28.422	17.664	3.132	1.372
Ceiba	Ceiba pentandra	46.295	326.940	128.815	102.702	148.647
Akasa	Chrysophyllum spp.	2.555	3.330	3.415	6.190	5.977
Bonsamdua/Ayan	Distemonanthus benthamianus	7.708	2.503	6.479	3.877	4.399
Candollei 1/	Entandophragma candollei	8.504	6.014	5.261	3.904	3.712
Guarea	Guarea spp	8.211	5.470	7.572	5.985	8.050
Niangon 1/	Heritiera utilis	13.713	22.701	25.790	33.437	24.074
Kaku/Ekki	Lophira alata	9.343	18.402	5.337	9.278	4.311
Walnut 1/	Lovoa trichilioides	6.380	5.484	4.713	4.811	4.014
Mansonia 1/	Mansonia altissima	3.964	3.967	3.856	3.833	2.751
Dahoma	Piptadenisatrum africanum	28.971	35.113	41.365	50.300	45.298
Bombax	Rhodognaphalon/Bombax brevicuspe	9.011	12.040	7.769	5.999	5.628
Emire 2/	Terminalia ivorensis	24.498	24.954	57.571	22.577	20.814
SUB TOTAL	**Red stars**	**197.857**	**669.629**	**569.952**	**314.128**	**321.323**
PINK STARS						
Albizia	Albizia adianthifolia	4	9	34	7	0
Albizia	Albizia zygia			277	501	284
Alstonia	Alstonia boonei	41	156	65	28	38
Yaya	Amphimas pterocarpoides	2.263	7.893	33.801	2.455	1.831
Kokote	Anopyxis klaineana	189	636	440	754	518
Berlinia	Berlinia confusa			72	0	0
Berlinia	Berlinia spp.			0	0	0
Celtis/Essa	Celtis mildbraedii/zenkeri	856	9.757	20.485	4.861	2.273
Pamprana	Coryanthe pachyceras			0	0	0
Denya	Cylicodiscus gabonensis	1.168	1.373	2.543	4.504	1.445
Ananta	Cynometra antana	0	2.660	820	504	220
Duabankye	Dialium aubrevillei		9	17	4	120
Potrodum	Erytrophleum sauaveolens	603	484	882	741	432
Nakwa	Holoptelea grandis	32	42	37	0	0
Kroma	Klaindoxa gabonensis		371	112	39	360
Kumanini	Lannea welwitschii	15		17	14	3
Bompagya	Mammea africana	443	490	286	304	871
Wonton	Morus mesozygia	13	74	47	7	125
Danta 2/	Nesogordonia papaverifera	9.594	5.765	9.397	11.915	9.138
Akwana	Ongokea gore	7		51	81	55
Afam	Parinari exelsa		50	0	0	0
Asoma	Parkia bicolor	25	665	46	61	124
Esia	Petersianthus macrocarpus	2.357	4.161	6.267	3.017	3.811
Otie	Pycnanthus angolensis	25.268	49.945	51.652	28.695	26.530
Bombax	Rhodognaphalon/Bombax buonopozense		15	58	457	46
Wama	Ricinodendron heudelotii		17	31	28	10
Sterculia	Sterculia spp.	2.372	2.301	1.257	5.817	7.922
Afena	Strombosia glaucescens		13	0	0	16
Ofram 2a/	Terminalia superba	36.216	80.103	167.285	39.349	42.177
Tanuronini	Trichilia tessmannii		15	0	23	0
Okure	Trilepisium madagascariense			0	0	0
Avodire 1/	Turreanthus africanus	1.271	3.806	1.990	4.102	3.999
SUB TOTAL	**Pink stars**	**82.737**	**170.810**	**297.969**	**108.268**	**102.348**
Subaha 2/, 4/	Hallea spp./Mitragyna spp.	4	11	17	8	15
Other species 5/		480	1.571	412	978	2.232
GRAND TOTAL		**1.318.404**	**1.682.288**	**1.681.857**	**1.194.415**	**1.166.452**

Notes:
1/ Log export banned in 1979
2/ Log export banned in 1988
2a/ Log export banned in 1988 but lifted in 1990
3/ Log export banned in 1993

Source:
FPIB LMC Totals (1986-96)

Appendix 5

GHANA
TOTAL RECORDED ON-RESERVE TIMBER EXTRACTION BY SPECIES 1986-93 (Cu.m. ROUNDWOOD)

SPECIES		YEAR							
Trade name	**Botanical name**	1986	1987	1988	1989	1990	1991	1992	1993
SCARLET STARS									
Albizia	Albizia ferruginea								
Asanfina 2/	Aningeria altissima/robusta	7.134	8.979	11.061	7.185	7.065	9.580		
Daniella	Daniella ogea/thurifera								
Edinam 1/	Entandophragma angolense	17.315	19.636	19.831	17.936	29.789	17.057		
Sapele 1/	Entandophragma cylindricum	28.528	28.352	30.249	26.373	51.598	27.496		
Utile 1/	Entandophragma utile	48.687	39.243	23.587	21.484	19.425	8.140		
Hyedua 1/	Guibourtia ehie	4.409	4.507	8.351	6.695	5.271	3.906		
Mahogany 1/	Khaya anthotheca/grandifoliola								
Mahogany 1/	Khaya ivorensis	37.989	41.889	43.708	49.571	67.286	38.479		
Odum 1/	Milicia excelsa/regia	38.882	33.188	55.924	77.708	80.481	35.242		
Kusia	Nauclea diderrichii								
Afrormosia 1/	Pericopsis elata	12.998	6.419	5.297	8.041	1.961	2.702		
Kyere	Pterygota macrocarpa	12.921	15.256	14.269	17.583	17.374	8.141		
Makore 1/	Tieghemella heckelii	12.070	16.804	13.872	22.025	18.224	9.160		
Wawa 3/	Triplochiton scleroxylon	70.967	77.800	99.089	149.583	203.499	148.032		
SUB TOTAL		**291.900**	**292.073**	**325.238**	**404.184**	**501.973**	**307.935**	**0**	**0**
RED STARS									
Afzelia/Papao	Afzelia africana/bellea	2.611	3.784	4.713	3.110	5.523	1.524		
Kyenkyen	Antiaris toxicaria	2.742	5.162	6.485	4.700	6.539	9.130		
Aprokuma	Antrocaryon micraster								
Canarium	Canarium schweinfurthii								
Ceiba	Ceiba pentandra	2.111	2.598	1.540	2.793	7.342	8.220		
Akasa	Chrysophyllum spp.								
Bonsamdua/Ayan	Distemonanthus benthamianus								
Candollei 1/	Entandophragma candollei	2.461	3.140	5.109	5.679	9.024	3.385		
Guarea	Guarea spp	3.049	4.285	5.281	4.046	10.427	7.219		
Niangon 1/	Heritiera utilis	693	3.507	3.722	4.738	8.956	5.405		
Kaku/Ekki	Lophira alata								
Walnut 1/	Lovoa trichilioides	1.597	2.372	2.994	4.006	6.516	4.814		
Mansonia 1/	Mansonia altissima	5.179	4.029	5.177	4.284	3.883	2.167		
Dahoma	Piptadenisatrum africanum	5.579	7.598	11.737	16.883	32.631	21.060		
Bombax	Rhodognaphalon/Bombax brevicuspe								
Emire 2/	Terminalia ivorensis	5.809	5.193	5.692	7.535	10.260	4.879		
SUB TOTAL		**31.831**	**41.668**	**52.450**	**57.774**	**101.101**	**67.803**	**0**	**0**
PINK STARS									
Albizia	Albizia adianthifolia								
Albizia	Albizia zygia								
Alstonia	Alstonia boonei								
Yaya	Amphimas pterocarpoides								
Kokote	Anopyxis klaineana								
Berlinia	Berlinia confusa								
Berlinia	Berlinia spp.								
Celtis/Essa	Celtis mildbraedii/zenkeri								
Pamprana	Coryanthe pachyceras								
Denya	Cylicodiscus gabonensis								
Ananta	Cynometra antana								
Duabankye	Dialium aubrevillei								
Potrodum	Erytrophleum sauaveolens								
Nakwa	Holoptelea grandis								
Kroma	Klaindoxa gabonensis								
Kumanini	Lannea welwitschii								
Bompagya	Mammea africana								
Wonton	Morus mesozygia								
Danta 2/	Nesogordonia papaverifera	6.713	8.853	8.351	3.356	8.181	6.273		
Akwana	Ongokea gore								
Afam	Parinari exelsa								
Asoma	Parkia bicolor								
Esia	Petersianthus macrocarpus								
Otie	Pycnanthus angolensis	4.306	5.383	7.140	6.410	6.988	4.213		
Bombax	Rhodognaphalon/Bombax buonopozense								
Wama	Ricinodendron heudelotii								
Sterculia	Sterculia spp.								
Afena	Strombosia glaucescens								
Ofram 2a/	Terminalia superba	6.166	2.960	2.062	5.959	3.022	8.374		
Tanuronini	Trichilia tessmannii								
Okure	Trilepisium madagascariense								
Avodire 1/	Turreanthus africanus								
SUB TOTAL		**17.185**	**17.196**	**17.553**	**15.725**	**18.191**	**18.860**	**0**	**0**
Subaha 2/, 4/	Hallea spp./Mitragyna spp.								
Other species 5/		8.080	9.051	10.749	9.156	16.762	10.386		
GRAND TOTAL		**348.996**	**359.988**	**405.990**	**486.839**	**638.027**	**404.984**	**0**	**0**

Notes:
1/ Log export banned in 1979
2/ Log export banned in 1988
2a/ Log export banned in 1988 but lifted in 1990
3/ Log export banned in 1993
4/ Harvest banned in 1993
Log exports was entirely banned in 1995
5/ For the period 1986-91 this group includes some of the listed species.

Source:
FPIB LMC Totals (1986-96)

Appendix 5

GHANA

TOTAL RECORDED ON-RESERVE TIMBER EXTRACTION BY SPECIES 1994-96 (Cu.m. ROUNDWOOD)

SPECIES Trade name	Botanical name	YEAR 1994 No. of trees	Volume	Vol./tree	1995 No. of trees	Volume	Vol./tree	1996 No. of trees	Volume	Vol./tree
SCARLET STARS										
Albizia	Albizia ferruginea	38	304	8,0	70	655	9,4	111	1.089	9,8
Asanfina 2/	Aningeria altissima/robusta	247	2.188	8,9	492	4.436	9,0	951	9.789	10,3
Daniella	Daniella ogea/thurifera	176	1.566	8,9	164	1.783	10,9	272	3.182	11,7
Edinam 1/	Entandophragma angolense	552	8.160	14,8	535	7.792	14,6	718	9.668	13,5
Sapele 1/	Entandophragma cylindricum	375	5.625	15,0	501	6.847	13,7	896	12.723	14,2
Utile 1/	Entandophragma utile	562	12.094	21,5	295	5.498	18,6	226	4.280	18,9
Hyedua 1/	Guibourtia ehie	22	121	5,5	86	466	5,4	207	1.460	7,1
Mahogany 1/	Khaya anthotheca/grandifoliola				166	1.717	10,3	128	1.207	9,4
Mahogany 1/	Khaya ivorensis	864	11.330	13,1	920	11.086	12,1	1.482	18.427	12,4
Odum 1/	Milicia excelsa/regia	422	6.394	15,2	538	7.367	13,7	1.602	20.598	12,9
Kusia	Nauclea diderrichii	94	852	9,1	74	827	11,2	97	1.000	10,3
Afrormosia 1/	Pericopsis elata	15	66	4,4	10	64	6,4	50	391	7,8
Kyere	Pterygota macrocarpa	257	2.244	8,7	448	4.316	9,6	624	6.391	10,2
Makore 1/	Tieghemella heckelii	364	7.716	21,2	241	4.817	20,0	240	5.962	24,8
Wawa 3/	Triplochiton scleroxylon	1.627	23.518	14,5	3.618	54.287	15,0	5.446	87.702	16,1
SUB TOTAL		**5.615**	**82.178**		**8.158**	**111.958**		**13.050**	**183.869**	
RED STARS										
Afzelia/Papao	Afzelia africana/bellea	8	48	6,0	73	330	4,5	200	1.043	5,2
Kyenkyen	Antiaris toxicaria	481	5.317	11,1	374	3.752	10,0	828	9.178	11,1
Aprokuma	Antrocaryon micraster	19	136	7,2	49	393	8,0	117	832	7,1
Canarium	Canarium schweinfurthii	89	1.158	13,0	42	484	11,5	57	598	10,5
Ceiba	Ceiba pentandra	235	3.093	13,2	178	2.600	14,6	437	7.723	17,7
Akasa	Chrysophyllum spp.	66	412	6,2	220	1.240	5,6	240	1.521	6,3
Bonsamdua/Ayan	Distemonanthus benthamianus	54	365	6,8	99	689	7,0	296	2.013	6,8
Candollei 1/	Entandophragma candollei	46	735	16,0	65	1.051	16,2	101	1.577	15,6
Guarea	Guarea spp	142	1.184	8,3	231	1.948	8,4	642	5.290	8,2
Niangon 1/	Heritiera utilis	564	3.428	6,1	558	3.568	6,4	1.235	7.745	6,3
Kaku/Ekki	Lophira alata	63	523	8,3	123	1.362	11,1	117	1.228	10,5
Walnut 1/	Lovoa trichilioides	154	1.468	9,5	134	1.419	10,6	221	2.405	10,9
Mansonia 1/	Mansonia altissima	219	1.498	6,8	424	2.651	6,3	315	1.753	5,6
Dahoma	Piptadenisatrum africanum	503	4.850	9,6	1.080	10.859	10,1	1.739	17.967	10,3
Bombax	Rhodognaphalon/Bombax brevicuspe	72	668	9,3	127	1.347	10,6	260	2.705	10,4
Emire 2/	Terminalia ivorensis	170	1.673	9,8	183	1.790	9,8	309	2.892	9,4
SUB TOTAL		**2.885**	**26.556**		**3.960**	**35.483**		**7.114**	**66.470**	
PINK STARS										
Albizia	Albizia adianthifolia									
Albizia	Albizia zygia	5	28	5,6	2	15	7,5	2	16	8,0
Alstonia	Alstonia boonei	1	6	6,0						
Yaya	Amphimas pterocarpoides	28	243	8,7	45	259	5,8	22	210	9,5
Kokote	Anopyxis klaineana	1	4	4,0	30	354	11,8	36	454	12,6
Berlinia	Berlinia confusa									
Berlinia	Berlinia spp.									
Celtis/Essa	Celtis mildbraedii/zenkeri	82	592	7,2	189	1.368	7,2	234	1.624	6,9
Pamprana	Coryanthe pachyceras									
Denya	Cylicodiscus gabonensis	14	161	11,5	40	497	12,4	79	936	11,8
Ananta	Cynometra antana	21	88	4,2				31	203	6,5
Duabankye	Dialium aubrevillei				1	4	4,0	18	91	5,1
Potrodum	Erytrophleum sauaveolens	27	154	5,7	43	217	5,0	23	138	6,0
Nakwa	Holoptelea grandis									
Kroma	Klaindoxa gabonensis				5	21	4,2	21	176	8,4
Kumanini	Lannea welwitschii	2	6	3,0	1	5	5,0			
Bompagya	Mammea africana				3	41	13,7	66	560	8,5
Wonton	Morus mesozygia									
Danta 2/	Nesogordonia papaverifera	209	1.173	5,6	521	2.851	5,5	774	4.514	5,8
Akwana	Ongokea gore									
Afam	Parinari exelsa									
Asoma	Parkia bicolor									
Esia	Petersianthus macrocarpus	26	263	10,1	137	944	6,9	230	1.701	7,4
Otie	Pycnanthus angolensis	266	1.535	5,8	702	4.448	6,3	1.149	7.316	6,4
Bombax	Rhodognaphalon/Bombax buonopozense	9	58	6,4	1	17	17,0	4	30	7,5
Wama	Ricinodendron heudelotii	6	31	5,2						
Sterculia	Sterculia spp.	17	95	5,6	462	1.980	4,3	568	2.810	4,9
Afena	Strombosia glaucescens							3	13	4,3
Ofram 2a/	Terminalia superba	403	3.291	8,2	578	4.287	7,4	1.149	9.043	7,9
Tanuronini	Trichilia tessmannii									
Okure	Trilepisium madagascariense									
Avodire 1/	Turreanthus africanus	160	459	2,9	687	2.447	3,6	771	3.585	4,6
SUB TOTAL		**1.277**	**8.187**		**3.447**	**19.755**		**5.180**	**33.420**	
Subaha 2/, 4/	Hallea spp./Mitragyna spp.									
Other species 5/		7	41		7	45		381	1.173	
GRAND TOTAL		**9.784**	**116.962**		**15.572**	**167.241**		**25.725**	**284.932**	

Notes:
1/ Log export banned in 1979
2/ Log export banned in 1988
2a/ Log export banned in 1988 but lifted in 1990
3/ Log export banned in 1993
4/ Harvest banned in 1993
Log exports was entirely banned in 1995
5/ For the period 1986-91 this group includes some of the listed species.

Source:
FPIB LMC Totals (1986-96)

Appendix 5

GHANA
TOTAL RECORDED OFF-RESERVE TIMBER EXTRACTION BY SPECIES 1986-93 (Cu.m. ROUNDWOOI

SPECIES		YEAR							
Trade name	**Botanical name**	**1986**	**1987**	**1988**	**1989**	**1990**	**1991**	**1992**	**1993**
SCARLET STARS									
Albizia	Albizia ferruginea								
Asanfina 2/	Aningeria altissima/robusta	15.855	27.388	24.675	1.800	6.675	15.990		
Daniella	Daniella ogea/thurifera								
Edinam 1/	Entandophragma angolense	7.989	14.516	8.304	1.895	9.627	8.618		
Sapele 1/	Entandophragma cylindricum	6.717	13.526	10.886	13.432	9.140	18.581		
Utile 1/	Entandophragma utile	7.989	14.516	8.304	1.895	9.627	8.618		
Hyedua 1/	Guibourtia ehie	1.636	4.761	2.269	975	4.476	5.084		
Mahogany 1/	Khaya anthotheca/grandifoliola								
Mahogany 1/	Khaya ivorensis	7.019	15.451	23.169	21.720	33.962	47.354		
Odum 1/	Milicia excelsa/regia	8.845	5.128	1.893	1.189	7.984	4.361		
Kusia	Nauclea diderrichii								
Afrormosia 1/	Pericopsis elata	8.845	5.128	1.893	1.189	7.984	4.361		
Kyere	Pterygota macrocarpa	16.124	27.911	35.287	22.202	24.046	33.949		
Makore 1/	Tieghemella heckelii	5.285	9.505	25.770	8.533	11.280	13.312		
Wawa 3/	Triplochiton scleroxylon	0	0	0	0	0	0		
SUB TOTAL		**86.304**	**137.830**	**142.450**	**74.830**	**124.801**	**160.228**	**0**	**0**
RED STARS									
Afzelia/Papao	Afzelia africana/bellea	334.089	494.333	544.160	430.247	527.465	662.734		
Kyenkyen	Antiaris toxicaria	0	0	281	221	553	1.768		
Aprokuma	Antrocaryon micraster								
Canarium	Canarium schweinfurthii								
Ceiba	Ceiba pentandra	3.779	25.250	24.698	7.111	11.849	29.635		
Akasa	Chrysophyllum spp.								
Bonsamdua/Ayan	Distemonanthus benthamianus								
Candollei 1/	Entandophragma candollei	2.232	2.162	3.113	2.244	953	5.373		
Guarea	Guarea spp	0	0	5.695	2.108	8.652	5.381		
Niangon 1/	Heritiera utilis	2.103	-180	-6	-376	-537	6.492		
Kaku/Ekki	Lophira alata								
Walnut 1/	Lovoa trichilioides	2.103	-180	-6	-376	-537	6.492		
Mansonia 1/	Mansonia altissima	0	0	3.004	5.968	1.196	11.814		
Dahoma	Piptadenisatrum africanum	50.112	44.838	36.550	12.560	23.771	15.812		
Bombax	Rhodognaphalon/Bombax brevicuspe								
Emire 2/	Terminalia ivorensis	50.112	44.838	36.550	12.560	23.771	15.812		
SUB TOTAL		**444.530**	**611.061**	**654.039**	**472.267**	**597.136**	**761.313**	**0**	**0**
PINK STARS									
Albizia	Albizia adianthifolia								
Albizia	Albizia zygia								
Alstonia	Alstonia boonei								
Yaya	Amphimas pterocarpoides								
Kokote	Anopyxis klaineana								
Berlinia	Berlinia confusa								
Berlinia	Berlinia spp.								
Celtis/Essa	Celtis mildbraedii/zenkeri								
Pamprana	Coryanthe pachyceras								
Denya	Cylicodiscus gabonensis								
Ananta	Cynometra antana								
Duabankye	Dialium aubrevillei								
Potrodum	Erytrophleum sauaveolens								
Nakwa	Holoptelea grandis								
Kroma	Klaindoxa gabonensis								
Kumanini	Lannea welwitschii								
Bompagya	Mammea africana								
Wonton	Morus mesozygia								
Danta 2/	Nesogordonia papaverifera	0	0	479	346	285	335		
Akwana	Ongokea gore								
Afam	Parinari exelsa								
Asoma	Parkia bicolor								
Esia	Petersianthus macrocarpus								
Otie	Pycnanthus angolensis	6.908	9.414	21.113	11.216	11.521	16.491		
Bombax	Rhodognaphalon/Bombax buonopozense								
Wama	Ricinodendron heudelotii								
Sterculia	Sterculia spp.								
Afena	Strombosia glaucescens								
Ofram 2a/	Terminalia superba	0	0	271	1.125	2.271	1.885		
Tanuronini	Trichilia tessmannii								
Okure	Trilepisium madagascariense								
Avodire 1/	Turreanthus africanus								
SUB TOTAL		**6.908**	**9.414**	**21.863**	**12.687**	**14.077**	**18.710**	**0**	**0**
Subaha 2/, 4/	Hallea spp./Mitragyna spp.								
Other species 5/		35.388	38.020	65.266	59.529	61.941	77.057		
GRAND TOTAL		**573.130**	**796.325**	**883.618**	**619.313**	**797.955**	**1.017.308**	**0**	**0**

Notes:
1/ Log export banned in 1979
2/ Log export banned in 1988
2a/ Log export banned in 1988 but lifted in 1990
3/ Log export banned in 1993
4/ Harvest banned in 1993
Log exports was entirely banned in 1995
5/ For the period 1986-91 this group includes some of the listed species.

Source:
FPIB LMC Totals (1986-96)

Appendix 5

GHANA
TOTAL RECORDED OFF-RESERVE TIMBER EXTRACTION BY SPECIES 1994-96 (Cu.m. ROUNDWOOD)

SPECIES		YEAR								
Trade name	Botanical name	1994			1995			1996		
		No. of trees	Volume	Vol./tree	No. of trees	Volume	Vol./tree	No. of trees	Volume	Vol./tree
SCARLET STARS										
Albizia	Albizia ferruginea	9	77	8,6	784	5.946	7,6	956	8.186	8,6
Asanfina 2/	Aningeria altissima/robusta	407	2.541	6,2	4.108	30.226	7,4	3.670	28.561	7,8
Daniella	Daniella ogea/thurifera	245	2.655	10,8	678	6.740	9,9	530	6.177	11,7
Edinam 1/	Entandophragma angolense	261	2.097	8,0	1.799	13.814	7,7	1.463	11.830	8,1
Sapele 1/	Entandophragma cylindricum	171	1.532	9,0	1.353	12.265	9,1	901	8.536	9,5
Utile 1/	Entandophragma utile	12	78	6,5	458	3.869	8,4	300	2.909	9,7
Hyedua 1/	Guibourtia ehie	14	97	6,9	145	788	5,4	151	1.015	6,7
Mahogany 1/	Khaya anthotheca/grandifoliola				610	4.283	7,0	1.259	8.508	6,8
Mahogany 1/	Khaya ivorensis	1.186	10.685	9,0	4.139	32.786	7,9	2.984	27.685	9,3
Odum 1/	Milicia excelsa/regia	212	1.684	7,9	7.837	68.908	8,8	4.490	43.335	9,7
Kusia	Nauclea diderrichii	230	2.272	9,9	1.132	10.580	9,3	588	5.733	9,8
Afrormosia 1/	Pericopsis elata				135	986	7,3	76	525	6,9
Kyere	Pterygota macrocarpa	173	1.167	6,7	3.190	25.437	8,0	4.121	33.369	8,1
Makore 1/	Tieghemella heckelii	210	4.153	19,8	559	6.949	12,4	445	6.356	14,3
Wawa 3/	Triplochiton scleroxylon	2.206	22.460	10,2	25.617	271.804	10,6	26.996	322.624	12,0
SUB TOTAL		**5.336**	**51.498**		**52.544**	**495.381**		**48.930**	**515.349**	
RED STARS										
Afzelia/Papao	Afzelia africana/bellea	57	425	7,5	2.178	10.437	4,8	1.541	6.456	4,2
Kyenkyen	Antiaris toxicaria	378	4.481	11,9	3.046	32.279	10,6	1.979	20.948	10,6
Aprokuma	Antrocaryon micraster	64	331	5,2	339	2.384	7,0	195	1.343	6,9
Canarium	Canarium schweinfurthii	160	1.467	9,2	263	2.302	8,8	72	709	9,8
Ceiba	Ceiba pentandra	275	4.434	16,1	4.320	70.715	16,4	7.566	134.590	17,8
Akasa	Chrysophyllum spp.	17	83	4,9	538	3.190	5,9	649	4.145	6,4
Bonsamdua/Ayan	Distemonanthus benthamianus	176	1.065	6,1	383	2.388	6,2	325	2.300	7,1
Candollei 1/	Entandophragma candollei	13	122	9,4	237	2.236	9,4	188	1.933	10,3
Guarea	Guarea spp	91	572	6,3	461	2.909	6,3	335	2.540	7,6
Niangon 1/	Heritiera utilis	1.159	6.161	5,3	4.386	24.864	5,7	2.485	15.171	6,1
Kaku/Ekki	Lophira alata	29	328	11,3	694	6.045	8,7	341	3.012	8,8
Walnut 1/	Lovoa trichilioides	66	489	7,4	347	2.756	7,9	197	1.499	7,6
Mansonia 1/	Mansonia altissima	2	7	3,5	147	765	5,2	184	926	5,0
Dahoma	Piptadenisatrum africanum	394	3.325	8,4	3.305	30.663	9,3	2.489	25.473	10,2
Bombax	Rhodognaphalon/Bombax brevicuspe	55	451	8,2	385	3.412	8,9	304	2.717	8,9
Emire 2/	Terminalia ivorensis	1.376	7.564	5,5	2.727	16.257	6,0	2.525	16.871	6,7
SUB TOTAL		**4.312**	**31.305**		**23.756**	**213.602**		**21.375**	**240.633**	
PINK STARS										
Albizia	Albizia adianthifolia									
Albizia	Albizia zygia	8	38	4,8	85	423	5,0	32	260	8,1
Alstonia	Alstonia boonei	4	28	7,0	5	28	5,6	7	38	5,4
Yaya	Amphimas pterocarpoides	139	1.064	7,7	222	1.793	8,1	164	1.505	9,2
Kokote	Anopyxis klaineana	1	8	8,0	29	266	9,2	9	64	7,1
Berlinia	Berlinia confusa	10	56	5,6						
Berlinia	Berlinia spp.									
Celtis/Essa	Celtis mildbraedii/zenkeri	97	713	7,4	342	2.433	7,1	79	564	7,1
Pamprana	Coryanthe pachyceras									
Denya	Cylicodiscus gabonensis	13	99	7,6	313	3.718	11,9	43	497	11,6
Ananta	Cynometra antana	3	15	5,0	105	504	4,8	4	17	4,3
Duabankye	Dialium aubrevillei							4	29	7,3
Potrodum	Erytrophleum sauaveolens				50	243	4,9	44	262	6,0
Nakwa	Holoptelea grandis									
Kroma	Klaindoxa gabonensis	2	29	14,5	2	13	6,5	24	174	7,3
Kumanini	Lannea welwitschii				1	9	9,0	1	3	3,0
Bompagya	Mammea africana	16	91	5,7	23	186	8,1	30	272	9,1
Wonton	Morus mesozygia							18	125	6,9
Danta 2/	Nesogordonia papaverifera	44	218	5,0	1.138	6.454	5,7	726	4.293	5,9
Akwana	Ongokea gore				13	81	6,2	7	55	7,9
Afam	Parinari exelsa									
Asoma	Parkia bicolor	7	31	4,4	7	58	8,3	15	124	8,3
Esia	Petersianthus macrocarpus	70	445	6,4	185	1.313	7,1	252	1.988	7,9
Otie	Pycnanthus angolensis	337	1.887	5,6	2.298	15.393	6,7	2.602	18.263	7,0
Bombax	Rhodognaphalon/Bombax buonopozense				38	413	10,9	3	16	5,3
Wama	Ricinodendron heudelotii							1	10	10,0
Sterculia	Sterculia spp.	43	237	5,5	494	2.897	5,9	953	4.784	5,0
Afena	Strombosia glaucescens									
Ofram 2a/	Terminalia superba	841	6.283	7,5	3.687	26.654	7,2	4.141	31.346	7,6
Tanuronini	Trichilia tessmannii									
Okure	Trilepisium madagascariense									
Avodire 1/	Turreanthus africanus	105	345	3,3	320	1.177	3,7	84	379	4,5
SUB TOTAL		**1.740**	**11.587**		**9.357**	**64.056**		**9.243**	**65.068**	
Subaha 2/, 4/	Hallea spp./Mitragyna spp.	1	5	5,0	1	4	4,0	3	15	5,0
Other species 5/		10	45		153	807		254	829	
GRAND TOTAL		**11.399**	**94.440**		**85.811**	**773.850**		**79.805**	**821.894**	

Notes:
1/ Log export banned in 1979
2/ Log export banned in 1988
2a/ Log export banned in 1988 but lifted in 1990
3/ Log export banned in 1993
4/ Harvest banned in 1993
Log exports was entirely banned in 1995
5/ For the period 1986-91 this group includes some of the listed species.

Source:
FPIB LMC Totals (1986-96)

Appendix 5

GHANA
TOTAL RECORDED EXTRACTION FROM UNSPECIFIED SOURCE
1986-93 (Cu.m. ROUNDWOOD)

SPECIES		YEAR							
Trade name	**Botanical name**	**1986**	**1987**	**1988**	**1989**	**1990**	**1991**	**1992**	**1993**
SCARLET STARS									
Albizia	Albizia ferruginea								
Asanfina 2/	Aningeria altissima/robusta								
Daniella	Daniella ogea/thurifera								
Edinam 1/	Entandophragma angolense								
Sapele 1/	Entandophragma cylindricum								
Utile 1/	Entandophragma utile								
Hyedua 1/	Guibourtia ehie								
Mahogany 1/	Khaya anthotheca/grandifoliola								
Mahogany 1/	Khaya ivorensis								
Odum 1/	Milicia excelsa/regia								
Kusia	Nauclea diderrichii								
Afrormosia 1/	Pericopsis elata								
Kyere	Pterygota macrocarpa								
Makore 1/	Tieghemella heckelii								
Wawa 3/	Triplochiton scleroxylon								
SUB TOTAL									
RED STARS									
Afzelia/Papao	Afzelia africana/bellea								
Kyenkyen	Antiaris toxicaria								
Aprokuma	Antrocaryon micraster								
Canarium	Canarium schweinfurthii								
Ceiba	Ceiba pentandra								
Akasa	Chrysophyllum spp.								
Bonsamdua/Ayan	Distemonanthus benthamianus								
Candollei 1/	Entandophragma candollei								
Guarea	Guarea spp								
Niangon 1/	Heritiera utilis								
Kaku/Ekki	Lophira alata								
Walnut 1/	Lovoa trichilioides								
Mansonia 1/	Mansonia altissima								
Dahoma	Piptadenisatrum africanum								
Bombax	Rhodognaphalon/Bombax brevicuspe								
Emire 2/	Terminalia ivorensis								
SUB TOTAL									
PINK STARS									
Albizia	Albizia adianthifolia								
Albizia	Albizia zygia								
Alstonia	Alstonia boonei								
Yaya	Amphimas pterocarpoides								
Kokote	Anopyxis klaineana								
Berlinia	Berlinia confusa								
Berlinia	Berlinia spp.								
Celtis/Essa	Celtis mildbraedii/zenkeri								
Pamprana	Coryanthe pachyceras								
Denya	Cylicodiscus gabonensis								
Ananta	Cynometra antana								
Duabankye	Dialium aubrevillei								
Potrodum	Erytrophleum sauaveolens								
Nakwa	Holoptelea grandis								
Kroma	Klaindoxa gabonensis								
Kumanini	Lannea welwitschii								
Bompagya	Mammea africana								
Wonton	Morus mesozygia								
Danta 2/	Nesogordonia papaverifera								
Akwana	Ongokea gore								
Afam	Parinari exelsa								
Asoma	Parkia bicolor								
Esia	Petersianthus macrocarpus								
Otie	Pycnanthus angolensis								
Bombax	Rhodognaphalon/Bombax buonopozense								
Wama	Ricinodendron heudelotii								
Sterculia	Sterculia spp.								
Afena	Strombosia glaucescens								
Ofram 2a/	Terminalia superba								
Tanuronini	Trichilia tessmannii								
Okure	Trilepisium madagascariense								
Avodire 1/	Turreanthus africanus								
SUB TOTAL									
Subaha 2/, 4/	Hallea spp./Mitragyna spp.								
Other species 5/		35.388	38.020	12.811	13.446	8.632	5.045	2.041	14.234
GRAND TOTAL		**35.388**	**38.020**	**12.811**	**13.446**	**8.632**	**5.045**	**2.041**	**14.234**

Notes:
1/ Log export banned in 1979
2/ Log export banned in 1988
2a/ Log export banned in 1988 but lifted in 1990
3/ Log export banned in 1993
4/ Harvest banned in 1993
Log exports was entirely banned in 1995
5/ For the period 1986-91 this group includes some of the listed species.

Source:
FPIB LMC Totals (1986-96)

Appendix 5

GHANA

TOTAL RECORDED EXTRACTION FROM UNSPECIFIED SOURCE 1994-96 (Cu.m. ROUNDWOOD)

SPECIES		YEAR							
Trade name	Botanical name	1994			1995			1996	
		No. of trees	Volume	Vol./tree	No. of trees	Volume	Vol./tree	No. of trees	Volume
SCARLET STARS									
Albizia	Albizia ferruginea	930	7.423	8,0	216	1.716	7,9	85	649
Asanfina 2/	Aningeria altissima/robusta	3.978	28.524	7,2	1.037	7.594	7,3	263	1.945
Daniella	Daniella ogea/thurifera	2.557	27.080	10,6	181	1.523	8,4	29	323
Edinam 1/	Entandophragma angolense	3.145	22.799	7,2	707	5.211	7,4	143	1.085
Sapele 1/	Entandophragma cylindricum	1.950	17.907	9,2	355	3.412	9,6	71	646
Utile 1/	Entandophragma utile	777	8.917	11,5	149	1.326	8,9	16	172
Hyedua 1/	Guibourtia ehie	196	1.204	6,1	38	315	8,3	6	64
Mahogany 1/	Khaya anthotheca/grandifoliola				154	1.115	7,2	49	264
Mahogany 1/	Khaya ivorensis	7.139	60.315	8,4	1.114	8.868	8,0	419	3.559
Odum 1/	Milicia excelsa/regia	12.860	126.630	9,8	1.360	13.345	9,8	325	2.837
Kusia	Nauclea diderrichii	1.097	10.868	9,9	269	2.750	10,2	12	104
Afrormosia 1/	Pericopsis elata	274	2.058	7,5	44	291	6,6	1	4
Kyere	Pterygota macrocarpa	4.667	37.496	8,0	1.338	10.881	8,1	392	2.909
Makore 1/	Tieghemella heckelii	770	9.345	12,1	105	1.241	11,8	28	291
Wawa 3/	Triplochiton scleroxylon	31.007	319.265	10,3	9.565	104.106	10,9	2.569	26.464
SUB TOTAL		**71.347**	**679.831**		**16.632**	**163.694**		**4.408**	**41.316**
RED STARS									
Afzelia/Papao	Afzelia africana/bellea	3.515	21.517	6,1	465	2.224	4,8	270	862
Kyenkyen	Antiaris toxicaria	18.633	217.438	11,7	478	5.073	10,6	137	1.490
Aprokuma	Antrocaryon micraster	728	4.652	6,4	177	1.231	7,0	14	124
Canarium	Canarium schweinfurthii	1.522	15.039	9,9	46	346	7,5	8	65
Ceiba	Ceiba pentandra	8.054	121.288	15,1	1.837	29.387	16,0	389	6.334
Akasa	Chrysophyllum spp.	518	2.920	5,6	279	1.760	6,3	49	311
Bonsamdua/Ayan	Distemonanthus benthamianus	809	5.049	6,2	134	800	6,0	16	86
Candollei 1/	Entandophragma candollei	463	4.404	9,5	78	617	7,9	29	202
Guarea	Guarea spp	851	5.816	6,8	176	1.128	6,4	29	220
Niangon 1/	Heritiera utilis	2.715	16.201	6,0	917	5.005	5,5	204	1.158
Kaku/Ekki	Lophira alata	482	4.486	9,3	240	1.871	7,8	7	71
Walnut 1/	Lovoa trichilioides	349	2.756	7,9	75	636	8,5	16	110
Mansonia 1/	Mansonia altissima	412	2.351	5,7	83	417	5,0	17	72
Dahoma	Piptadenisatrum africanum	3.484	33.190	9,5	937	8.778	9,4	198	1.858
Bombax	Rhodognaphalon/Bombax brevicuspe	772	6.650	8,6	143	1.240	8,7	25	206
Emire 2/	Terminalia ivorensis	7.817	48.334	6,2	712	4.530	6,4	168	1.051
SUB TOTAL		**51.124**	**512.091**		**6.777**	**65.043**		**1.576**	**14.220**
PINK STARS									
Albizia	Albizia adianthifolia	5	34	6,8	1	7	7,0		
Albizia	Albizia zygia	37	211	5,7	11	63	5,7	2	8
Alstonia	Alstonia boonei	4	31	7,8					
Yaya	Amphimas pterocarpoides	3.451	32.494	9,4	51	403	7,9	12	116
Kokote	Anopyxis klaineana	51	428	8,4	8	134	16,8		
Berlinia	Berlinia confusa	2	16	8,0					
Berlinia	Berlinia spp.								
Celtis/Essa	Celtis mildbraedii/zenkeri	2.629	19.180	7,3	165	1.060	6,4	13	85
Pamprana	Coryanthe pachyceras								
Denya	Cylicodiscus gabonensis	193	2.283	11,8	30	289	9,6	1	12
Ananta	Cynometra antana	115	717	6,2					
Duabankye	Dialium aubrevillei	1	17	17,0					
Potrodum	Erytrophleum sauaveolens	128	728	5,7	48	281	5,9	5	32
Nakwa	Holoptelea grandis	6	37	6,2					
Kroma	Klaindoxa gabonensis	12	83	6,9	1	5	5,0	1	10
Kumanini	Lannea welwitschii	2	11	5,5					
Bompagya	Mammea africana	29	195	6,7	13	77	5,9	5	39
Wonton	Morus mesozygia	8	47	5,9	1	7	7,0		
Danta 2/	Nesogordonia papaverifera	1.528	8.006	5,2	483	2.610	5,4	64	331
Akwana	Ongokea gore	9	51	5,7					
Afam	Parinari exelsa								
Asoma	Parkia bicolor	2	15	7,5	1	3	3,0		
Esia	Petersianthus macrocarpus	800	5.559	6,9	130	760	5,8	21	122
Otie	Pycnanthus angolensis	7.290	48.230	6,6	1.349	8.854	6,6	155	951
Bombax	Rhodognaphalon/Bombax buonopozense				3	27	9,0		
Wama	Ricinodendron heudelotii				8	28	3,5		
Sterculia	Sterculia spp.	177	925	5,2	195	940	4,8	68	328
Afena	Strombosia glaucescens							1	3
Ofram 2a/	Terminalia superba	19.076	157.711	8,3	1.095	8.408	7,7	242	1.788
Tanuronini	Trichilia tessmannii				2	23	11,5		
Okure	Trilepisium madagascariense								
Avodire 1/	Turreanthus africanus	332	1.186	3,6	114	478	4,2	11	35
SUB TOTAL		**35.887**	**278.195**		**3.709**	**24.457**		**601**	**3.860**
Subaha 2/, 4/	Hallea spp./Mitragyna spp.	2	12	6,0	1	4	4,0		
Other species 5/		62	326		30	126		80	230
GRAND TOTAL		**158.422**	**1.470.455**		**27.149**	**253.324**		**6.665**	**59.626**

Notes:
1/ Log export banned in 1979
2/ Log export banned in 1988
2a/ Log export banned in 1988 but lifted in 1990
3/ Log export banned in 1993
4/ Harvest banned in 1993
Log exports was entirely banned in 1995
5/ For the period 1986-91 this group includes some of the listed species.

Source:
FPIB LMC Totals (1986-96)

Appendix 6

GHANA

ESTIMATED LOG HARVEST ON- AND OFF-RESERVES (1,000 Cu.m.)

Year	TOTAL (Cu.m.)	On-reserves 1/ (Cu.m.)	Off-reserves 2/ (Cu.m.)	Not specified /3 (Cu.m.)
1960	1.650	240	1.410	
1961	1.500	260	1.240	
1962	1.680	500	1.180	
1963	1.490	350	1.140	
1964	1.740	640	1.100	
1965	1.650	660	990	
1966	1.400	590	810	
1967	1.350	520	830	
1968	1.420	480	940	
1969	1.650	680	970	
1970	1.600	810	790	
1971	1.640	860	780	
1972	1.810	1.150	660	
1973	2.050	1.386	664	
1974	1.420	808	612	
1975	1.680	828	852	
1976	1.590	644	946	
1977	1.420	666	754	
1978	1.100	650	450	
1979	880	366	514	
1980	660	276	384	
1981	420	322	98	
1982	420	218	202	
1983	445	245	200	
1984	510	297	213	
1985	626	387	239	
1986	803	349	454	
1987	1.025	360	665	
1988	1.137	406	731	
1989	996	487	509	
1990	1.290	638	652	
1991	1.229	405	824	
1992	1.318	280	1.038	
1993	1.683	189	1.494	
1994	1.682	407	1.275	1.470
1995	1.194	221	973	253
1996	1.166	298	869	60

Sources: FIMP, 1995b [1960-72 & 1992]
IIED, 1993; App.1 [1973-91]
FPIB LMC Totals (1986-96)

Notes:

1/ A 15% under-reporting of the harvest on-reserves is, according to Kakrada and Maginnis (1992), assumed for the period 1973-91.

2/ Off-reserve harvest is for the period 1973-91 calculated as the residual between LMC-totals and the estimated harvest on-reserves. 1992 figures are from FIMP (1995b). For 1993 data does, in fact, not allow for a reliable distinction between on- and off-reserve harvest. For the period 1994-96 volumes in the column "not specified" are allocated to on- and off-reserves by the ratios of 1992.

3/ The column "not specified" absorbs LMCs which do not indicate the source of the logs (on- or off-reserves).

Appendix 7

GHANA
WOOD EXPORTS BY VOLUME 1972-96 (Cu.m. PRODUCTS)

Commodity	1972	1973	1974	1975	1976	1977	1978	1979	1980	1981	1982	1983	1984
Logs	1.020.937	1.007.040	481.668	602.303	531.022	453.906	311.610	198.381	104.951	54.156	53.233	61.460	69.963
Poles													
Sub total	**1.020.937**	**1.007.040**	**481.668**	**602.303**	**531.022**	**453.906**	**311.610**	**198.381**	**104.951**	**54.156**	**53.233**	**61.460**	**69.963**
Lumber (Air dried)	188.443	234.829	215.373	173.999	149.289	73.075	77.176	77.619	69.445	52.863	39.652	44.753	56.409
Lumber (Kiln dried)													
Lumber (Overland)													
Sub total	**188.443**	**234.829**	**215.373**	**173.999**	**149.289**	**73.075**	**77.176**	**77.619**	**69.445**	**52.863**	**39.652**	**44.753**	**56.409**
Sliced Veneer													
Rotary Veneer													
Reconst. Veneer													
Furniture Parts													
Boules													
Processed L/Moulding													
Plywood													
Curls													
Floorings													
Layons													
Profile Boards													
Door Lippings													
Door Panels													
Doors													
Toys													
Dowels													
Broomsticks													
Block Boards													
Flitches													
Rectng. Poles													
Pallets													
Raw Canes													
Sleepers													
Hockey sticks													
Briquet													
Sub total													
GRAND TOTAL	1.209.380	1.241.869	697.041	776.302	680.311	526.981	388.786	276.000	174.396	107.019	92.885	106.213	126.372

Commodity	1985	1986	1987	1988	1989	1990	1991	1992	1993	1994	1995	1996
Logs	129.872	220.983	318.627	338.928	200.659	197.874	215.438	177.037	495.613	572.374	80.617	0
Poles						1.554	2.605	4.460	24.311	41.883	69.285	1.087
Sub total	**129.872**	**220.983**	**318.627**	**338.928**	**200.659**	**199.428**	**218.043**	**181.497**	**519.924**	**614.257**	**149.902**	**1.087**
Lumber (Air dried)	79.641	103.818	172.930	169.641	154.054	200.517	146.320	182.456	188.896	188.624	194.068	140.297
Lumber (Kiln dried)							14.934	28.972	35.018	60.217	80.464	84.919
Lumber (Overland)							21.390	21.025	15.279	9.670	11.573	13.387
Sub total	**79.641**	**103.818**	**172.930**	**169.641**	**154.054**	**200.517**	**182.644**	**232.453**	**239.193**	**258.511**	**286.105**	**238.603**
Sliced Veneer		12.075	15.026	14.460	11.937	13.032	13.034	15.726	15.544	20.858	27.486	27.882
Rotary Veneer		2.312	5.991	6.391	2.940	4.308	6.325	7.903	9.580	14.181	18.429	25.524
Reconst. Veneer			130	140	158		115			49	16	
Furniture Parts			2.790	2.976	3.258	2.754	1.984	1.981	2.668	2.400	2.874	1.472
Boules			647	696				3.274	11.669	39.046	44.351	35.772
Processed L/Moulding		503	1.027	742	967	1.936	1.950	3.574	3.023	4.509	7.476	9.610
Plywood .		721	825	1.057	1.011	1.801	1.419	1.502	2.409	964	3.600	18.888
Curls		144	252	443	223	458	472	640	873	955	565	166
Floorings		286	86	139	167	960	1.134	2.136	2.776	5.493	4.892	3.413
Layons							88	259	527	399	291	338
Profile Boards		402	617	473	344	500	473	642	429	835	270	178
Door Lippings		24		8								
Door Panels				20								
Doors			19				6					
Toys												
Dowels										757	796	1.227
Broomsticks		91	80		21	88	70	155	67	218	231	232
Block Boards						57	157	27		221	150	184
Flitches								44				
Rectng. Poles								35				
Pallets					13							
Raw Canes												
Sleepers											77	192
Hockey sticks											90	
Briquet											29	
Sub total		**4.483**	**12.464**	**13.085**	**9.102**	**12.862**	**14.193**	**22.172**	**34.021**	**70.027**	**84.137**	**97.196**
GRAND TOTAL	209.513	341.359	519.047	536.114	375.752	425.839	427.914	451.848	808.682	963.653	547.630	364.768

Sources:
General Woods 1992; WD. no. 3, App. 3 (1972-85)
FPIB Export Permits (1986-96)

Appendix 7

GHANA
ESTIMATED WOOD EXPORTS AND APPARENT DOMESTIC CONSUMPTION IN ROUNDWOOD EQUIVALENTS (Cu.m.)

Assumptions about Recovery Rates and Use of Residues

Product Category	Recovery rates	Residues	Residues re-processed for domestic use other than firewood
-Lumber	40,00%	60,00%	55,00%
-Sliced Veneer	30,00%	70,00%	55,00%
-All Others	35,00%	65,00%	55,00%

YEAR	1972	1973	1974	1975	1976	1077	1978	1979	1980	1981	1982	1983
Export Commodities:												
Logs 1/	1.020.937	1.007.040	481.668	602.303	531.022	453.906	311.610	198.381	104.951	54.156	53.233	61.460
Lumber	471.108	587.073	538.433	434.998	373.223	182.688	192.940	194.048	173.613	132.158	99.130	111.883
Sliced Veneer												
All Others												
Total (The Export Harvest)	**1.492.045**	**1.594.113**	**1.020.101**	**1.037.301**	**904.245**	**636.594**	**504.550**	**392.429**	**278.564**	**186.314**	**152.363**	**173.343**

YEAR	1984	1985	1986	1987	1988	1989	1990	1991	1992	1993	1994	1995	1996
Export Commodities:													
Logs 1/	69.963	129.872	220.983	318.627	338.928	200.659	197.874	215.438	177.037	495.613	572.374	80.617	0
Lumber	141.023	199.103	259.545	432.325	424.103	385.135	501.293	456.610	581.133	597.983	646.278	715.263	596.508
Sliced Veneer			40.250	50.087	48.200	39.790	43.440	43.447	52.420	51.813	69.527	91.620	92.940
All Others			12.809	35.611	37.386	26.006	36.749	40.551	63.349	97.203	200.077	240.391	277.703
Total (The Export Harvest)	**210.986**	**328.975**	**533.587**	**836.650**	**848.616**	**651.590**	**779.355**	**756.046**	**873.938**	**1.242.612**	**1.488.255**	**1.127.891**	**967.150**
Total Recorded Extraction			**804.091**	**1.024.600**	**1.136.737**	**996.207**	**1.289.623**	**1.229.433**	**1.318.404**	**1.682.288**	**1.681.872**	**1.194.412**	**1.166.436**
Apparent Domestic Consumption			**376.230**	**362.632**	**459.997**	**496.328**	**705.556**	**655.292**	**679.068**	**691.709**	**505.184**	**423.771**	**531.194**

Notes:
1/ Poles are not included as they consist of plantation Teak for which reliable data on growth the total volume harvested is not available.
Exports for the period 1972-85 are probably on the low side as only logs and lumber are included.
Apparent domestic consumption is a crude estimate associated with some factors of uncertainty:
- Changes in stocks are not specified which means that stock increases are taken as apparent domestic consumption.
- The assumed recovery rates, especially the one for 'All others', are rough estimates.
- It is assumed that 55% of residues are re-processed for domestic markets other then firewood and charcoal.
- In addition to recorded extraction some illegal harvest takes place of which most is likely to be consumed domestically.
Harvest of plantation wood, where Teak is the major species, is not included as reliable extraction figures are not available.
However, Teak exports are insignificant for most years which means that 'apparent domestic consumption' should be increased by the plantation harvest.

Sources:
General Woods 1992; WD. no. 3, App. 3 (1972-85)
FPIB Export Permits (1986-96)
FPIB LMC Totals (1986-96)

Appendix 7

GHANA
LOG EXPORTS BY VOLUME (Cu.m. ROUNDWOOD)

SPECIES Trade name	Botanical name	Year 1986	1987	1988	1989	1990	1991	1992	1993	1994	1995
SCARLET STARS											
Albizia	Albizia ferruginea	12		5.149	9.447	5.465	11.396	1.278	209	1.941	
Asanfina 2/	Aningeria altissima/robusta	11.868	24.682	18.783							
Daniella	Daniella ogea/thurifera		356	1.038				794	16.267	15.749	
Edinam 1/	Entandophragma angolense										
Sapele 1/	Entandophragma cylindricum										
Utile 1/	Entandophragma utile										
Hyedua 1/	Guibourtia ehie										
Mahogany 1/	Khaya anthotheca/grandifolia/ivorensis										
Odum 1/	Milicia excelsa/regia										
Kusia	Nauclea diderrichii	1.286	2.690	1.461	7.483	829	2.292	1.298	6.858	4.961	
Afrormosia 1/	Pericopsis elata										
Kyere	Pterygota macrocarpa	16.993	27.005	31.878	23.690	18.662	15.708	8.728	9.580	4.623	
Makore 1/	Tieghemelia heckelii										
Wawa 3/	Triplochiton scleroxylon	110.673	134.250	147.829	113.473	123.441	119.063	112.008	55.931		
SUB TOTAL	**Scarlet stars**	**140.832**	**188.983**	**206.138**	**154.093**	**148.397**	**148.459**	**124.106**	**88.845**	**27.273**	
RED STARS											
Afzelia/Papao	Afzelia africana/bellea	270	767	401	499	9.809	5.839		686	5.390	
Kyenkyen	Antiaris toxicaria	46	19.499	10.572	6.801	5.849	4.952	2.364	98.528	224.277	
Aprokuma	Antrocaryon micraster	117	51	76			47	56	557	1.053	
Canarium	Canarium schweinfurthii		54				708		21.735	15.127	
Ceiba	Ceiba pentandra	3.891	9.847	20.490	3.737	2.960	10.553	5.397	167.485	79.404	
Akasa	Chrysophyllum spp.	16	1.427	2.307							
Bonsamdua/Ayan	Distemonanthus benthamianus	3.945	5.373	2.790	653	6.529	3.562	2.145	422	2.569	
Candollei 1/	Entandophragma candollei										
Guarea	Guarea spp	4.037	6.137	3.842	1.307	5.583	5.705	1.213	1.144	2.447	
Niangon 1/	Heritiera utilis										
Kaku/Ekki	Lophira alata	306	352	2.558	5.152	718	8.553	8.333	13.608	1.966	
Walnut 1/	Lovoa trichilioides										
Mansonia 1/	Mansonia altissima										
Dahoma	Piptadenisatrum africanum				11.432	1.242			8.828	10.372	
Bombax	Rhodognaphalon/Bombax brevicuspe	4.483	3.741	5.668	2.882	4.687	2.479	2.479	6.246	2.398	
Emire 2/	Terminalia ivorensis	33.449	28.024	22.245							
SUB TOTAL	**Red stars**	**50.560**	**75.272**	**70.949**	**32.463**	**37.377**	**42.398**	**21.987**	**319.239**	**345.004**	
PINK STARS											
Albizia	Albizia adianthifolia								67		
Albizia	Albizia zygia										
Alstonia	Alstonia boonei										
Yaya	Amphimas pterocarpoides			8					3.703	23.006	
Kokote	Anopyxis klaineana								423	205	
Berlinia	Berlinia confusa								26		
Berlinia	Berlinia spp.										
Celtis/Essa	Celtis mildbraedii/zenkeri	16	113	993	1.670	1.990	178	811	6.016	15.100	
Pamprana	Coryanthe pachyceras										
Denya	Cylicodiscus gabonensis				470		815	426	772	1.641	
Ananta	Cynometra antana				1.492				2.035	829	
Duabankye	Dialium aubrevillei										
Potrodum	Erytrophleum sauaveolens			2.262	360			18	171	1.030	
Nakwa	Holoptelea grandis								6		
Kroma	Klaindoxa gabonensis								803	29	
Kumanini	Lannea welwitschii										
Bompagya	Mammea africana								8		
Wonton	Morus mesozygia								24		
Danta 2/	Nesogordonia papaverifera	9.276	22.719	21.484							
Akwana	Ongokea gore										
Afam	Parinari exelsa								36		
Asoma	Parkia bicolor								463	15	
Esia	Petersianthus macrocarpus								1.075	2.235	
Otie	Pycnanthus angolensis	6.777	11.323	23.236	8.561	9.455	12.018	11.181	30.509	35.731	
Bombax	Rhodognaphalon/Bombax buonopozense										
Wama	Ricinodendron heudelotii								4		
Sterculia	Sterculia spp.						34		212	778	
Afena	Strombosia glaucescens										
Ofram 2a/	Terminalia superba	11.987	17.375	7.085		152	11.028	18.455	40.717	119.316	
Tanuronini	Trichilia tessmannii										
Okure	Trilepisium madagascariense										
Avodire 1/	Turreanthus africanus										
SUB TOTAL	**Pink stars**	**28.056**	**51.530**	**55.068**	**12.553**	**11.597**	**24.073**	**30.891**	**87.070**	**199.913**	
Subaha 2/, 4/	Hallea spp./Mitragyna spp.	1.453	2.344	6.620	1.499	57					
Other species		82	498	153	0	446	508	53	459	184	
GRAND TOTAL		**220.983**	**318.627**	**338.928**	**200.608**	**197.874**	**215.438**	**177.037**	**495.613**	**572.374**	**80.617**

Notes:
1/ Log export banned in 1979
2/ Log export banned in 1988
2a/ Log export banned in 1988 but lifted in 1990
3/ Log export banned in 1993
4/ Harvest banned in 1993
Log exports was entirely banned during 1995. Species specific data on the export of 80,617 cu.m logs in 1995 has not been available.

Source:
FPIB Export permits (1986-96)

Appendix 7

GHANA

AIR-DRIED, KILN-DRIED AND OVERLAND LUMBER EXPORTS BY VOLUME (Cu.m. ROUNDWOOD EQUIVALENTS)

Assumed Recovery Rate: 40,00%

SPECIES Trade name	Botanical name	Year 1986	1987	1988	1989	1990	1991	1992	1993	1994	1995	1996
SCARLET STARS												
Albizia	Albizia ferruginea					478	1.040	3.040	2.135	2.109	5.735	5.749
Asanfina 2/	Aningeria altissima/robusta		165	405	410	503	1.833	993	863	1.462	2.025	2.200
Daniella	Daniella ogea/thurifera								105			
Edinam 1/	Entandophragma angolense	2.308	10.275	11.343	12.888	18.008	7.528	16.430	16.903	12.964	10.981	11.288
Sapele 1/	Entandophragma cylindricum	8.398	14.750	15.325	16.873	26.165	13.718	14.968	13.313	10.162	8.549	7.374
Utile 1/	Entandophragma utile	36.990	53.608	24.595	17.658	18.620	7.705	9.545	14.318	17.414	8.200	6.534
Hyedua 1/	Guibourtia ehie	1.245	3.078	3.120	2.555	1.520	1.085	535	580	298	160	434
Mahogany 1/	Khaya anthotheca/grandifolia/ivorensis	14.358	35.538	24.213	26.158	50.870	32.605	39.203	47.735	43.068	38.059	41.826
Odum 1/	Milicia excelsa/regia	82.998	123.905	129.773	94.368	88.908	65.713	93.703	113.100	118.350	93.397	76.879
Kusia	Nauclea diderrichii		648	2.240	1.685	2.533	4.723	5.758	4.298	3.564	3.330	3.431
Afrormosia 1/	Pericopsis elata	12.443	9.070	4.943	5.510	3.995	2.933	3.815	2.143	133	271	136
Kyere	Pterygota macrocarpa	215	1.108	1.550	80	43	2.180	7.120	8.505	13.961	23.243	16.528
Makore 1/	Tieghemella heckelii	3.630	11.473	17.845	14.170	13.500	9.740	12.653	10.223	5.218	7.980	4.491
Wawa 3/	Triplochiton scleroxylon	78.475	127.768	146.840	162.045	215.710	246.265	303.465	271.955	328.398	401.330	311.715
SUB TOTAL	**Scarlet stars**	**241.058**	**391.383**	**382.190**	**354.398**	**440.850**	**397.065**	**511.225**	**506.173**	**557.102**	**603.261**	**488.585**
RED STARS												
Afzelia/Papao	Afzelia africana/bellea	1.370	3.105	8.130	2.015	2.248	2.958	6.808	10.098	6.382	11.175	4.312
Kyenkyen	Antiaris toxicaria	200	118	13		43			20	503	52	14.312
Aprokuma	Antrocaryon micraster									101		
Canarium	Canarium schweinfurthii									1		261
Ceiba	Ceiba pentandra								75	1.365	3.878	1.409
Akasa	Chrysophyllum spp.			1.975	643	2.860	2.383	1.825	473	1.135	2.558	1.205
Bonsamdua/Ayan	Distemonanthus benthamianus	190	685	513	103	908	510	2.423	1.923	831	2.220	2.521
Candollei 1/	Entandophragma candollei	368	915	1.028	2.018	3.773	1.203	1.490	1.383	1.072	1.214	177
Guarea	Guarea spp	55	530	1.003	1.328	2.220	1.125	3.033	2.790	1.765	3.082	3.021
Niangon 1/	Heritiera utilis	388	3.948	4.610	4.060	6.845	7.060	8.303	12.693	10.530	12.887	8.647
Kaku/Ekki	Lophira alata						1.005		753	298	1.697	3.747
Walnut 1/	Lovoa trichilioides	245	1.245	1.848	2.075	3.238	2.713	2.018	2.590	2.003	1.731	1.269
Mansonia 1/	Mansonia altissima	1.100	3.218	2.913	1.945	1.250	955	1.800	2.225	1.902	1.646	1.468
Dahoma	Piptadenisatrum africanum	448	1.108	463	188	2.903	8.355	7.355	14.323	6.965	2.947	1.766
Bombax	Rhodognaphalon/Bombax brevicuspe											99
Emire 2/	Terminalia ivorensis	8.095	12.828	9.543	11.743	22.983	12.655	12.258	15.498	29.488	11.848	9.533
SUB TOTAL	**Red stars**	**12.458**	**27.698**	**32.035**	**26.115**	**49.268**	**40.920**	**47.310**	**64.840**	**64.341**	**56.934**	**53.745**
PINK STARS												
Yaya	Amphimas pterocarpoides						3			9		
Celtis/Essa	Celtis mildbraedii/zenkeri						60		193	450	159	167
Denya	Cylicodiscus gabonensis		83					480	33	158	392	211
Ananta	Cynometra antana											109
Potrodum	Erytrophleum sauaveolens				180	360	325	373	110			434
Wonton	Morus mesozygia											
Danta 2/	Nesogordonia papaverifera	1.690	1.263	1.290	1.070	3.945	2.435	2.568	2.305	2.581	3.809	1.396
Asoma	Parkia bicolor											
Esia	Petersianthus macrocarpus								163			53
Otie	Pycnanthus angolensis	188			270	125	250	728	795	255	1.380	1.474
Bombax	Rhodognaphalon/Bombax buonopozense					85						
Sterculia	Sterculia spp.										2.971	3.064
Ofram 2a/	Terminalia superba	988	1.338	1.263	80	1.165	1.135	7.123	17.423	17.517	8.531	17.474
Avodire 1/	Turreanthus africanus	245	960	635	138	1.070	1.483	413	1.413	616	943	244
SUB TOTAL	**Pink stars**	**3.110**	**3.643**	**3.188**	**1.738**	**6.750**	**5.690**	**11.683**	**22.433**	**21.586**	**18.185**	**24.625**
Subaha 2/, 4/	Hallea spp./Mitragyna spp.	70	208	430	238	128	23					
Other Species		2.850	9.395	6.260	4.873	4.298	12.913	10.915	4.538	3.246	36.883	29.554
GRAND TOTAL		**259.545**	**432.325**	**424.103**	**387.360**	**501.293**	**456.610**	**581.133**	**597.983**	**646.276**	**715.262**	**596.510**

Notes:
1/ Log export banned in 1979
2/ Log export banned in 1988
2a/ Log export banned in 1988 but lifted in 1990
3/ Log export banned in 1993
4/ Harvest banned in 1993
Log exports was entirely banned in 1995

Source:
FPIB Export permits (1986-96)

Appendix 7

GHANA

SLICED VENEER EXPORTS BY VOLUME IN ROUNDWOOD EQUIVALENTS (Cu.m.)

Assumed Recovery Rate: **30,00%**

SPECIES Trade name	Botanical name	Year 1986	1987	1988	1989	1990	1991	1992	1993	1994	1995	1996
SCARLET STARS												
Albizia	Albizia ferruginea						9					4
Asanfina 2/	Aningeria altissima/robusta	3.124	4.937	9.708	10.667	12.238	19.924	25.782	26.043	39.775	61.236	63.150
Daniella	Daniella ogea/thurifera								5			
Edinam 1/	Entandophragma angolense	3.890	4.657	4.174	1.767	1.924	1.523	1.338	2.215	2.003	2.301	1.864
Sapele 1/	Entandophragma cylindricum	9.492	9.635	7.702	4.101	7.504	4.550	6.805	4.331	3.806	4.104	6.074
Utile 1/	Entandophragma utile	3.124	667	11	113	13	205	438	161	230	392	751
Hyedua 1/	Guibourtia ehie	126	101	50	59		12	47	15	26	57	76
Mahogany 1/	Khaya anthotheca/grandifolia/ivorensis	4.107	6.435	5.576	2.982	3.890	2.591	2.819	3.921	5.725	4.139	3.571
Odum 1/	Milicia excelsa/regia	202		8	17	40	31	41	189	45	150	99
Kusia	Nauclea diderrichii											
Afrormosia 1/	Pericopsis elata	31	3						69			
Kyere	Pterygota macrocarpa	3.702	5.931	5.054	3.389	2.973	5.488	4.067	3.684	3.325	5.905	4.066
Makore 1/	Tieghemella heckelii	8.965	13.316	11.927	8.540	10.320	6.267	7.878	8.015	10.700	4.723	5.779
Wawa 3/	Triplochiton scleroxylon			2		7	143	30	109	136	121	352
SUB TOTAL		**36.763**	**45.683**	**44.211**	**31.635**	**38.910**	**40.743**	**49.246**	**48.757**	**65.771**	**83.129**	**85.786**
RED STARS												
Afzelia/Papao	Afzelia africana/bellea						5		7			
Kyenkyen	Antiaris toxicaria	82	146	50	81	897	13	10	301	447	254	1.658
Aprokuma	Antrocaryon micraster						11					
Canarium	Canarium schweinfurthii									10		544
Ceiba	Ceiba pentandra					7		37			174	61
Akasa	Chrysophyllum spp.	27		382	445	452	313	31	646	1.542	5.992	2.635
Bonsamdua/Ayan	Distemonanthus benthamianus		12									
Candollei 1/	Entandophragma candollei	2.747	2.212	2.884	1.497	2.132	1.975	2.049	1.571	1.165	1.286	356
Guarea	Guarea spp				14		58	32	9	30	150	138
Niangon 1/	Heritiera utilis											
Kaku/Ekki	Lophira alata											
Walnut 1/	Lovoa trichilioides	136	277	207	88	78	13	43	94	163	124	297
Mansonia 1/	Mansonia altissima	183	198	51	112	59	1		101	119	107	156
Dahoma	Piptadenisatrum africanum									14		
Bombax	Rhodognaphalon/Bombax brevicuspe									8		
Emire 2/	Terminalia ivorensis					8	13	7			29	99
SUB TOTAL		**3.175**	**2.845**	**3.575**	**2.236**	**3.634**	**2.402**	**2.209**	**2.728**	**3.499**	**8.116**	**5.945**
PINK STARS												
Yaya	Amphimas pterocarpoides										26	
Celtis/Essa	Celtis mildbraedii/zenkeri								22	11	22	
Denya	Cylicodiscus gabonensis											
Ananta	Cynometra antana											
Potrodum	Erytrophleum sauaveolens											
Wonton	Morus mesozygia											
Danta 2/	Nesogordonia papaverifera									12		
Asoma	Parkia bicolor											
Esia	Petersianthus macrocarpus									3		
Otie	Pycnanthus angolensis										35	648
Bombax	Rhodognaphalon/Bombax buonopozense											
Sterculia	Sterculia spp.									4		
Ofram 2a/	Terminalia superba	18		53	16	79	94	226	203	204	282	425
Avodire 1/	Turreanthus africanus	9	3					15	83	21	3	70
SUB TOTAL		**28**	**3**	**53**	**16**	**79**	**94**	**241**	**307**	**256**	**368**	**1.143**
Subaha 2/, 4/	Hallea spp./Mitragyna spp.											
Other Species		286	1.554	363	5.903	814	436	725	20	26	7	64
GRAND TOTAL		**40.251**	**50.085**	**48.201**	**39.790**	**43.436**	**43.675**	**52.420**	**51.812**	**69.526**	**91.619**	**92.938**

Notes:
1/ Log export banned in 1979
2/ Log export banned in 1988
2a/ Log export banned in 1988 but lifted in 1990
3/ Log export banned in 1993
4/ Harvest banned in 1993
Log exports was entirely banned in 1995

Source:
FPIB Export permits (1986-96)

Appendix 7

GHANA
ESTIMATED ON-RESERVE AACs COMPARED WITH THE 1996 ESTIMATED HARVEST FOR LUMBER & SLICED VENEER EXPORT (Cu.m. ROUNDWOOD EQUIVALENTS)

Assumed Recovery Rates

Product Category	Recovery rates
-Lumber	40,00%
-Sliced Veneer	30,00%

SPECIES		Large-grained protection only (production forest = 762,400 ha)					Large- and fine-grained protection (production forest = 645,080 ha)				
		AAC total volume	1996 Export harvest in Round Wood Equivalents				AAC total volume	1996 Export harvest in Round Wood Equivalents			
			Lumber	Sliced Veneer	Total	Export harvest as % of AAC		Lumber	Sliced Veneer	Total	Export harvest as % of AAC
Trade name	**Botanical name**	(Cu.m.)	(Cu.m.)	(Cu.m.)	(Cu.m.)	(%)	(Cu.m.)	(Cu.m.)	(Cu.m.)	(Cu.m.)	(%)
SCARLET STARS											
Albizia	Albizia ferruginea	4.628	5.749	4	5.753	124	4.264	5.749	4	5.753	135
Asanfina 2/	Aningeria altissima/robusta	3.450	2.200	63.150	65.350	1.894	3.098	2.200	63.150	65.350	2.110
Daniella	Daniella ogea/thurifera	7.167					5.642				
Edinam 1/	Entandophragma angolense	3.564	11.288	1.864	13.152	369	3.269	11.288	1.864	13.152	402
Sapele 1/	Entandophragma cylindricum	5.175	7.374	6.074	13.448	260	4.433	7.374	6.074	13.448	303
Utile 1/	Entandophragma utile	2.397	6.534	751	7.285	304	2.321	6.534	751	7.285	314
Hyedua 1/	Guibourtia ehie	2.847	434	76	511	18	2.556	434	76	511	20
Mahogany 1/	Khaya anthotheca/grandifolia/ivorensis	6.696	41.826	3.571	45.397	678	5.904	41.826	3.571	45.397	769
Odum 1/	Milicia excelsa/regia	6.820	76.879	99	76.978	1.129	6.659	76.879	99	76.978	1.156
Kusia	Nauclea diderrichii	2.342	3.431		3.431	147	2.074	3.431		3.431	165
Afrormosia 1/	Pericopsis elata	616	136		136	22	583	136		136	23
Kyere	Pterygota macrocarpa	4.888	16.528	4.066	20.594	421	4.525	16.528	4.066	20.594	455
Makore 1/	Tieghemella heckelii	3.293	4.491	5.779	10.270	312	2.764	4.491	5.779	10.270	372
Wawa 3/	Triplochiton scleroxylon	78.555	311.715	352	312.067	397	72.727	311.715	352	312.067	429
SUB TOTAL	**Scarlet stars**	**132.437**	**488.585**	**85.786**	**574.371**	**434**	**120.817**	**488.585**	**85.786**	**574.371**	**475**
RED STARS											
Afzelia/Papao	Afzelia africana/bellea	3.175	4.312		4.312	136	2.618	4.312		4.312	165
Kyenkyen	Antiaris toxicaria	41.405	14.312	1.658	15.970	39	38.087	14.312	1.658	15.970	42
Aprokuma	Antrocaryon micraster	6.610					5.962				
Canarium	Canarium schweinfurthii	4.207	261	544	805	19	3.454	261	544	805	23
Ceiba	Ceiba pentandra	80.906	1.409	61	1.470	2	73.292	1.409	61	1.470	2
Akasa	Chrysophyllum spp.	4.176	1.205	2.635	3.840	92	3.712	1.205	2.635	3.840	103
Bonsamdua/Ayan	Distemonanthus benthamianus	5.892	2.521		2.521	43	5.171	2.521		2.521	49
Candollei 1/	Entandophragma candollei	638	177	356	533	84	638	177	356	533	84
Guarea	Guarea spp	6.440	3.021	138	3.159	49	5.605	3.021	138	3.159	56
Niangon 1/	Heritiera utilis	10.222	8.647		8.647	85	8.444	8.647		8.647	102
Kaku/Ekki	Lophira alata	9.815	3.747		3.747	38	8.918	3.747		3.747	42
Walnut 1/	Lovoa trichilioides	3.758	1.269	297	1.566	42	3.175	1.269	297	1.566	49
Mansonia 1/	Mansonia altissima	5.280	1.468	156	1.623	31	5.040	1.468	156	1.623	32
Dahoma	Piptadenisatrum africanum	27.522	1.766		1.766	6	24.680	1.766		1.766	7
Bombax	Rhodognaphalon/Bombax brevicuspe	7.843	99		99	1	6.864	99		99	1
Emire 2/	Terminalia ivorensis	6.307	9.533	99	9.632	153	5.692	9.533	99	9.632	169
SUB TOTAL	**Red stars**	**224.195**	**53.745**	**5.945**	**59.690**	**27**	**201.353**	**53.745**	**5.945**	**59.690**	**30**
PINK STARS											
Albizia	Albizia adianthifolia	3.629					3.437				
Albizia	Albizia zygia	6.653					6.132				
Alstonia	Alstonia boonei	15.115					13.166				
Yaya	Amphimas pterocarpoides	5.650					5.108				
Kokote	Anopyxis klaineana	5.975					5.131				
Berlinia	Berlinia confusa	4.682					4.019				
Berlinia	Berlinia spp.	14.086					11.629				
Celtis/Essa	Celtis mildbraedii/zenkeri	138.200	167		167	0	127.028	167		167	0
Pamprana	Coryanthe pachyceras	9.520					8.738				
Denya	Cylicodiscus gabonensis	33.627	211		211	1	29.096	211		211	1
Ananta	Cynometra antana	18.131	109		109	1	15.445	109		109	1
Duabankye	Dialium aubrevillei	13.448					10.742				
Potrodum	Erytrophleum sauaveolens	7.465	434		434	6	6.374	434		434	7
Nakwa	Holoptelea grandis	3.898					3.756				
Kroma	Klaindoxa gabonensis	8.466					7.068				
Kumanini	Lannea welwitschii	5.100					4.575				
Bompagya	Mammea africana	4.893					3.933				
Wonton	Morus mesozygia	6.029					5.780				
Danta 2/	Nesogordonia papaverifera	19.465	1.396		1.396	7	17.680	1.396		1.396	8
Akwana	Ongokea gore	4.147					3.696				
Afam	Parinari exelsa	7.402					6.541				
Asoma	Parkia bicolor	7.642					6.874				
Esia	Petersianthus macrocarpus	37.391	53		53	0	33.317	53		53	0
Otie	Pycnanthus angolensis	32.070	1.474	648	2.122	7	28.395	1.474	648	2.122	7
Bombax	Rhodognaphalon/Bombax buonopozense	5.868					5.106				
Wama	Ricinodendron heudelotii	9.065					8.523				
Sterculia	Sterculia spp.	8.880	3.064		3.064	34	8.288	3.064		3.064	37
Afena	Strombosia glaucescens	11.560					10.200				
Ofram 2a/	Terminalia superba	56.055	17.474	425	17.899	32	51.404	17.474	425	17.899	35
Tanuronini	Trichilia tessmannii	6.491					5.925				
Okure	Trilepisium madagascariense	4.176					3.816				
Avodire 1/	Turreanthus africanus	20.100	244	70	314	2	18.600	244	70	314	2
SUB TOTAL	**Pink stars**	**534.877**	**24.625**	**1.143**	**25.769**	**5**	**479.518**	**24.625**	**1.143**	**25.769**	**5**
Subaha 2/, 4/	Hallea spp./Mitragyna spp.										
Other species			29.554	64	29.618			29.554	64	29.618	
GRAND TOTAL		**891.509**	**596.510**	**92.938**	**689.448**	**77**	**801.688**	**596.510**	**92.938**	**689.448**	**86**

Notes:
1/ Log export banned in 1979
2/ Log export banned in 1988
2a/ Log export banned in 1988 but lifted in 1990
3/ Log export banned in 1993
4/ Harvest banned in 1993
Log exports was entirely banned during 1995.

Sources:
FPIB Export Permits (1996)
FIMP (1995b)

Appendix 7

GHANA
ESTIMATED ON-RESERVE AACs COMPARED WITH THE ESTIMATED AVERAGE 1986-96 EXPORT HARVEST OF LOGS, LUMBER AND SLICED VENEER (Cu.m. ROUNDWOOD EQUIVALENTS)

Assumed Recovery Rates:	
-Lumber	40,00%
-Sliced Veneer	30,00%

SPECIES		Export harvest							
Trade name	Botanical name	1986	1987	1988	1989	1990	1991	1992	1993
SCARLET STARS									
Albizia	Albizia ferruginea	12		5.149	9.447	5.943	12.445	4.318	2.344
Asanfina 2/	Aningeria altissima/robusta	14.992	29.784	28.896	11.077	12.740	21.757	26.775	26.906
Daniella	Daniella ogea/thurifera		356	1.038				794	16.377
Edinam 1/	Entandophragma angolense	6.197	14.932	15.516	14.654	19.931	9.050	17.768	19.117
Sapele 1/	Entandophragma cylindricum	17.889	24.385	23.027	20.974	33.669	18.267	21.773	17.643
Utile 1/	Entandophragma utile	40.114	54.275	24.606	17.771	18.633	7.910	9.983	14.479
Hyedua 1/	Guibourtia ehie	1.371	3.178	3.170	2.614	1.520	1.097	582	595
Mahogany 1/	Khaya anthotheca/grandifolia/ivorensis	18.465	41.972	29.789	29.139	54.760	35.196	42.022	51.656
Odum 1/	Milicia excelsa/regia	83.200	123.905	129.780	94.385	88.948	65.743	93.743	113.289
Kusia	Nauclea diderrichii	1.286	3.338	3.701	9.168	3.362	7.015	7.056	11.156
Afrormosia 1/	Pericopsis elata	12.473	9.073	4.943	5.510	3.995	2.933	3.815	2.211
Kyere	Pterygota macrocarpa	20.910	34.044	38.482	27.159	21.677	23.376	19.915	21.769
Makore 1/	Tieghemella heckelii	12.595	24.789	29.772	22.710	23.820	16.007	20.530	18.238
Wawa 3/	Triplochiton scleroxylon	189.148	262.018	294.671	275.518	339.158	365.471	415.503	327.995
SUB TOTAL	**Scarlet stars**	**418.652**	**626.048**	**632.539**	**540.125**	**628.157**	**586.267**	**684.577**	**643.774**
RED STARS									
Afzelia/Papao	Afzelia africana/bellea	1.640	3.872	8.531	2.514	12.057	8.802	6.808	10.790
Kyenkyen	Antiaris toxicaria	328	19.762	10.635	6.882	6.789	4.965	2.374	98.849
Aprokuma	Antrocaryon micraster	117	51	76			58	56	557
Canarium	Canarium schweinfurthii		54				708		21.735
Ceiba	Ceiba pentandra	3.891	9.847	20.490	3.737	2.967	10.553	5.434	167.560
Akasa	Chrysophyllum spp.	43	1.427	4.664	1.087	3.312	2.695	1.856	1.118
Bonsamdua/Ayan	Distemonanthus benthamianus	4.135	6.070	3.303	756	7.437	4.072	4.568	2.345
Candollei 1/	Entandophragma candollei	3.115	3.127	3.911	3.514	5.904	3.177	3.539	2.954
Guarea	Guarea spp	4.092	6.667	4.845	2.648	7.803	6.888	4.278	3.943
Niangon 1/	Heritiera utilis	388	3.948	4.610	4.060	6.845	7.060	8.303	12.693
Kaku/Ekki	Lophira alata	306	352	2.558	5.152	718	9.558	8.333	14.361
Walnut 1/	Lovoa trichilioides	381	1.522	2.054	2.163	3.315	2.725	2.060	2.684
Mansonia 1/	Mansonia altissima	1.283	3.416	2.964	2.057	1.309	956	1.800	2.326
Dahoma	Piptadenisatrum africanum	448	1.108	463	11.620	4.145	8.355	7.355	23.151
Bombax	Rhodognaphalon/Bombax brevicuspe	4.483	3.741	5.668	2.882	4.687	2.479	2.479	6.246
Emire 2/	Terminalia ivorensis	41.544	40.852	31.788	11.743	22.991	12.668	12.265	15.498
SUB TOTAL	**Red stars**	**66.192**	**105.814**	**106.559**	**60.814**	**90.278**	**85.720**	**71.506**	**386.807**
PINK STARS									
Albizia	Albizia adianthifolia								67
Albizia	Albizia zygia								
Alstonia	Alstonia boonei								
Yaya	Amphimas pterocarpoides			8			3		3.703
Kokote	Anopyxis klaineana								423
Berlinia	Berlinia confusa								26
Berlinia	Berlinia spp.								
Celtis/Essa	Celtis mildbraedii/zenkeri	16	113	993	1.670	1.990	238	811	6.230
Pamprana	Coryanthe pachyceras								
Denya	Cylicodiscus gabonensis		83		470		815	906	805
Ananta	Cynometra antana				1.492				2.035
Duabankye	Dialium aubrevillei								
Potrodum	Erytrophleum sauaveolens			2.262	540	360	325	391	281
Nakwa	Holoptelea grandis								6
Kroma	Klaindoxa gabonensis								803
Kumanini	Lannea welwitschii								
Bompagya	Mammea africana								8
Wonton	Morus mesozygia								24
Danta 2/	Nesogordonia papaverifera	10.966	23.982	22.774	1.070	3.945	2.435	2.568	2.305
Akwana	Ongokea gore								
Afam	Parinari exelsa								36
Asoma	Parkia bicolor								463
Esia	Petersianthus macrocarpus								1.238
Otie	Pycnanthus angolensis	6.965	11.323	23.236	8.831	9.580	12.268	11.909	31.304
Bombax	Rhodognaphalon/Bombax buonopozense					85			
Wama	Ricinodendron heudelotii								4
Sterculia	Sterculia spp.						34		212
Afena	Strombosia glaucescens								
Ofram 2a/	Terminalia superba	12.993	18.713	8.401	96	1.396	12.257	25.803	58.342
Tanuronini	Trichilia tessmannii								
Okure	Trilepisium madagascariense								
Avodire 1/	Turreanthus africanus	254	963	635	138	1.070	1.483	428	1.495
SUB TOTAL	**Pink stars**	**31.194**	**55.176**	**58.309**	**14.307**	**18.426**	**29.857**	**42.815**	**109.810**
Subaha 2/, 4/	Hallea spp./Mitragyna spp.	1.523	2.552	7.050	1.737	185	23		
Other species		3.218	11.447	6.776	10.776	5.557	13.857	11.693	5.016
GRAND TOTAL		**520.779**	**801.037**	**811.231**	**627.758**	**742.602**	**715.723**	**810.590**	**1.145.408**

Notes:
1/ Log export banned in 1979
2/ Log export banned in 1988
2a/ Log export banned in 1988 but lifted in 1990
3/ Log export banned in 1993
4/ Harvest banned in 1993
Log exports was entirely banned during 1995.

Sources:
FPIB Export Permits (1996)
FIMP (1995b)

Appendix 7

GHANA
ESTIMATED ON-RESERVE AACs COMPARED WITH THE ESTIMATED AVERAGE 1986-96 EXPORT HARVEST OF LOGS, LUMBER AND SLICED VENEER (Cu.m. ROUNDWOOD EQUIVALENTS)

		Assumed Recovery Rates: -Lumber 40,00% -Sliced Veneer 30,00%				Large-grained protection only (prod. forest = 762,400 ha)		Large- and fine-grained protection (prod. forest = 645,080 ha)	
SPECIES		Export harvest					Avg. 1986-96		Avg. 1986-96
Trade name	Botanical name	1994	1995	1996	Avg. 1986-96	AAC	hvst. as % of AAC	AAC	hvst. as % of AAC
SCARLET STARS									
Albizia	Albizia ferruginea	4.051	5.735	5.753	5.018	4.628	108	4.264	118
Asanfina 2/	Aningeria altissima/robusta	41.237	63.261	65.350	31.161	3.450	903	3.098	1.006
Daniella	Daniella ogea/thurifera	15.749			3.119	7.167	44	5.642	55
Edinam 1/	Entandophragma angolense	14.968	13.282	13.152	14.415	3.564	404	3.269	441
Sapele 1/	Entandophragma cylindricum	13.968	12.653	13.448	19.791	5.175	382	4.433	446
Utile 1/	Entandophragma utile	17.644	8.593	7.285	20.117	2.397	839	2.321	867
Hyedua 1/	Guibourtia ehie	324	217	511	1.380	2.847	48	2.556	54
Mahogany 1/	Khaya anthotheca/grandifolia/ivorensis	48.793	42.198	45.397	39.944	6.696	597	5.904	677
Odum 1/	Milicia excelsa/regia	118.396	93.548	76.978	98.356	6.820	1.442	6.659	1.477
Kusia	Nauclea diderrichii	8.524	3.330	3.431	5.579	2.342	238	2.074	269
Afrormosia 1/	Pericopsis elata	133	271	136	4.136	616	672	583	709
Kyere	Pterygota macrocarpa	21.908	29.148	20.594	25.362	4.888	519	4.525	560
Makore 1/	Tieghemella heckelii	15.917	12.703	10.270	18.850	3.293	572	2.764	682
Wawa 3/	Triplochiton scleroxylon	328.534	401.451	312.067	319.230	78.555	406	72.727	439
SUB TOTAL	**Scarlet stars**	**650.147**	**686.389**	**574.371**	**606.459**	**132.437**	**458**	**120.817**	**502**
RED STARS									
Afzelia/Papao	Afzelia africana/bellea	11.771	11.175	4.312	7.479	3.175	236	2.618	286
Kyenkyen	Antiaris toxicaria	225.228	305	15.970	35.644	41.405	86	38.087	94
Aprokuma	Antrocaryon micraster	1.154			188	6.610	3	5.962	3
Canarium	Canarium schweinfurthii	15.138		805	3.495	4.207	83	3.454	101
Ceiba	Ceiba pentandra	80.769	4.052	1.470	28.252	80.906	35	73.292	39
Akasa	Chrysophyllum spp.	2.677	8.550	3.840	2.843	4.176	68	3.712	77
Bonsamdua/Ayan	Distemonanthus benthamianus	3.401	2.220	2.521	3.711	5.892	63	5.171	72
Candollei 1/	Entandophragma candollei	2.238	2.500	533	3.137	638	492	638	492
Guarea	Guarea spp	4.243	3.232	3.159	4.709	6.440	73	5.605	84
Niangon 1/	Heritiera utilis	10.530	12.887	8.647	7.270	10.222	71	8.444	86
Kaku/Ekki	Lophira alata	2.264	1.697	3.747	4.459	9.815	45	8.918	50
Walnut 1/	Lovoa trichilioides	2.166	1.855	1.566	2.045	3.758	54	3.175	64
Mansonia 1/	Mansonia altissima	2.022	1.752	1.623	1.955	5.280	37	5.040	39
Dahoma	Piptadenisatrum africanum	17.350	2.947	1.766	7.155	27.522	26	24.680	29
Bombax	Rhodognaphalon/Bombax brevicuspe	2.406		99	3.197	7.843	41	6.864	47
Emire 2/	Terminalia ivorensis	29.488	11.877	9.632	21.849	6.307	346	5.692	384
SUB TOTAL	**Red stars**	**412.845**	**65.050**	**59.690**	**137.389**	**224.195**	**61**	**201.353**	**68**
PINK STARS									
Albizia	Albizia adianthifolia				6	3.629	0	3.437	0
Albizia	Albizia zygia					6.653		6.132	
Alstonia	Alstonia boonei					15.115		13.166	
Yaya	Amphimas pterocarpoides	23.015	26		2.432	5.650	43	5.108	48
Kokote	Anopyxis klaineana	205			57	5.975	1	5.131	1
Berlinia	Berlinia confusa				2	4.682	0	4.019	0
Berlinia	Berlinia spp.					14.086		11.629	
Celtis/Essa	Celtis mildbraedii/zenkeri	15.561	181	167	2.543	138.200	2	127.028	2
Pamprana	Coryanthe pachyceras					9.520		8.738	
Denya	Cylicodiscus gabonensis	1.799	392	211	498	33.627	1	29.096	2
Ananta	Cynometra antana	829		109	406	18.131	2	15.445	3
Duabankye	Dialium aubrevillei					13.448		10.742	
Potrodum	Erytrophleum sauaveolens	1.030		434	511	7.465	7	6.374	8
Nakwa	Holoptelea grandis				1	3.898	0	3.756	0
Kroma	Klaindoxa gabonensis	29			76	8.466	1	7.068	1
Kumanini	Lannea welwitschii					5.100		4.575	
Bompagya	Mammea africana				1	4.893	0	3.933	0
Wonton	Morus mesozygia				2	6.029	0	5.780	0
Danta 2/	Nesogordonia papaverifera	2.593	3.809	1.396	7.077	19.465	36	17.680	40
Akwana	Ongokea gore					4.147		3.696	
Afam	Parinari exelsa				3	7.402	0	6.541	0
Asoma	Parkia bicolor	15			43	7.642	1	6.874	1
Esia	Petersianthus macrocarpus	2.238		53	321	37.391	1	33.317	1
Otie	Pycnanthus angolensis	35.986	1.415	2.122	14.085	32.070	44	28.395	50
Bombax	Rhodognaphalon/Bombax buonopozense				8	5.868	0	5.106	0
Wama	Ricinodendron heudelotii				0	9.065	0	8.523	0
Sterculia	Sterculia spp.	782	2.971	3.064	642	8.880	7	8.288	8
Afena	Strombosia glaucescens					11.560		10.200	
Ofram 2a/	Terminalia superba	137.037	8.813	17.899	27.432	56.055	49	51.404	53
Tanuronini	Trichilia tessmannii					6.491		5.925	
Okure	Trilepisium madagascariense					4.176		3.816	
Avodire 1/	Turreanthus africanus	637	946	314	760	20.100	4	18.600	4
SUB TOTAL	**Pink stars**	**221.755**	**18.552**	**25.769**	**56.906**	**534.877**	**11**	**479.518**	**12**
Subaha 2/, 4/	Hallea spp./Mitragyna spp.				1.188				
Other species		3.456	36.890	29.618	12.573				
GRAND TOTAL		**1.288.202**	**806.882**	**689.448**	**814.515**	**891.509**	**91**	**801.688**	**102**

Notes:
1/ Log export banned in 1979
2/ Log export banned in 1988
2a/ Log export banned in 1988 but lifted in 1990
3/ Log export banned in 1993
4/ Harvest banned in 1993
Log exports was entirely banned during 1995.

Sources:
FPIB Export Permits (1996)
FIMP (1995b)

Appendix 8

GHANA
WOOD EXPORTS IN FOB VALUE PER VOLUME OF PRODUCTS (DM/Cu.m. Products)

	Year										
Commodity	**1986**	**1987**	**1988**	**1989**	**1990**	**1991**	**1992**	**1993**	**1994**	**1995**	**1996**
Logs	231	222	235	223	240	223	210	189	180	134	
Poles					359	373	329	327	358	344	1.060
WEIGHTED AVERAGE	**231**	**222**	**235**	**223**	**241**	**225**	**213**	**196**	**192**	**231**	**1.060**
Weighted average in 1996 DM	*298*	*287*	*299*	*275*	*291*	*262*	*236*	*207*	*198*	*234*	*1.060*
Lumber (Air dried)	464	453	469	533	600	505	490	521	646	562	536
Lumber (Kiln dried)						727	647	582	575	577	549
Lumber (Overland)						276	250	220	145	131	131
WEIGHTED AVERAGE	**464**	**453**	**469**	**533**	**600**	**497**	**488**	**510**	**611**	**549**	**518**
Weighted average in 1996 DM	*599*	*584*	*596*	*659*	*723*	*578*	*540*	*541*	*630*	*557*	*518*
Sliced Veneer	647	739	872	1.261	1.431	1.315	1.384	1.313	1.400	1.434	1.371
Sliced Veneer in 1996 DM											
Rotary Veneer	404	373	406	504	546	541	535	535	537	549	447
Reconst. Veneer		1.571	1.747	2.189		1.588			1.053	2.256	
Furniture Parts		1.204	1.400	1.620	1.952	3.163	3.014	2.524	2.954	2.454	3.174
Boules		621	540				509	500	508	478	452
Processed L/Moulding	1.252	930	899	1.261	1.308	869	928	1.000	1.022	888	888
Plywood	361	529	487	562	591	671	744	609	615	542	591
Curls	1.532	1.606	1.794	1.720	1.885	1.765	1.952	1.906	1.908	1.899	1.988
Floorings	1.041	833	533	1.328	1.272	1.586	1.586	1.390	1.019	1.304	1.303
Layons						2.225	2.094	1.766	1.981	1.440	1.794
Profile Boards	1.213	1.185	1.175	1.298	1.183	1.265	1.140	1.264	518	1.368	1.157
Door Lippings	795		641								
Door Panels			1.461								
Doors		261				626					
Toys											
Dowels									720	809	975
Broomsticks	307	373		695	812	834	784	819	762	801	798
Block Boards					286	537	551		468	266	312
Flitches							913				
Rectng. Poles							795				
Pallets				60							
Raw Canes											
Sleepers										492	438
Hockey sticks										831	
Briquet										401	
WEIGHTED AVERAGE	**1.124**	**721**	**784**	**1.097**	**1.095**	**1.134**	**1.011**	**859**	**703**	**668**	**608**
Weighted average in 1996 DM	*1.451*	*930*	*997*	*1.358*	*1.320*	*1.320*	*1.119*	*910*	*725*	*678*	*608*
WEIGHTED AVERAGE	**328**	**326**	**339**	**404**	**472**	**404**	**434**	**338**	**368**	**525**	**609**
Weighted average in 1996 DM	*424*	*420*	*432*	*500*	*569*	*470*	*481*	*358*	*379*	*532*	*609*

Sources:
FPIB Export permits (1986-96)
Statistisches Bundesamt Deutschland (1999)

Appendix 8

GHANA
ESTIMATED WEIGHTED AVERAGE FOB VALUES OF WOOD EXPORTS IN DM PER ROUNDWOOD EQUIVALENT (DM/Cu.m. RWE)

Assumptions about Recovery Rates	
-Lumber	40,00%
-Sliced Veneer	30,00%
-All Others	35,00%

	YEAR										
	1986	**1987**	**1988**	**1989**	**1990**	**1991**	**1992**	**1993**	**1994**	**1995**	**1996**
Export Commodities:											
Logs (current prices)	231	222	235	223	240	223	210	189	180	134	
Logs (1996 prices)	*298*	*287*	*299*	*275*	*290*	*260*	*233*	*200*	*185*	*136*	
Poles 1/ (current prices)					359	373	329	327	358	344	1.060
Poles 1/ (1996 prices)					*432*	*434*	*365*	*346*	*369*	*349*	*1.060*
Lumber (current prices)	186	181	187	213	240	199	195	204	244	220	207
Lumber (1996 prices)	*240*	*233*	*239*	*264*	*289*	*231*	*216*	*216*	*252*	*223*	*207*
Sliced Veneer (current prices)	194	222	262	378	429	394	415	394	420	430	411
Sliced Veneer (1996 prices)	*251*	*286*	*333*	*468*	*517*	*459*	*460*	*417*	*433*	*436*	*411*
All Others (current prices)	393	252	274	384	383	397	354	301	246	234	213
All Others (1996 prices)	*508*	*325*	*349*	*475*	*462*	*462*	*392*	*318*	*254*	*237*	*213*
Total Wood Exports (current prices)	**210**	**202**	**214**	**233**	**258**	**228**	**223**	**216**	**232**	**240**	**229**
Total Wood Exports (1996 prices)	***271***	***261***	***273***	***288***	***310***	***265***	***247***	***229***	***239***	***243***	***229***

Note:
1/ Poles are exclusively Teak. The unusually high price per cubic metre in 1996 probably reflects an error in FPIB's data. However, the effect of this on the value per RWE of "Total Wood Exports" is negligible as the export volume of Teak poles in 1996 was only 1,087 cubic metres.

Sources:
FPIB Export permits (1986-96)
Statistisches Bundesamt Deutschland (1999)

Appendix 9

GHANA
TIMBER ROYALTIES 1976-93

SPECIES Trade name	Botanical name	Forestry Department vol. per tree	1976-82 Royalty per tree	1976-82 Royalty per Cu.m.	1983-85 Royalty per tree	1983-85 Royalty per Cu.m.	1986-88 Royalty per tree	1986-88 Royalty per Cu.m.	1989-93 Royalty per tree	1989-93 Royalty per Cu.m.
Scarlet Stars		(Cu.m.)	(Cedis)	(Cedis)	(Cedis)	(Cedis)	(Cedis)	(Cedis)	(Cedis)	(Cedis)
Albizia	Albizia ferruginea	7	15	2,14	200	28,57	2.500	357,14	2.300	328,57
Asantina 2/	Aningeria altissima/robusta	9,5	12	1,26	300	31,58	2.500	263,16	6.900	726,32
Daniella	Daniella ogea/thurifera	10	22	2,20	200	20,00	1.000	100,00	3.900	390,00
Edinam 1/	Entandophragma angolense	15	30	2,00	400	26,67	4.500	300,00	12.200	813,33
Sapele 1/	Entandophragma cylindricum	15	45	3,00	450	30,00	6.000	400,00	16.600	1.106,67
Utile 1/	Entandophragma utile	18	54	3,00	500	27,78	9.000	500,00	23.600	1.311,11
Hyedua 1/	Guibourtia ehie	8,5	22	2,59	400	47,06	5.500	647,06	11.200	1.317,65
Mahogany 1/	Khaya anthotheca/grandifolia	15	35	2,33	400	26,67	5.500	366,67	16.900	1.126,67
Mahogany 1/	Khaya ivorensis	15	35	2,33	400	26,67	5.500	366,67	16.900	1.126,67
Odum 1/	Milicia excelsa/regia	15,5	35	2,26	400	25,81	5.000	322,58	13.400	864,52
Kusia	Nauclea diderrichii	10	28	2,80	350	35,00	3.500	350,00	6.100	610,00
Afrormosia 1/	Pericopsis elata	8,5	52	6,12	500	58,82	4.500	529,41	24.000	2.823,53
Kyere	Pterygota macrocarpa	10	12	1,20	300	30,00	3.000	300,00	7.100	710,00
Makore 1/	Tieghemella heckelii	20	48	2,40	450	22,50	9.000	450,00	17.700	885,00
Wawa 3/	Triplochiton scleroxylon	13	15	1,15	200	15,38	2.000	153,85	6.400	492,31
Average	**Scarlet stars**			**2,45**		**30,17**		**360,44**		**975,49**
Red Stars										
Afzelia/Papao	Afzelia africana/bellea	7	18	2,57	250	35,71	1.000	142,86	3.500	500,00
Kyenkyen	Antiaris toxicaria	20	15	0,75	200	10,00	2.000	100,00	4.600	230,00
Aprokuma	Antrocaryon micraster	12	6	0,50	150	12,50	1.500	125,00	6.000	500,00
Canarium	Canarium schweinfurthii	10	9	0,90	150	15,00	1.500	150,00	2.500	250,00
Ceiba	Ceiba pentandra	10	12	1,20	150	15,00	1.500	150,00	1.500	150,00
Akasa	Chrysophyllum spp.	10	6	0,60	100	10,00	1.000	100,00	2.500	250,00
Bonsamdua/Ayan	Distemonanthus benthamianus	7	10	1,43	150	21,43	1.500	214,29	6.900	985,71
Candollei 1/	Entandophragma candollei	15	35	2,33	400	26,67	6.000	400,00	12.500	833,33
Guarea	Guarea spp	8	15	1,88	200	25,00	3.000	375,00	7.600	950,00
Niangon 1/	Heritiera utilis	7	14	2,00	300	42,86	2.500	357,14	6.300	900,00
Kaku/Ekki	Lophira alata	9	28	3,11	350	38,89	2.500	277,78	6.100	677,78
Walnut 1/	Lovoa trichilioides	10	24	2,40	350	35,00	3.000	300,00	14.500	1.450,00
Mansonia 1/	Mansonia altissima	6	18	3,00	350	58,33	3.000	500,00	10.200	1.700,00
Dahoma	Piptadenisatrum africanum	14	10	0,71	200	14,29	1.500	107,14	7.200	514,29
Bombax	Rhodognaphalon/Bombax brevicuspe	10,5	6	0,57	200	19,05	1.500	142,86	5.500	523,81
Emire 2/	Terminalia ivorensis	8	21	2,63	350	43,75	3.500	437,50	7.400	925,00
Average	**Red stars**			**1,66**		**26,47**		**242,47**		**708,75**
Pink Stars										
Albizia	Albizia adianthifolia	7	15	2,14	200	28,57	2.000	285,71	2.500	357,14
Albizia	Albizia zygia	7	15	2,14	200	28,57	2.000	285,71	2.500	357,14
Alstonia	Alstonia boonei	10	6	0,60	100	10,00	1.000	100,00	2.500	250,00
Yaya	Amphimas pterocarpoides	11	6	0,55	100	9,09	1.000	90,91	2.500	227,27
Kokote	Anopyxis klaineana	10	15	1,50	200	20,00	2.500	250,00	2.500	250,00
Berlinia	Berlinia confusa	10	6	0,60	100	10,00	1.000	100,00	2.500	250,00
Berlinia	Berlinia spp.	10	6	0,60	100	10,00	1.000	100,00	2.500	250,00
Celtis/Essa	Celtis mildbraedii/zenkeri	8	8	1,00	100	12,50	3.000	375,00	3.400	425,00
Pamprana	Coryanthe pachyceras	10	6	0,60	100	10,00	1.000	100,00	2.500	250,00
Denya	Cylicodiscus gabonensis	10	16	1,60	300	30,00	3.000	300,00	2.500	250,00
Ananta	Cynometra antana	10	6	0,60	100	10,00	1.000	100,00	2.500	250,00
Duabankye	Dialium aubrevillei	10	6	0,60	100	10,00	1.000	100,00	2.500	250,00
Potrodum	Erytrophleum sauaveolens	10	6	0,60	150	15,00	1.500	150,00	2.500	250,00
Nakwa	Holoptelea grandis	10	6	0,60	100	10,00	1.000	100,00	2.500	250,00
Kroma	Klaindoxa gabonensis	10	6	0,60	100	10,00	1.000	100,00	2.500	250,00
Kumanini	Lannea welwitschii	10	6	0,60	100	10,00	1.000	100,00	2.500	250,00
Bompagya	Mammea africana	10	6	0,60	100	10,00	1.000	100,00	2.500	250,00
Wonton	Morus mesozygia	8	12	1,50	100	12,50	1.500	187,50	2.500	312,50
Danta 2/	Nesogordonia papaverifera	10	14	1,40	200	20,00	3.000	300,00	4.600	460,00
Akwana	Ongokea gore	10	6	0,60	100	10,00	1.000	100,00	2.500	250,00
Afam	Parinari exelsa	10	6	0,60	100	10,00	1.000	100,00	2.500	250,00
Asoma	Parkia bicolor	10	6	0,60	100	10,00	2.000	200,00	2.500	250,00
Esia	Petersianthus macrocarpus	10	6	0,60	100	10,00	1.000	100,00	2.500	250,00
Otie	Pycnanthus angolensis	10	6	0,60	150	15,00	1.500	150,00	2.300	230,00
Bombax	Rhodognaphalon/Bombax buonopozense	10,5	6	0,57	200	19,05	1.500	142,86	5.500	523,81
Wama	Ricinodendron heudelotii	10	6	0,60	100	10,00	1.000	100,00	2.500	250,00
Sterculia	Sterculia spp.	10	6	0,60	100	10,00	1.000	100,00	2.500	250,00
Afena	Strombosia glaucescens	10	6	0,60	100	10,00	1.000	100,00	2.500	250,00
Ofram 2a/	Terminalia superba	9	18	2,00	300	33,33	3.000	333,33	4.200	466,67
Tanuronini	Trichilia tessmannii	10	6	0,60	100	10,00	1.000	100,00	2.500	250,00
Okure	Trilepisium madagascariense	10	6	0,60	100	10,00	1.000	100,00	2.500	250,00
Avodire 1/	Turreanthus africanus	8	14	1,75	300	37,50	3.000	375,00	6.400	800,00
Average	**Pink stars**			**0,90**		**14,72**		**163,31**		**301,86**
Subaha 2/, 4/	Hallea spp./Mitragyna spp.	10	14	1,40	150	15,00	2.000	200,00	2.000	200,00
AVERAGE	**ALL SPECIES**			**1,46**		**21,38**		**230,35**		**565,58**

Notes:
1/ Log export banned in 1979
2/ Log export banned in 1988
2a/ Log export banned in 1988 but lifted in 1990
3/ Log export banned in 1993
4/ Harvest banned in 1993
Log exports was entirely banned in 1995
5/ Calculated on the basis of 1996 air-dried lumber prices using average star group prices when other prices are not available.

Sources:
Forest Fees Regulations (1976)
Forest Fees (Amendment) Regulations (1983)
Forest Fees (Amendment) Regulations (1986)
Forest Fees (Amendment) Regulations (1989)
Forest Fees (Amendment) Regulations (1993)
Timber Resources Management Regulations (1998) (came into force on 16 December, 1998).

Appendix 9

GHANA
TIMBER ROYALTIES 1994-99

SPECIES Trade name	Botanical name	Forestry Department vol. per tree (Cu.m.)	FOB Price of air-dried lumber per Cu.m. 5/ (DM)	35% of FOB price on air-dried lumber (DM/Cu.m.)	1999- Stumpage rate	Stumpage fee per Cu.m. (DM)	Stumpage fee per Cu.m. using avg. star group prices for species where prices are not available (DM)
Scarlet Stars							
Albizia	Albizia ferruginea	7	436,47	152,76	10,00%	15,28	15,28
Asanfina 2/	Aningeria altissima/robusta	9,5	851,95	298,18	20,00%	59,64	59,64
Daniella	Daniella ogea/thurifera	10			10,00%		22,82
Edinam 1/	Entandophragma angolense	15	529,43	185,30	10,00%	18,53	18,53
Sapele 1/	Entandophragma cylindricum	15	741,93	259,68	20,00%	51,94	51,94
Utile 1/	Entandophragma utile	18	899,46	314,81	20,00%	62,96	62,96
Hyedua 1/	Guibourtia ehie	8,5	867,45	303,61	20,00%	60,72	60,72
Mahogany 1/	Khaya anthotheca/grandifolia	15	648,59	227,01	20,00%	45,40	45,40
Mahogany 1/	Khaya ivorensis	15	648,59	227,01	20,00%	45,40	45,40
Odum 1/	Milicia excelsa/regia	15,5	618,24	216,39	20,00%	43,28	43,28
Kusia	Nauclea diderrichii	10	459,50	160,82	20,00%	32,16	32,16
Afrormosia 1/	Pericopsis elata	8,5			20,00%		45,64
Kyere	Pterygota macrocarpa	10			10,00%		22,82
Makore 1/	Tieghemella heckelii	20	733,41	256,69	20,00%	51,34	51,34
Wawa 3/	Triplochiton scleroxylon	13	388,24	135,88	10,00%	13,59	13,59
Average	**Scarlet stars**			**228,18**		**41,69**	**39,43**
Red Stars							
Afzelia/Papao	Afzelia africana/bellea	7	686,12	240,14	10,00%	24,01	24,01
Kyenkyen	Antiaris toxicaria	20	143,94	50,38	5,00%	2,52	2,52
Aprokuma	Antrocaryon micraster	12			10,00%		19,96
Canarium	Canarium schweinfurthii	10	671,88	235,16	10,00%	23,52	23,52
Ceiba	Ceiba pentandra	10			10,00%		19,96
Akasa	Chrysophyllum spp.	10			10,00%		19,96
Bonsamdua/Ayan	Distemonanthus benthamianus	7	559,80	195,93	10,00%	19,59'	19,59
Candollei 1/	Entandophragma candollei	15	583,11	204,09	20,00%	40,82	40,82
Guarea	Guarea spp	8	630,56	220,70	10,00%	22,07	22,07
Niangon 1/	Heritiera utilis	7	749,12	262,19	10,00%	26,22	26,22
Kaku/Ekki	Lophira alata	9	516,25	180,69	10,00%	18,07	18,07
Walnut 1/	Lovoa trichilioides	10	581,59	203,56	20,00%	40,71	40,71
Mansonia 1/	Mansonia altissima	6	757,20	265,02	10,00%	26,50	26,50
Dahoma	Piptadenisatrum africanum	14	426,61	149,31	5,00%	7,47	7,47
Bombax	Rhodognaphalon/Bombax brevicuspe	10,5	459,98	160,99	10,00%	16,10	16,10
Emire 2/	Terminalia ivorensis	8	648,44	226,96	10,00%	22,70	22,70
Average	**Red stars**			**199,62**		**22,33**	**21,89**
Pink Stars							
Albizia	Albizia adianthifolia	7			5,00%		7,64
Albizia	Albizia zygia	7			5,00%		7,64
Alstonia	Alstonia boonei	10			5,00%		7,64
Yaya	Amphimas pterocarpoides	11			5,00%		7,64
Kokote	Anopyxis klaineana	10			5,00%		7,64
Berlinia	Berlinia confusa	10			5,00%		7,64
Berlinia	Berlinia spp.	10			5,00%		7,64
Celtis/Essa	Celtis mildbraedii/zenkeri	8	517,25	181,04	5,00%	9,05	9,05
Pamprana	Coryanthe pachyceras	10			5,00%		7,64
Denya	Cylicodiscus gabonensis	10	374,19	130,97	5,00%	6,55	6,55
Ananta	Cynometra antana	10	340,02	119,01	5,00%	5,95	5,95
Duabankye	Dialium aubrevillei	10			5,00%		7,64
Potrodum	Erytrophleum sauaveolens	10	378,81	132,58	5,00%	6,63	6,63
Nakwa	Holoptelea grandis	10			5,00%		7,64
Kroma	Klaindoxa gabonensis	10			5,00%		7,64
Kumanini	Lannea welwitschii	10			5,00%		7,64
Bompagya	Mammea africana	10	339,48	118,82	5,00%	5,94	5,94
Wonton	Morus mesozygia	8			5,00%		7,64
Danta 2/	Nesogordonia papaverifera	10	521,82	182,64	5,00%	9,13	9,13
Akwana	Ongokea gore	10			5,00%		7,64
Afam	Parinari exelsa	10			5,00%		7,64
Asoma	Parkia bicolor	10			5,00%		7,64
Esia	Petersianthus macrocarpus	10	340,00	119,00	5,00%	5,95	5,95
Otie	Pycnanthus angolensis	10	455,79	159,53	5,00%	7,98	7,98
Bombax	Rhodognaphalon/Bombax buonopozense	10,5			5,00%		7,64
Wama	Ricinodendron heudelotii	10			5,00%		7,64
Sterculia	Sterculia spp.	10	453,11	158,59	5,00%	7,93	7,93
Afena	Strombosia glaucescens	10			5,00%		7,64
Ofram 2a/	Terminalia superba	9	364,54	127,59	10,00%	12,76	12,76
Tanuronini	Trichilia tessmannii	10			5,00%		7,64
Okure	Trilepisium madagascariense	10			5,00%		7,64
Avodire 1/	Turreanthus africanus	8	716,32	250,71	10,00%	25,07	25,07
Average	**Pink stars**			**152,77**		**9,36**	**8,23**
Subaha 2/, 4/	Hallea spp./Mitragyna spp.	10					
AVERAGE	**ALL SPECIES**					**24,82**	**19,13**

Notes:
1/ Log export banned in 1979
2/ Log export banned in 1988
2a/ Log export banned in 1988 but lifted in 1990
3/ Log export banned in 1993
4/ Harvest banned in 1993
Log exports was entirely banned in 1995
5/ Calculated on the basis of 1996 air-dried lumber prices using average star group prices when other prices are not available.

Sources:
Forest Fees Regulations (1976)
Forest Fees (Amendment) Regulations (1983)
Forest Fees (Amendment) Regulations (1986)
Forest Fees (Amendment) Regulations (1989)
Forest Fees (Amendment) Regulations (1993)
Timber Resources Management Regulations (1998) (came into force on 16 December, 1998).

Appendix 9

GHANA
ROYALTIES CONVERTED TO CURRENT DM/Cu.m. AT BANK EXCHANGE RATES (1976-87)

SPECIES		Year											
Trade name	**Botanical name**	**1976**	**1977**	**1978**	**1979**	**1980**	**1981**	**1982**	**1983**	**1984**	**1985**	**1986**	**1987**
Scarlet Stars													
Albizia	Albizia ferruginea	4,73	4,45	2,85	1,45	1,44	1,64	1,85	3,47	2,10	1,71	8,98	3,80
Asanfina 2/	Aningeria altissima/robusta	2,79	2,63	1,68	0,85	0,85	0,97	1,09	3,83	2,32	1,89	6,62	2,80
Daniella	Daniella ogea/thurifera	4,86	4,57	2,92	1,49	1,48	1,69	1,90	2,43	1,47	1,20	2,52	1,06
Edinam 1/	Entandophragma angolense	4,42	4,16	2,66	1,35	1,34	1,53	1,73	3,23	1,96	1,60	7,55	3,19
Sapele 1/	Entandophragma cylindricum	6,63	6,23	3,99	2,03	2,02	2,30	2,60	3,64	2,20	1,80	10,06	4,26
Utile 1/	Entandophragma utile	6,63	6,23	3,99	2,03	2,02	2,30	2,60	3,37	2,04	1,66	12,58	5,32
Hyedua 1/	Guibourtia ehie	5,72	5,38	3,44	1,75	1,74	1,98	2,24	5,71	3,46	2,82	16,28	6,89
Mahogany 1/	Khaya anthotheca/grandifolia	5,15	4,85	3,10	1,58	1,57	1,79	2,02	3,23	1,96	1,60	9,22	3,90
Mahogany 1/	Khaya ivorensis	5,15	4,85	3,10	1,58	1,57	1,79	2,02	3,23	1,96	1,60	9,22	3,90
Odum 1/	Milicia excelsa/regia	4,99	4,69	3,00	1,53	1,52	1,73	1,95	3,13	1,90	1,55	8,11	3,43
Kusia	Nauclea diderrichii	6,18	5,82	3,72	1,89	1,88	2,15	2,42	4,25	2,57	2,10	8,80	3,73
Afrormosia 1/	Pericopsis elata	13,51	12,71	8,13	4,14	4,11	4,69	5,29	7,14	4,32	3,52	13,32	5,64
Kyere	Pterygota macrocarpa	2,65	2,49	1,59	0,81	0,81	0,92	1,04	3,64	2,20	1,80	7,55	3,19
Makore 1/	Tieghemella heckelii	5,30	4,99	3,19	1,62	1,61	1,84	2,08	2,73	1,65	1,35	11,32	4,79
Wawa 3/	Triplochiton scleroxylon	2,55	2,40	1,53	0,78	0,78	0,88	1,00	1,87	1,13	0,92	3,87	1,64
Average	**Scarlet stars**	**5,42**	**5,10**	**3,26**	**1,66**	**1,65**	**1,88**	**2,12**	**3,66**	**2,22**	**1,81**	**9,07**	**3,84**
Red Stars													
Afzelia/Papao	Afzelia africana/bellea	5,68	5,34	3,42	1,74	1,73	1,97	2,22	4,33	2,62	2,14	3,59	1,52
Kyenkyen	Antiaris toxicaria	1,66	1,56	1,00	0,51	0,50	0,58	0,65	1,21	0,73	0,60	2,52	1,06
Aprokuma	Antrocaryon micraster	1,10	1,04	0,66	0,34	0,34	0,38	0,43	1,52	0,92	0,75	3,14	1,33
Canarium	Canarium schweinfurthii	1,99	1,87	1,20	0,61	0,61	0,69	0,78	1,82	1,10	0,90	3,77	1,60
Ceiba	Ceiba pentandra	2,65	2,49	1,59	0,81	0,81	0,92	1,04	1,82	1,10	0,90	3,77	1,60
Akasa	Chrysophyllum spp.	1,33	1,25	0,80	0,41	0,40	0,46	0,52	1,21	0,73	0,60	2,52	1,06
Bonsamdua/Ayan	Distemonanthus benthamianus	3,16	2,97	1,90	0,97	0,96	1,10	1,24	2,60	1,57	1,28	5,39	2,28
Candollei 1/	Entandophragma candollei	5,15	4,85	3,10	1,58	1,57	1,79	2,02	3,23	1,96	1,60	10,06	4,26
Guarea	Guarea spp	4,14	3,90	2,49	1,27	1,26	1,44	1,62	3,03	1,84	1,50	9,43	3,99
Niangon 1/	Heritiera utilis	4,42	4,16	2,66	1,35	1,34	1,53	1,73	5,20	3,15	2,57	8,98	3,80
Kaku/Ekki	Lophira alata	6,87	6,47	4,13	2,10	2,09	2,39	2,69	4,72	2,86	2,33	6,99	2,96
Walnut 1/	Lovoa trichilioides	5,30	4,99	3,19	1,62	1,61	1,84	2,08	4,25	2,57	2,10	7,55	3,19
Mansonia 1/	Mansonia altissima	6,63	6,23	3,99	2,03	2,02	2,30	2,60	7,08	4,29	3,49	12,58	5,32
Dahoma	Piptadenisatrum africanum	1,58	1,48	0,95	0,48	0,48	0,55	0,62	1,73	1,05	0,86	2,70	1,14
Bombax	Rhodognaphalon/Bombax brevicuspe	1,26	1,19	0,76	0,39	0,38	0,44	0,49	2,31	1,40	1,14	3,59	1,52
Emire 2/	Terminalia ivorensis	5,80	5,46	3,49	1,78	1,76	2,01	2,27	5,31	3,21	2,62	11,01	4,66
Average	**Red stars**	**3,67**	**3,45**	**2,21**	**1,12**	**1,12**	**1,27**	**1,44**	**3,21**	**1,94**	**1,59**	**6,10**	**2,58**
Pink Stars													
Albizia	Albizia adianthifolia	4,73	4,45	2,85	1,45	1,44	1,64	1,85	3,47	2,10	1,71	7,19	3,04
Albizia	Albizia zygia	4,73	4,45	2,85	1,45	1,44	1,64	1,85	3,47	2,10	1,71	7,19	3,04
Alstonia	Alstonia boonei	1,33	1,25	0,80	0,41	0,40	0,46	0,52	1,21	0,73	0,60	2,52	1,06
Yaya	Amphimas pterocarpoides	1,20	1,13	0,72	0,37	0,37	0,42	0,47	1,10	0,67	0,54	2,29	0,97
Kokote	Anopyxis klaineana	3,31	3,12	1,99	1,01	1,01	1,15	1,30	2,43	1,47	1,20	6,29	2,66
Berlinia	Berlinia confusa	1,33	1,25	0,80	0,41	0,40	0,46	0,52	1,21	0,73	0,60	2,52	1,06
Berlinia	Berlinia spp.	1,33	1,25	0,80	0,41	0,40	0,46	0,52	1,21	0,73	0,60	2,52	1,06
Celtis/Essa	Celtis mildbraedii/zenkeri	2,21	2,08	1,33	0,68	0,67	0,77	0,87	1,52	0,92	0,75	9,43	3,99
Pamprana	Coryanthe pachyceras	1,33	1,25	0,80	0,41	0,40	0,46	0,52	1,21	0,73	0,60	2,52	1,06
Denya	Cylicodiscus gabonensis	3,53	3,33	2,13	1,08	1,08	1,23	1,38	3,64	2,20	1,80	7,55	3,19
Ananta	Cynometra antana	1,33	1,25	0,80	0,41	0,40	0,46	0,52	1,21	0,73	0,60	2,52	1,06
Duabankye	Dialium aubrevillei	1,33	1,25	0,80	0,41	0,40	0,46	0,52	1,21	0,73	0,60	2,52	1,06
Potrodum	Erytrophleum sauaveolens	1,33	1,25	0,80	0,41	0,40	0,46	0,52	1,82	1,10	0,90	3,77	1,60
Nakwa	Holoptelea grandis	1,33	1,25	0,80	0,41	0,40	0,46	0,52	1,21	0,73	0,60	2,52	1,06
Kroma	Klaindoxa gabonensis	1,33	1,25	0,80	0,41	0,40	0,46	0,52	1,21	0,73	0,60	2,52	1,06
Kumanini	Lannea welwitschii	1,33	1,25	0,80	0,41	0,40	0,46	0,52	1,21	0,73	0,60	2,52	1,06
Bompagya	Mammea africana	1,33	1,25	0,80	0,41	0,40	0,46	0,52	1,21	0,73	0,60	2,52	1,06
Wonton	Morus mesozygia	3,31	3,12	1,99	1,01	1,01	1,15	1,30	1,52	0,92	0,75	4,72	2,00
Danta 2/	Nesogordonia papaverifera	3,09	2,91	1,86	0,95	0,94	1,07	1,21	2,43	1,47	1,20	7,55	3,19
Akwana	Ongokea gore	1,33	1,25	0,80	0,41	0,40	0,46	0,52	1,21	0,73	0,60	2,52	1,06
Afam	Parinari exelsa	1,33	1,25	0,80	0,41	0,40	0,46	0,52	1,21	0,73	0,60	2,52	1,06
Asoma	Parkia bicolor	1,33	1,25	0,80	0,41	0,40	0,46	0,52	1,21	0,73	0,60	5,03	2,13
Esia	Petersianthus macrocarpus	1,33	1,25	0,80	0,41	0,40	0,46	0,52	1,21	0,73	0,60	2,52	1,06
Otie	Pycnanthus angolensis	1,33	1,25	0,80	0,41	0,40	0,46	0,52	1,82	1,10	0,90	3,77	1,60
Bombax	Rhodognaphalon/Bombax buonopozense	1,26	1,19	0,76	0,39	0,38	0,44	0,49	2,31	1,40	1,14	3,59	1,52
Wama	Ricinodendron heudelotii	1,33	1,25	0,80	0,41	0,40	0,46	0,52	1,21	0,73	0,60	2,52	1,06
Sterculia	Sterculia spp.	1,33	1,25	0,80	0,41	0,40	0,46	0,52	1,21	0,73	0,60	2,52	1,06
Afena	Strombosia glaucescens	1,33	1,25	0,80	0,41	0,40	0,46	0,52	1,21	0,73	0,60	2,52	1,06
Ofram 2a/	Terminalia superba	4,42	4,16	2,66	1,35	1,34	1,53	1,73	4,04	2,45	2,00	8,39	3,55
Tanuronini	Trichilia tessmannii	1,33	1,25	0,80	0,41	0,40	0,46	0,52	1,21	0,73	0,60	2,52	1,06
Okure	Trilepisium madagascariense	1,33	1,25	0,80	0,41	0,40	0,46	0,52	1,21	0,73	0,60	2,52	1,06
Avodire 1/	Turreanthus africanus	3,87	3,64	2,33	1,18	1,18	1,34	1,51	4,55	2,75	2,25	9,43	3,99
Average	**Pink stars**	**1,98**	**1,87**	**1,19**	**0,61**	**0,60**	**0,69**	**0,78**	**1,79**	**1,08**	**0,88**	**4,11**	**1,74**
Subaha 2/, 4/	Hallea spp./Mitragyna spp.	3,09	2,91	1,86	0,95	0,94	1,07	1,21	1,82	1,10	0,90	5,03	2,13
AVERAGE	ALL SPECIES	3,23	3,04	1,94	0,99	0,98	1,12	1,26	2,59	1,57	1,28	5,79	2,45

Notes:
1/ Log export banned in 1979
2/ Log export banned in 1988
2a/ Log export banned in 1988 but lifted in 1990
3/ Log export banned in 1993
4/ Harvest banned in 1993
Log exports was entirely banned in 1995

Sources:
Forest Fees Regulations (1976)
Forest Fees (Amendment) Regulations (1983)
Forest Fees (Amendment) Regulations (1986)
Forest Fees (Amendment) Regulations (1989)
Forest Fees (Amendment) Regulations (1993)

Appendix 9

GHANA
ROYALTIES CONVERTED TO CURRENT DM/Cu.m. AT BANK EXCHANGE RATES (1988-99)

SPECIES Trade name	Botanical name	1988	1989	1990	1991	1992	1993	1994	1995	1996	1997	1998	1999
Scarlet Stars													5/
Albizia	Albizia ferruginea	2,80	2,20	1,66	1,34	1,18	0,89	6,63	5,23	3,83	3,61	2,84	15,28
Asanfina 2/	Aningeria altissima/robusta	2,07	4,87	3,68	2,97	2,60	1,97	21,14	16,67	12,21	11,49	9,05	59,64
Daniella	Daniella ogea/thurifera	0,79	2,62	1,97	1,59	1,40	1,06	7,49	5,91	4,33	4,07	3,21	
Edinam 1/	Entandophragma angolense	2,36	5,46	4,12	3,32	2,91	2,20	9,76	7,69	5,63	5,31	4,18	18,53
Sapele 1/	Entandophragma cylindricum	3,14	7,42	5,60	4,52	3,96	3,00	16,59	13,08	9,58	9,02	7,10	51,94
Utile 1/	Entandophragma utile	3,93	8,80	6,64	5,36	4,69	3,55	17,76	14,00	10,25	9,66	7,60	62,96
Hyedua 1/	Guibourtia ehie	5,08	8,84	6,67	5,39	4,71	3,57	17,89	14,10	10,33	9,73	7,66	60,72
Mahogany 1/	Khaya anthotheca/grandifolia	2,88	7,56	5,70	4,61	4,03	3,05	11,71	9,23	6,76	6,37	5,01	45,40
Mahogany 1/	Khaya ivorensis	2,88	7,56	5,70	4,61	4,03	3,05	11,71	9,23	6,76	6,37	5,01	45,40
Odum 1/	Milicia excelsa/regia	2,53	5,80	4,38	3,53	3,09	2,34	10,57	8,33	6,10	5,75	4,53	43,28
Kusia	Nauclea diderrichii	2,75	4,09	3,09	2,49	2,18	1,65	6,63	5,23	3,83	3,61	2,84	32,16
Afrormosia 1/	Pericopsis elata	4,16	18,94	14,29	11,54	10,10	7,65	21,50	16,95	12,41	11,69	9,21	
Kyere	Pterygota macrocarpa	2,36	4,76	3,59	2,90	2,54	1,92	13,01	10,26	7,51	7,07	5,57	
Makore 1/	Tieghemella heckelii	3,53	5,94	4,48	3,62	3,17	2,40	10,70	8,44	6,18	5,82	4,58	51,34
Wawa 3/	Triplochiton scleroxylon	1,21	3,30	2,49	2,01	1,76	1,33	7,48	5,90	4,32	4,07	3,20	13,59
Average	**Scarlet stars**	**2,83**	**6,54**	**4,94**	**3,99**	**3,49**	**2,64**	**12,70**	**10,02**	**7,34**	**6,91**	**5,44**	**41,69**
Red Stars													
Afzelia/Papao	Afzelia africana/bellea	1,12	3,35	2,53	2,04	1,79	1,35	13,01	10,26	7,51	7,07	5,57	24,01
Kyenkyen	Antiaris toxicaria	0,79	1,54	1,16	0,94	0,82	0,62	7,49	5,91	4,33	4,07	3,21	2,52
Aprokuma	Antrocaryon micraster	0,98	3,35	2,53	2,04	1,79	1,35	7,49	5,91	4,33	4,07	3,21	
Canarium	Canarium schweinfurthii	1,18	1,68	1,27	1,02	0,89	0,68	7,49	5,91	4,33	4,07	3,21	23,52
Ceiba	Ceiba pentandra	1,18	1,01	0,76	0,61	0,54	0,41	7,49	5,91	4,33	4,07	3,21	
Akasa	Chrysophyllum spp.	0,79	1,68	1,27	1,02	0,89	0,68	11,38	8,97	6,57	6,19	4,87	
Bonsamdua/Ayan	Distemonanthus benthamianus	1,68	6,61	4,99	4,03	3,53	2,67	11,06	8,72	6,38	6,01	4,74	19,59
Candollei 1/	Entandophragma candollei	3,14	5,59	4,22	3,41	2,98	2,26	9,76	7,69	5,63	5,31	4,18	40,82
Guarea	Guarea spp	2,94	6,37	4,81	3,88	3,40	2,57	10,96	8,64	6,33	5,96	4,69	22,07
Niangon 1/	Heritiera utilis	2,80	6,04	4,56	3,68	3,22	2,44	10,41	8,21	6,01	5,66	4,46	26,22
Kaku/Ekki	Lophira alata	2,18	4,55	3,43	2,77	2,42	1,84	6,63	5,23	3,83	3,61	2,84	18,07
Walnut 1/	Lovoa trichilioides	2,36	9,73	7,34	5,93	5,19	3,93	11,51	9,08	6,65	6,26	4,93	40,71
Mansonia 1/	Mansonia altissima	3,93	11,41	8,61	6,95	6,08	4,61	14,54	11,46	8,39	7,90	6,23	26,50
Dahoma	Piptadenisatrum africanum	0,84	3,45	2,60	2,10	1,84	1,39	6,50	5,13	3,76	3,54	2,79	7,47
Bombax	Rhodognaphalon/Bombax brevicuspe	1,12	3,51	2,65	2,14	1,87	1,42	6,50	5,13	3,76	3,54	2,79	16,10
Emire 2/	Terminalia ivorensis	3,43	6,21	4,68	3,78	3,31	2,51	10,41	8,21	6,01	5,66	4,46	22,70
Average	**Red stars**	**1,90**	**4,76**	**3,59**	**2,90**	**2,54**	**1,92**	**9,54**	**7,52**	**5,51**	**5,19**	**4,09**	**22,33**
Pink Stars													
Albizia	Albizia adianthifolia	2,24	2,40	1,81	1,46	1,28	0,97	6,50	5,13	3,76	3,54	2,79	
Albizia	Albizia zygia	2,24	2,40	1,81	1,46	1,28	0,97	6,50	5,13	3,76	3,54	2,79	
Alstonia	Alstonia boonei	0,79	1,68	1,27	1,02	0,89	0,68	6,50	5,13	3,76	3,54	2,79	
Yaya	Amphimas pterocarpoides	0,71	1,52	1,15	0,93	0,81	0,62	6,50	5,13	3,76	3,54	2,79	
Kokote	Anopyxis klaineana	1,96	1,68	1,27	1,02	0,89	0,68	6,50	5,13	3,76	3,54	2,79	
Berlinia	Berlinia confusa	0,79	1,68	1,27	1,02	0,89	0,68	6,50	5,13	3,76	3,54	2,79	
Berlinia	Berlinia spp.	0,79	1,68	1,27	1,02	0,89	0,68	6,50	5,13	3,76	3,54	2,79	
Celtis/Essa	Celtis mildbraedii/zenkeri	2,94	2,85	2,15	1,74	1,52	1,15	7,49	5,91	4,33	4,07	3,21	9,05
Pamprana	Coryanthe pachyceras	0,79	1,68	1,27	1,02	0,89	0,68	6,50	5,13	3,76	3,54	2,79	
Denya	Cylicodiscus gabonensis	2,36	1,68	1,27	1,02	0,89	0,68	6,50	5,13	3,76	3,54	2,79	6,55
Ananta	Cynometra antana	0,79	1,68	1,27	1,02	0,89	0,68	6,50	5,13	3,76	3,54	2,79	5,95
Duabankye	Dialium aubrevillei	0,79	1,68	1,27	1,02	0,89	0,68	6,50	5,13	3,76	3,54	2,79	
Potrodum	Erytrophleum sauaveolens	1,18	1,68	1,27	1,02	0,89	0,68	6,50	5,13	3,76	3,54	2,79	6,63
Nakwa	Holoptelea grandis	0,79	1,68	1,27	1,02	0,89	0,68	6,50	5,13	3,76	3,54	2,79	
Kroma	Klaindoxa gabonensis	0,79	1,68	1,27	1,02	0,89	0,68	6,50	5,13	3,76	3,54	2,79	
Kumanini	Lannea welwitschii	0,79	1,68	1,27	1,02	0,89	0,68	6,50	5,13	3,76	3,54	2,79	
Bompagya	Mammea africana	0,79	1,68	1,27	1,02	0,89	0,68	6,50	5,13	3,76	3,54	2,79	5,94
Wonton	Morus mesozygia	1,47	2,10	1,58	1,28	1,12	0,85	11,06	8,72	6,38	6,01	4,74	
Danta 2/	Nesogordonia papaverifera	2,36	3,09	2,33	1,88	1,65	1,25	9,11	7,18	5,26	4,95	3,90	9,13
Akwana	Ongokea gore	0,79	1,68	1,27	1,02	0,89	0,68	6,50	5,13	3,76	3,54	2,79	
Afam	Parinari exelsa	0,79	1,68	1,27	1,02	0,89	0,68	6,50	5,13	3,76	3,54	2,79	
Asoma	Parkia bicolor	1,57	1,68	1,27	1,02	0,89	0,68	6,50	5,13	3,76	3,54	2,79	
Esia	Petersianthus macrocarpus	0,79	1,68	1,27	1,02	0,89	0,68	6,50	5,13	3,76	3,54	2,79	5,95
Otie	Pycnanthus angolensis	1,18	1,54	1,16	0,94	0,82	0,62	7,49	5,91	4,33	4,07	3,21	7,98
Bombax	Rhodognaphalon/Bombax buonopozense	1,12	3,51	2,65	2,14	1,87	1,42	6,50	5,13	3,76	3,54	2,79	
Wama	Ricinodendron heudelotii	0,79	1,68	1,27	1,02	0,89	0,68	6,50	5,13	3,76	3,54	2,79	
Sterculia	Sterculia spp.	0,79	1,68	1,27	1,02	0,89	0,68	7,49	5,91	4,33	4,07	3,21	7,93
Afena	Strombosia glaucescens	0,79	1,68	1,27	1,02	0,89	0,68	6,50	5,13	3,76	3,54	2,79	
Ofram 2a/	Terminalia superba	2,62	3,13	2,36	1,91	1,67	1,26	7,15	5,64	4,13	3,89	3,06	12,76
Tanuronini	Trichilia tessmannii	0,79	1,68	1,27	1,02	0,89	0,68	6,50	5,13	3,76	3,54	2,79	
Okure	Trilepisium madagascariense	0,79	1,68	1,27	1,02	0,89	0,68	6,50	5,13	3,76	3,54	2,79	
Avodire 1/	Turreanthus africanus	2,94	5,37	4,05	3,27	2,86	2,17	9,11	7,18	5,26	4,95	3,90	25,07
Average	**Pink stars**	**1,28**	**2,03**	**1,53**	**1,23**	**1,08**	**0,82**	**6,92**	**5,46**	**4,00**	**3,76**	**2,96**	**9,36**
Subaha 2/, 4/	Hallea spp./Mitragyna spp.	1,57	1,34	1,01	0,82	0,72	0,54	11,06	8,72	6,38	6,01		
AVERAGE	**ALL SPECIES**	**1,81**	**3,79**	**2,86**	**2,31**	**2,02**	**1,53**	**8,96**	**7,07**	**5,18**	**4,87**	**3,84**	**24,82**

Notes:
1/ Log export banned in 1979
2/ Log export banned in 1988
2a/ Log export banned in 1988 but lifted in 1990
3/ Log export banned in 1993
4/ Harvest banned in 1993
Log exports was entirely banned in 1995
5/ 1996 prices are used as a proxy due to data constraints

Sources:
Forest Fees Regulations (1976)
Forest Fees (Amendment) Regulations (1983)
Forest Fees (Amendment) Regulations (1986)
Forest Fees (Amendment) Regulations (1989)
Forest Fees (Amendment) Regulations (1993)
Timber Resources Management Regulations (1998) (came into force on 16 December, 1998).

Appendix 10

GHANA
ROYALTIES AS PER CENT OF AVAILABLE FOB LOG PRICES (1976-85)

SPECIES		Year									
Trade name	**Botanical name**	**1976**	**1977**	**1978**	**1979**	**1980**	**1981**	**1982**	**1983**	**1984**	**1985**
Scarlet Stars											
Albizia	Albizia ferruginea										
Asanfina 2/	Aningeria altissima/robusta										
Daniella	Daniella ogea/thurifera										
Edinam 1/	Entandophragma angolense	1,29	1,09	1,30	0,72	0,52					
Sapele 1/	Entandophragma cylindricum	1,44		1,29	0,78	0,47	0,56	0,67	0,91	0,50	0,33
Utile 1/	Entandophragma utile	1,28	1,16	0,76	0,49	0,35	0,46	0,54	0,74	0,44	0,29
Hyedua 1/	Guibourtia ehie										
Mahogany 1/	Khaya anthotheca/grandifolia/ivorensis	1,36	1,28	0,97		0,46	0,54	0,62	1,10	0,70	0,46
Odum 1/	Milicia excelsa/regia	1,38		1,40							
Kusia	Nauclea didichii	3,03									
Afrormosia 1/	Pericopsis elata	1,94	1,81	1,37	0,86	0,89	1,11	1,34	1,86		
Kyere	Pterygota macrocarpa										
Makore 1/	Tieghemella heckelii	1,15	1,23		0,53	0,39	0,41	0,53			
Wawa 3/	Triplochiton scleroxylon	1,21	1,14	0,81	0,60	0,37	0,42	0,47	0,92	0,55	0,35
Average	**Scarlet stars**	**1,56**	**1,29**	**1,13**	**0,66**	**0,49**	**0,58**	**0,69**	**1,10**	**0,55**	**0,36**
Red Stars											
Afzelia/Papao	Afzelia africana/bellea					0,50					
Kyenkyen	Antiaris toxicaria	0,70	0,99	0,74	0,38	0,25	0,44				
Aprokuma	Antrocaryon micraster										
Canarium	Canarium schweinfurthii	1,88									
Ceiba	Ceiba pentandra		2,32		1,10	0,83		0,61	1,43	0,93	0,61
Akasa	Chrysophyllum spp.										
Bonsamdua/Ayan	Distemonanthus benthamianus										
Candollei 1/	Entandophragma candollei	1,18			0,68	0,45	0,58	0,68	1,21	0,73	0,46
Guarea	Guarea spp	2,00	1,32	1,52	0,85		0,41	0,53	1,41		
Niangon 1/	Heritiera utilis	1,40	1,86	2,07	0,91	0,52	0,53	0,65	1,74	0,96	0,60
Kaku/Ekki	Lophira alata	2,80									
Walnut 1/	Lovoa trichilioides	1,62	1,50	1,44	0,81	0,89	0,68	0,75	1,61	0,99	0,64
Mansonia 1/	Mansonia altissima	1,87	2,03	2,11	1,92	0,82					
Dahoma	Piptadenisatrum africanum				0,49	0,30	0,22				
Bombax	Rhodognaphalon/Bombax brevicuspe										
Emire 2/	Terminalia ivorensis	2,40	2,27		1,00	0,58	0,60	0,82	1,85	1,13	
Average	**Red stars**	**1,76**	**1,76**	**1,58**	**0,91**	**0,57**	**0,49**	**0,67**	**1,54**	**0,95**	**0,58**
Pink Stars											
Albizia	Albizia adianthifolia										
Albizia	Albizia zygia										
Alstonia	Alstonia boonei										
Yaya	Amphimas pterocarpoides										
Kokote	Anopyxis klaineana										
Berlinia	Berlinia confusa										
Berlinia	Berlinia spp.										
Celtis/Essa	Celtis mildbraedii/zenkeri										
Pamprana	Coryanthe pachyceras										
Denya	Cylicodiscus gabonensis	1,62									
Ananta	Cynometra antana										
Duabankye	Dialium aubrevillei										
Potrodum	Erytrophleum sauaveolens										
Nakwa	Holoptelea grandis										
Kroma	Klaindoxa gabonensis										
Kumanini	Lannea welwitschii										
Bompagya	Mammea africana										
Wonton	Morus mesozygia										
Danta 2/	Nesogordonia papaverifera	1,31		1,86	1,13	0,63		0,48			
Akwana	Ongokea gore										
Afam	Parinari exelsa										
Asoma	Parkia bicolor										
Esia	Petersianthus macrocarpus										
Otie	Pycnanthus angolensis	0,64	0,86			0,38		0,32	1,65	0,77	0,50
Bombax	Rhodognaphalon/Bombax buonopozense										
Wama	Ricinodendron heudelotii										
Sterculia	Sterculia spp.										
Afena	Strombosia glaucescens										
Ofram 2a/	Terminalia superba	3,67	3,74	3,28	1,61	0,62	0,78	0,57	1,88	1,18	0,73
Tanuronini	Trichilia tessmannii										
Okure	Trilepisium madagascariense										
Avodire 1/	Turreanthus africanus	1,19	1,15	1,12	0,50	0,59					
Average	**Pink stars**	**1,69**	**1,92**	**2,09**	**1,08**	**0,56**	**0,78**	**0,46**	**1,76**	**0,98**	**0,62**
Subaha 2/, 4/	Hallea spp./Mitragyna spp.	1,73									
AVERAGE	**ALL SPECIES**	**1,67**	**1,61**	**1,47**	**0,85**	**0,54**	**0,55**	**0,64**	**1,41**	**0,81**	**0,50**

Notes:
1/ Log export banned in 1979
2/ Log export banned in 1988
2a/ Log export banned in 1988 but lifted in 1990
3/ Log export banned in 1993
4/ Harvest banned in 1993
Log exports was entirely banned in 1995
Sources:
-FAO Forestry Paper 104 (1992). For the period 1980-90 average prices from Central and West African countries are included where Ghanaian prices are few or absent (italicised and bold print in the table).
-Forestry Department Annual Reports (1971-85). Actual prices obtained during the period 1971-85 (normal print in the table).
-FPIB Export Permits (1986-94). Actual prices obtained during the period 1986-94 (normal print in the table).
-TEDB (1993). Fob log prices from West Africa in 1993 (italicised and bold print for 1993 in the table).
Forest Fees Regulations (1976)
Forest Fees (Amendment) Regulations (1983)
Forest Fees (Amendment) Regulations (1986)
Forest Fees (Amendment) Regulations (1989)
Forest Fees (Amendment) Regulations (1993)

Appendix 10

GHANA
ROYALTIES AS PER CENT OF AVAILABLE FOB LOG PRICES (1986-94)

SPECIES		Year								
Trade name	**Botanical name**	**1986**	**1987**	**1988**	**1989**	**1990**	**1991**	**1992**	**1993**	**1994**
Scarlet Stars										
Albizia	Albizia ferruginea	4,62		1,69	1,10	0,84	0,68	0,58	0,49	3,10
Asanfina 2/	Aningeria altissima/robusta	2,40	0,97	0,53					0,30	
Daniella	Daniella ogea/thurifera			0,57				0,93	0,63	4,33
Edinam 1/	Entandophragma angolense								0,73	
Sapele 1/	Entandophragma cylindricum	2,64	1,16	0,88	1,76	1,16			0,59	
Utile 1/	Entandophragma utile	2,56	1,11	0,83	1,65	1,15			0,65	
Hyedua 1/	Guibourtia ehie								0,65	
Mahogany 1/	Khaya anthotheca/grandifolia/ivorensis	2,39	1,07	0,90	1,93	1,31			0,85	
Odum 1/	Milicia excelsa/regia								0,72	
Kusia	Nauclea didichii	4,32	2,41	1,53	2,48	1,58	1,55	1,13	0,81	3,20
Afrormosia 1/	Pericopsis elata								1,16	
Kyere	Pterygota macrocarpa	2,87	1,14	0,75	1,42	1,03	0,85	0,79	0,47	2,61
Makore 1/	Tieghemella heckelii								0,73	
Wawa 3/	Triplochiton scleroxylon	1,86	0,81	0,58	1,51	1,12	0,92	0,83	0,63	
Average	**Scarlet stars**	**1,69**	**1,24**	**0,92**	**1,69**	**1,17**	**1,00**	**0,85**	**0,67**	**3,31**
Red Stars										
Afzelia/Papao	Afzelia africana/bellea	1,52	0,64	0,49	1,07	0,73	0,59		0,32	3,00
Kyenkyen	Antiaris toxicaria	1,78	0,82	0,51	0,97	0,68	0,57	0,51	0,34	4,38
Aprokuma	Antrocaryon micraster	1,37	0,59	0,61			1,15	0,94	0,72	4,16
Canarium	Canarium schweinfurthii		1,07				0,68		0,39	4,49
Ceiba	Ceiba pentandra	3,12	1,19	0,84	0,62	0,49	0,41	0,38	0,23	4,49
Akasa	Chrysophyllum spp.	1,24	0,32	0,21					0,19	
Bonsamdua/Ayan	Distemonanthus benthamianus	1,97	0,83	0,56	2,06	1,91	1,32	1,13	0,83	3,44
Candollei 1/	Entandophragma candollei	2,70	1,21	0,96	1,61	1,13			0,75	
Guarea	Guarea spp	3,17	1,32	0,93	1,89	1,28	1,25	1,05	0,80	3,22
Niangon 1/	Heritiera utilis	2,33	1,02	0,83	1,74	1,12			0,76	
Kaku/Ekki	Lophira alata	4,48	1,61	1,18	2,44	1,86	1,58	1,37	0,95	3,28
Walnut 1/	Lovoa trichilioides	2,23	1,01	0,76	2,98	2,03			1,11	
Mansonia 1/	Mansonia altissima								1,03	
Dahoma	Piptadenisatrum africanum				2,04	1,48			0,79	3,11
Bombax	Rhodognaphalon/Bombax brevicuspe	2,13	0,87	0,62	1,80	1,31	1,08	0,95	0,83	3,11
Emire 2/	Terminalia ivorensis	3,61	1,57	1,06	1,79	1,21			0,78	
Average	**Red stars**	**2,43**	**1,00**	**0,74**	**1,75**	**1,27**	**0,96**	**0,90**	**0,68**	**3,67**
Pink Stars										
Albizia	Albizia adianthifolia								0,51	
Albizia	Albizia zygia									
Alstonia	Alstonia boonei									
Yaya	Amphimas pterocarpoides			0,45					0,38	3,89
Kokote	Anopyxis klaineana								0,34	3,97
Berlinia	Berlinia confusa									
Berlinia	Berlinia spp.									
Celtis/Essa	Celtis mildbraedii/zenkeri	5,42	5,02	1,96	1,70	1,18	0,91	0,81	0,58	4,23
Pamprana	Coryanthe pachyceras									
Denya	Cylicodiscus gabonensis				0,94		0,57	0,53	0,38	2,99
Ananta	Cynometra antana				1,13				0,41	3,82
Duabankye	Dialium aubrevillei									
Potrodum	Erytrophleum sauaveolens			0,74	0,56			0,18	0,31	3,32
Nakwa	Holoptelea grandis								0,50	
Kroma	Klaindoxa gabonensis								0,98	4,31
Kumanini	Lannea welwitschii									
Bompagya	Mammea africana								0,32	
Wonton	Morus mesozygia								0,63	
Danta 2/	Nesogordonia papaverifera	2,91	1,18	0,76					0,45	
Akwana	Ongokea gore									
Afam	Parinari exelsa								0,53	
Asoma	Parkia bicolor								0,39	4,27
Esia	Petersianthus macrocarpus								0,42	3,90
Otie	Pycnanthus angolensis	2,25	1,05	0,77	0,84	0,64	0,53	0,45	0,33	4,16
Bombax	Rhodognaphalon/Bombax buonopozense									
Wama	Ricinodendron heudelotii								0,52	
Sterculia	Sterculia spp.								0,36	4,44
Afena	Strombosia glaucescens									
Ofram 2a/	Terminalia superba	4,63	1,92	1,27		1,11	0,94	0,85	0,68	4,11
Tanuronini	Trichilia tessmannii									
Okure	Trilepisium madagascariense									
Avodire 1/	Turreanthus africanus								0,77	
Average	**Pink stars**	**3,80**	**2,29**	**0,99**	**1,03**	**0,98**	**0,74**	**0,56**	**0,49**	**3,95**
Subaha 2/, 4/	Hallea spp./Mitragyna spp.	2,16	0,81	0,56	0,42	0,28				
AVERAGE	**ALL SPECIES**	**2,13**	**1,28**	**0,85**	**1,58**	**1,20**	**0,92**	**0,79**	**0,60**	**3,74**

Notes:
1/ Log export banned in 1979
2/ Log export banned in 1988
2a/ Log export banned in 1988 but lifted in 1990
3/ Log export banned in 1993
4/ Harvest banned in 1993
Log exports was entirely banned in 1995
Sources:
-FAO Forestry Paper 104 (1992). For the period 1980-90 average prices from Central and West African countries are included where Ghanaian prices are few or absent (italicised and bold print in the table).
-Forestry Department Annual Reports (1971-85). Actual prices obtained during the period 1971-85 (normal print in the table).
-FPIB Export Permits (1986-94). Actual prices obtained during the period 1986-94 (normal print in the table).
-TEDB (1993). Fob log prices from West Africa in 1993 (italicised and bold print for 1993 in the table).
Forest Fees Regulations (1976)
Forest Fees (Amendment) Regulations (1983)
Forest Fees (Amendment) Regulations (1986)
Forest Fees (Amendment) Regulations (1989)
Forest Fees (Amendment) Regulations (1993)

Appendix 10

GHANA
ROYALTIES AS PER CENT OF AVAILABLE AIR-DRIED, KILN-DRIED AND OVER-LAND FOB LUMBER EXPORT PRICES (1986-99)

SPECIES		Year													
Trade name	**Botanical name**	**1986**	**1987**	**1988**	**1989**	**1990**	**1991**	**1992**	**1993**	**1994**	**1995**	**1996**	**1997** 5/	**1998** 5/	**1999** 5/
Scarlet Stars															
Albizia	Albizia fuginea					0,30	0,29	0,27	0,19	1,58	1,09	0,87	0,82	0,64	3,45
Asanfina 2/	Aningeria altissima/robusta		0,45	0,26	0,67	0,52	0,36	0,33	0,27	2,55	2,13	1,42	1,34	1,05	6,93
Daniella	Daniella ogea/thurifera								0,20						
Edinam 1/	Entandophragma angolense	1,54	0,62	0,41	0,87	0,62	0,28	0,52	0,38	1,39	1,22	1,06	1,00	0,79	3,49
Sapele 1/	Entandophragma cylindricum	1,50	0,67	0,41	0,92	0,60	0,60	0,55	0,42	1,90	1,62	1,26	1,18	0,93	6,82
Utile 1/	Entandophragma utile	1,71	0,76	0,48	0,96	0,59	0,52	0,49	0,36	1,56	1,38	1,09	1,03	0,81	6,71
Hyedua 1/	Guibourtia ehie	2,23	0,93	0,64	1,02	0,62	0,44	0,44	0,37	1,66	1,34	1,15	1,08	0,85	6,77
Mahogany 1/	Khaya anthotheca/grandifolia/ivorensis	1,67	0,68	0,45	1,09	0,81	0,69	0,64	0,47	1,52	1,37	1,04	0,98	0,77	6,99
Odum 1/	Milicia excelsa/regia	1,94	0,80	0,56	1,09	0,57	0,56	0,50	0,38	1,19	1,05	0,94	0,89	0,70	6,69
Kusia	Nauclea didichii		1,07	0,78	1,03	0,62	0,49	0,48	0,39	1,28	1,05	0,78	0,74	0,58	6,57
Afrormosia 1/	Pericopsis elata	1,64	0,75	0,54	1,71	1,03	0,97	0,85	0,71	1,87	1,39	1,03	0,97	0,77	3,80
Kyere	Pterygota macrocarpa	1,15	0,72	0,44	1,04	0,66	0,32	0,30	0,22	1,61	1,16	0,82	0,77	0,61	2,49
Makore 1/	Tieghemella heckelii	2,17	0,88	0,58	0,84	0,59	0,48	0,45	0,34	1,36	1,15	0,84	0,79	0,62	6,97
Wawa 3/	Triplochiton scleroxylon	1,33	0,57	0,39	0,91	0,68	0,57	0,50	0,37	1,76	1,34	1,02	0,96	0,76	3,21
Average	**Scarlet stars**	**0,93**	**0,74**	**0,50**	**1,01**	**0,63**	**0,51**	**0,49**	**0,36**	**1,63**	**1,33**	**1,03**	**0,97**	**0,76**	**5,45**
Red Stars															
Afzelia/Papao	Afzelia africana/bellea	0,88	0,38	0,28	0,42	0,30	0,25	0,24	0,19	1,48	1,12	1,09	1,03	0,81	3,50
Kyenkyen	Antiaris toxicaria	0,78	0,33	0,25		0,23			0,16	6,56	1,04	3,26	3,07	2,42	1,90
Aprokuma	Antrocaryon micraster									10,21					
Canarium	Canarium schweinfurthii									1,75		0,64	0,61	0,48	3,50
Ceiba	Ceiba pentandra								0,09	4,06	4,84	1,65	1,56	1,22	7,62
Akasa	Chrysophyllum spp.			0,13	0,21	0,14	0,12	0,11	0,09	1,33	1,02	0,68	0,64	0,51	2,07
Bonsamdua/Ayan	Distemonanthus benthamianus	0,89	0,52	0,28	1,05	0,71	0,60	0,62	0,48	1,88	1,49	1,14	1,07	0,85	3,50
Candollei 1/	Entandophragma candollei	2,05	0,77	0,54	0,89	0,62	0,52	0,51	0,39	1,45	1,17	0,97	0,91	0,72	7,00
Guarea	Guarea spp	1,96	1,10	0,60	1,01	0,80	0,62	0,55	0,44	1,67	1,35	1,00	0,94	0,74	3,49
Niangon 1/	Heritiera utilis	2,40	0,78	0,61	0,94	0,61	0,52	0,44	0,33	1,16	0,93	0,80	0,76	0,59	3,50
Kaku/Ekki	Lophira alata						0,59		0,40	1,29	1,02	0,74	0,70	0,55	3,50
Walnut 1/	Lovoa trichilioides	1,39	0,62	0,36	1,45	1,08	0,88	0,80	0,59	1,59	1,35	1,14	1,07	0,84	6,97
Mansonia 1/	Mansonia altissima	2,11	0,86	0,64	1,49	1,06	0,70	0,76	0,58	1,68	1,41	1,11	1,04	0,82	3,50
Dahoma	Piptadenisatrum africanum	1,65	0,51	0,23	0,73	0,63	0,60	0,44	0,37	1,47	1,23	0,81	0,77	0,60	1,62
Bombax	Rhodognaphalon/Bombax brevicuspe											0,82	0,77	0,61	3,50
Emire 2/	Terminalia ivorensis	2,24	0,96	0,62	0,91	0,64	0,54	0,47	0,37	1,37	1,16	0,90	0,85	0,67	3,40
Average	**Red stars**	**1,64**	**0,68**	**0,41**	**0,91**	**0,62**	**0,54**	**0,49**	**0,35**	**2,60**	**1,47**	**1,12**	**1,05**	**0,83**	**3,90**
Pink Stars															
Albizia	Albizia adianthifolia														
Albizia	Albizia zygia														
Alstonia	Alstonia boonei														
Yaya	Amphimas pterocarpoides						0,23			1,55					
Kokote	Anopyxis klaineana														
Berlinia	Berlinia confusa														
Berlinia	Berlinia spp.														
Celtis/Essa	Celtis mildbraedii/zenkeri						0,30		0,41	1,29	1,00	0,69	0,65	0,51	1,44
Pamprana	Coryanthe pachyceras														
Denya	Cylicodiscus gabonensis		1,69					0,20	0,18	1,24	1,10	0,94	0,88	0,69	1,63
Ananta	Cynometra antana											1,10	1,04	0,82	1,75
Duabankye	Dialium aubrevillei														
Potrodum	Erytrophleum sauaveolens				0,32	0,32	0,27	0,22	0,16			0,99			
Nakwa	Holoptelea grandis														
Kroma	Klaindoxa gabonensis														
Kumanini	Lannea welwitschii														
Bompagya	Mammea africana														
Wonton	Morus mesozygia														
Danta 2/	Nesogordonia papaverifera	2,42	0,66	0,44	0,51	0,38	0,33	0,27	0,22	1,40	1,13	0,93	0,87	0,69	1,61
Akwana	Ongokea gore														
Afam	Parinari exelsa														
Asoma	Parkia bicolor														
Esia	Petersianthus macrocarpus								0,16			1,10	1,04	0,82	1,75
Otie	Pycnanthus angolensis	1,27			0,23	0,23	0,17	0,14	0,08	1,36	1,18	0,90	0,85	0,67	1,66
Bombax	Rhodognaphalon/Bombax buonopozense					0,52									
Wama	Ricinodendron heudelotii														
Sterculia	Sterculia spp.										1,27	0,95	0,90	0,71	1,75
Afena	Strombosia glaucescens														
Ofram 2a/	Terminalia superba	2,93	1,14	1,20	0,85	0,51	0,44	0,47	0,34	1,79	1,35	1,11	1,04	0,82	3,42
Tanuronini	Trichilia tessmannii														
Okure	Trilepisium madagascariense														
Avodire 1/	Turreanthus africanus	2,60	1,21	0,65	0,93	0,60	0,61	0,50	0,35	1,39	1,06	0,68	0,64	0,51	3,26
Average	**Pink stars**	**2,31**	**1,18**	**0,76**	**0,57**	**0,43**	**0,34**	**0,30**	**0,24**	**1,43**	**1,16**	**0,94**	**0,88**	**0,69**	**2,03**
Subaha 2/, 4/	Hallea spp./Mitragyna spp.	1,98	0,55	0,31	0,23	0,14	0,15								
AVERAGE	**ALL SPECIES**	**1,78**	**0,78**	**0,48**	**0,87**	**0,57**	**0,47**	**0,44**	**0,32**	**1,95**	**1,31**	**1,01**	**0,95**	**0,75**	**3,89**

Notes:
1/ Log export banned in 1979
2/ Log export banned in 1988
2a/ Log export banned in 1988 but lifted in 1990
3/ Log export banned in 1993
4/ Harvest banned in 1993
Log exports was entirely banned in 1995
5/ 1996 prices are used as a proxy due to data constraints

Sources:
FPIB Export Permits (1986-96). Actual volumes and prices obtained for air-dried, kiln-dried and over-land lumber exports are lumped together.
Forest Fees Regulations (1976)
Forest Fees (Amendment) Regulations (1983)
Forest Fees (Amendment) Regulations (1986)
Forest Fees (Amendment) Regulations (1989)
Forest Fees (Amendment) Regulations (1993)
Timber Resources Management Regulations (1998) (came into force on 16 December, 1998).

Appendix 11

GHANA
DOMESTIC AND FOB LOG PRICES

SPECIES Trade name	Botanical name	FOB log prices 1993 (DM/Cu.m.)	1993 Domestic log prices (Cedis/Cu.m.)	(DM/Cu.m.)	Per cent of 1993 FOB price
SCARLET STARS					
Albizia	Albizia ferruginea	190	35.000	95	50
Asanfina 2/	Aningeria altissima/robusta	650	65.000	176	27
Daniella	Daniella ogea/thurifera	168	30.000	81	48
Edinam 1/	Entandophragma angolense	300	65.000	176	59
Sapele 1/	Entandophragma cylindricum	512	90.000	244	48
Utile 1/	Entandophragma utile	546	90.000	244	45
Hyedua 1/	Guibourtia ehie	550			
Mahogany 1/	Khaya anthotheca/grandifolia/ivorensis	360	65.000	176	49
Odum 1/	Milicia excelsa/regia	325	80.000	217	67
Kusia	Nauclea diderrichii	195	30.000	81	42
Afrormosia 1/	Pericopsis elata	661	120.000	325	49
Kyere	Pterygota macrocarpa	320	55.000	149	47
Makore 1/	Tieghemella heckelii	329	90.000	244	74
Wawa 3/	Triplochiton scleroxylon	230	35.000	95	41
AVERAGE	**Scarlet Stars**	**381**	**65.385**	**177**	**46**
RED STARS					
Afzelia/Papao	Afzelia africana/bellea	400	65.000	176	44
Kyenkyen	Antiaris toxicaria	181	35.000	95	52
Aprokuma	Antrocaryon micraster	180	30.000	81	45
Canarium	Canarium schweinfurthii	190	35.000	95	50
Ceiba	Ceiba pentandra	178	30.000	81	46
Akasa	Chrysophyllum spp.	350	30.000	81	23
Bonsamdua/Ayan	Distemonanthus benthamianus	323	30.000	81	25
Candollei 1/	Entandophragma candollei	300	65.000	176	59
Guarea	Guarea spp	337	55.000	149	44
Niangon 1/	Heritiera utilis	320	70.000	190	59
Kaku/Ekki	Lophira alata	200	30.000	81	41
Walnut 1/	Lovoa trichilioides	354	55.000	149	42
Mansonia 1/	Mansonia altissima	447	55.000	149	33
Dahoma	Piptadenisatrum africanum	176	35.000	95	54
Bombax	Rhodognaphalon/Bombax brevicuspe				
Emire 2/	Terminalia ivorensis	320	60.000	163	51
AVERAGE	**Red Stars**	**284**	**45.333**	**123**	**43**
PINK STARS					
Albizia	Albizia adianthifolia	180	30.000	81	45
Albizia	Albizia zygia	180	30.000	81	45
Alstonia	Alstonia boonei	180	30.000	81	45
Yaya	Amphimas pterocarpoides	200	35.000	95	47
Kokote	Anopyxis klaineana	180	30.000	81	45
Berlinia	Berlinia confusa	180	30.000	81	45
Berlinia	Berlinia spp.	180	30.000	81	45
Celtis/Essa	Celtis mildbraedii/zenkeri	190	30.000	81	43
Pamprana	Coryanthe pachyceras	180	30.000	81	45
Denya	Cylicodiscus gabonensis	177	35.000	95	54
Ananta	Cynometra antana	166	30.000	81	49
Duabankye	Dialium aubrevillei	180	30.000	81	45
Potrodum	Erytrophleum sauaveolens	218	30.000	81	37
Nakwa	Holoptelea grandis	180	30.000	81	45
Kroma	Klaindoxa gabonensis	180	30.000	81	45
Kumanini	Lannea welwitschii	180	30.000	81	45
Bompagya	Mammea africana	180	30.000	81	45
Wonton	Morus mesozygia	135	30.000	81	60
Danta 2/	Nesogordonia papaverifera	280	30.000	81	29
Akwana	Ongokea gore	180	30.000	81	45
Afam	Parinari exelsa	180	30.000	81	45
Asoma	Parkia bicolor	174	30.000	81	47
Esia	Petersianthus macrocarpus	163	30.000	81	50
Otie	Pycnanthus angolensis	180	35.000	95	53
Bombax	Rhodognaphalon/Bombax buonopozense	200	30.000	81	41
Wama	Ricinodendron heudelotii	180	30.000	81	45
Sterculia	Sterculia spp.	196	30.000	81	41
Afena	Strombosia glaucescens	180	30.000	81	45
Ofram 2a/	Terminalia superba	220	35.000	95	43
Tanuronini	Trichilia tessmannii	180	30.000	81	45
Okure	Trilepisium madagascariense	180	30.000	81	45
Avodire 1/	Turreanthus africanus	280	45.000	122	44
AVERAGE	**Pink Stars**	**188**	**31.094**	**84**	**45**
Subaha 2/, 4/	Hallea spp./Mitragyna spp.				
Other species		180	30.000	81	45

Notes:
1/ Log export banned in 1979
2/ Log export banned in 1988
2a/ Log export banned in 1988 but lifted in 1990
3/ Log export banned in 1993
4/ Harvest banned in 1993
Log exports was entirely banned in 1995

Sources:
FPIB Export permits (1993) [FOB log prices, 1993]
TEDB (1993) [FOB log prices from West Africa, 1993 for species which may not be exported from Ghana in log form]
GTMO (1994) [domestic log prices, 1993]
Treue and Armstrong-Mensah (1995) [domestic log prices, 1994]
Log sales contracts from 14 companies obtained by T.Treue [domestic log prices, 1996]

Appendix 11

GHANA
DOMESTIC AND FOB LOG PRICES

SPECIES Trade name	Botanical name	FOB log prices 1993 (DM/Cu.m.)	1994 Domestic log prices (Cedis/Cu.m.)	(DM/Cu.m.)	Per cent of 1993 FOB price
SCARLET STARS					
Albizia	Albizia ferruginea	190	30.000	49	26
Asanfina 2/	Aningeria altissima/robusta	650	65.000	106	16
Daniella	Daniella ogea/thurifera	168			
Edinam 1/	Entandophragma angolense	300	80.000	130	43
Sapele 1/	Entandophragma cylindricum	512	100.000	163	32
Utile 1/	Entandophragma utile	546	110.000	179	33
Hyedua 1/	Guibourtia ehie	550	60.000	98	18
Mahogany 1/	Khaya anthotheca/grandifolia/ivorensis	360	70.000	114	32
Odum 1/	Milicia excelsa/regia	325	80.000	130	40
Kusia	Nauclea diderrichii	195	30.000	49	25
Afrormosia 1/	Pericopsis elata	661	120.000	195	30
Kyere	Pterygota macrocarpa	320	60.000	98	30
Makore 1/	Tieghemella heckelii	329	70.000	114	35
Wawa 3/	Triplochiton scleroxylon	230	35.000	57	25
AVERAGE	**Scarlet Stars**	**381**	**70.000**	**114**	**30**
RED STARS					
Afzelia/Papao	Afzelia africana/bellea	400	60.000	98	24
Kyenkyen	Antiaris toxicaria	181	30.000	49	27
Aprokuma	Antrocaryon micraster	180			
Canarium	Canarium schweinfurthii	190			
Ceiba	Ceiba pentandra	178	25.000	41	23
Akasa	Chrysophyllum spp.	350	30.000	49	14
Bonsamdua/Ayan	Distemonanthus benthamianus	323	30.000	49	15
Candollei 1/	Entandophragma candollei	300	70.000	114	38
Guarea	Guarea spp	337	35.000	57	17
Niangon 1/	Heritiera utilis	320	60.000	98	30
Kaku/Ekki	Lophira alata	200			
Walnut 1/	Lovoa trichilioides	354	60.000	98	28
Mansonia 1/	Mansonia altissima	447	35.000	57	13
Dahoma	Piptadenisatrum africanum	176			
Bombax	Rhodognaphalon/Bombax brevicuspe				
Emire 2/	Terminalia ivorensis	320	35.000	57	18
AVERAGE	**Red Stars**	**284**	**42.727**	**69**	**24**
PINK STARS					
Albizia	Albizia adianthifolia	180			
Albizia	Albizia zygia	180			
Alstonia	Alstonia boonei	180			
Yaya	Amphimas pterocarpoides	200			
Kokote	Anopyxis klaineana	180			
Berlinia	Berlinia confusa	180			
Berlinia	Berlinia spp.	180			
Celtis/Essa	Celtis mildbraedii/zenkeri	190	25.000	41	21
Pamprana	Coryanthe pachyceras	180			
Denya	Cylicodiscus gabonensis	177			
Ananta	Cynometra antana	166			
Duabankye	Dialium aubrevillei	180			
Potrodum	Erytrophleum sauaveolens	218	35.000	57	26
Nakwa	Holoptelea grandis	180			
Kroma	Klaindoxa gabonensis	180			
Kumanini	Lannea welwitschii	180			
Bompagya	Mammea africana	180			
Wonton	Morus mesozygia	135			
Danta 2/	Nesogordonia papaverifera	280	30.000	49	17
Akwana	Ongokea gore	180			
Afam	Parinari exelsa	180			
Asoma	Parkia bicolor	174			
Esia	Petersianthus macrocarpus	163			
Otie	Pycnanthus angolensis	180			
Bombax	Rhodognaphalon/Bombax buonopozense	200			
Wama	Ricinodendron heudelotii	180			
Sterculia	Sterculia spp.	196			
Afena	Strombosia glaucescens	180			
Ofram 2a/	Terminalia superba	220			
Tanuronini	Trichilia tessmannii	180			
Okure	Trilepisium madagascariense	180			
Avodire 1/	Turreanthus africanus	280	50.000	81	29
AVERAGE	**Pink Stars**	**188**	**35.000**	**57**	**30**
Subaha 2/, 4/	Hallea spp./Mitragyna spp.				
Other species		180			

Notes:
1/ Log export banned in 1979
2/ Log export banned in 1988
2a/ Log export banned in 1988 but lifted in 1990
3/ Log export banned in 1993
4/ Harvest banned in 1993
Log exports was entirely banned in 1995

Sources:
FPIB Export permits (1993) [FOB log prices, 1993]
TEDB (1993) [FOB log prices from West Africa, 1993 for species which may not be exported from Ghana in log form]
GTMO (1994) [domestic log prices, 1993]
Treue and Armstrong-Mensah (1995) [domestic log prices, 1994]
Log sales contracts from 14 companies obtained by T.Treue [domestic log prices, 1996]

Appendix 11

GHANA
DOMESTIC AND FOB LOG PRICES

SPECIES Trade name	Botanical name	FOB log prices 1993 (DM/Cu.m.)	1996 Domestic log prices (Cedis/Cu.m.)	(DM/Cu.m.)	Per cent of 1993 FOB price
SCARLET STARS					
Albizia	Albizia ferruginea	190	40.000	38	20
Asanfina 2/	Aningeria altissima/robusta	650	110.000	103	16
Daniella	Daniella ogea/thurifera	168			
Edinam 1/	Entandophragma angolense	300			
Sapele 1/	Entandophragma cylindricum	512	200.000	188	37
Utile 1/	Entandophragma utile	546	200.000	188	34
Hyedua 1/	Guibourtia ehie	550			
Mahogany 1/	Khaya anthotheca/grandifolia/ivorensis	360	80.000	75	21
Odum 1/	Milicia excelsa/regia	325	170.000	160	49
Kusia	Nauclea diderrichii	195	45.000	42	22
Afrormosia 1/	Pericopsis elata	661			
Kyere	Pterygota macrocarpa	320	90.000	85	26
Makore 1/	Tieghemella heckelii	329	100.000	94	29
Wawa 3/	Triplochiton scleroxylon	230	60.000	56	24
AVERAGE	**Scarlet Stars**	**381**	**109.500**	**103**	**27**
RED STARS					
Afzelia/Papao	Afzelia africana/bellea	400			
Kyenkyen	Antiaris toxicaria	181	45.000	42	23
Aprokuma	Antrocaryon micraster	180			
Canarium	Canarium schweinfurthii	190			
Ceiba	Ceiba pentandra	178	45.000	42	24
Akasa	Chrysophyllum spp.	350	45.000	42	12
Bonsamdua/Ayan	Distemonanthus benthamianus	323	40.000	38	12
Candollei 1/	Entandophragma candollei	300			
Guarea	Guarea spp	337	60.000	56	17
Niangon 1/	Heritiera utilis	320	100.000	94	29
Kaku/Ekki	Lophira alata	200			
Walnut 1/	Lovoa trichilioides	354			
Mansonia 1/	Mansonia altissima	447	45.000	42	9
Dahoma	Piptadenisatrum africanum	176			
Bombax	Rhodognaphalon/Bombax brevicuspe				
Emire 2/	Terminalia ivorensis	320	50.000	47	15
AVERAGE	**Red Stars**	**284**	**53.750**	**50**	**18**
PINK STARS					
Albizia	Albizia adianthifolia	180			
Albizia	Albizia zygia	180			
Alstonia	Alstonia boonei	180			
Yaya	Amphimas pterocarpoides	200			
Kokote	Anopyxis klaineana	180			
Berlinia	Berlinia confusa	180			
Berlinia	Berlinia spp.	180			
Celtis/Essa	Celtis mildbraedii/zenkeri	190			
Pamprana	Coryanthe pachyceras	180			
Denya	Cylicodiscus gabonensis	177			
Ananta	Cynometra antana	166			
Duabankye	Dialium aubrevillei	180			
Potrodum	Erytrophleum sauaveolens	218			
Nakwa	Holoptelea grandis	180			
Kroma	Klaindoxa gabonensis	180			
Kumanini	Lannea welwitschii	180			
Bompagya	Mammea africana	180			
Wonton	Morus mesozygia	135			
Danta 2/	Nesogordonia papaverifera	280	45.000	42	15
Akwana	Ongokea gore	180			
Afam	Parinari exelsa	180			
Asoma	Parkia bicolor	174			
Esia	Petersianthus macrocarpus	163			
Otie	Pycnanthus angolensis	180			
Bombax	Rhodognaphalon/Bombax buonopozense	200			
Wama	Ricinodendron heudelotii	180			
Sterculia	Sterculia spp.	196			
Afena	Strombosia glaucescens	180			
Ofram 2a/	Terminalia superba	220	70.000	66	30
Tanuronini	Trichilia tessmannii	180			
Okure	Trilepisium madagascariense	180			
Avodire 1/	Turreanthus africanus	280			
AVERAGE	**Pink Stars**	**188**	**57.500**	**54**	**29**
Subaha 2/, 4/	Hallea spp./Mitragyna spp.				
Other species		180			

Notes:
1/ Log export banned in 1979
2/ Log export banned in 1988
2a/ Log export banned in 1988 but lifted in 1990
3/ Log export banned in 1993
4/ Harvest banned in 1993
Log exports was entirely banned in 1995

Sources:
FPIB Export permits (1993) [FOB log prices, 1993]
TEDB (1993) [FOB log prices from West Africa, 1993 for species which may not be exported from Ghana in log form]
GTMO (1994) [domestic log prices, 1993]
Treue and Armstrong-Mensah (1995) [domestic log prices, 1994]
Log sales contracts from 14 companies obtained by T.Treue [domestic log prices, 1996]

Appendix 12

GHANA

OBSERVED WILLINGNESS TO PAY FOR STANDING TIMBER PER Cu.m.

SPECIES		Year									
Trade name	Botanical name	1986		1991		1993		1994		1996	
SCARLET STARS		(Cedis)	(DM)	(Cedis)	(DM)	(Cedis)	(DM)	(Cedis)	(DM)	(Cedis)	(DM)
Albizia	Albizia ferruginea	550	14			32.500	88	32.500	53	55.000	52
Asanfina 2/	Aningeria altissima/robusta	813	20	6.000	25	15.000	41				
Daniella	Daniella ogea/thurifera	550	14			32.500	88	40.000	65		
Edinam 1/	Entandophragma angolense	1.500	38	6.000	25	45.000	122	50.000	81	100.000	94
Sapele 1/	Entandophragma cylindricum	1.500	38	8.000	33	45.000	122	55.000	89	100.000	94
Utile 1/	Entandophragma utile	1.500	38	8.000	33			30.000	49		
Hyedua 1/	Guibourtia ehie	550	14	4.000	16	32.500	88	35.000	57	40.000	38
Mahogany 1/	Khaya anthotheca/grandifolia/ivorensis	1.500	38	6.000	25	40.000	108	40.000	65	85.000	80
Odum 1/	Milicia excelsa/regia	950	24	4.000	16	15.000	41	15.000	24	22.500	21
Kusia	Nauclea diderrichii	550	14			60.000	163	60.000	98		
Afrormosia 1/	Pericopsis elata			8.000	33	27.500	75	30.000	49	45.000	42
Kyere	Pterygota macrocarpa	550	14			45.000	122	35.000	57	50.000	47
Makore 1/	Tieghemella heckelii	1.500	38	6.000	25	17.500	47	17.500	28	30.000	28
Wawa 3/	Triplochiton scleroxylon	663	17	3.000	12	32.692	89	35.000	57	54.750	51
AVERAGE	**Scarlet Stars**	**975**	**25**	**5.900**	**24**	**33.861**	**92**	**36.538**	**59**	**58.225**	**55**
RED STARS											
Afzelia/Papao	Afzelia africana/bellea	550	14	3.000	12	17.500	47	15.000	24	22.500	21
Kyenkyen	Antiaris toxicaria			3.000	12	15.000	41				
Aprokuma	Antrocaryon micraster					17.500	47				
Canarium	Canarium schweinfurthii					15.000	41	12.500	20	22.500	21
Ceiba	Ceiba pentandra					15.000	41	15.000	24	22.500	21
Akasa	Chrysophyllum spp.					15.000	41	15.000	24	20.000	19
Bonsamdua/Ayan	Distemonanthus benthamianus					32.500	88	35.000	57		
Candollei 1/	Entandophragma candollei	1.375	35	6.000	25	27.500	75	17.500	28	30.000	28
Guarea	Guarea spp	813	20	4.000	16	35.000	95	30.000	49	50.000	47
Niangon 1/	Heritiera utilis	550	14			15.000	41				
Kaku/Ekki	Lophira alata					27.500	75	30.000	49		
Walnut 1/	Lovoa trichilioides					27.500	75	17.500	28	22.500	21
Mansonia 1/	Mansonia altissima	550	14	6.000	25	17.500	47				
Dahoma	Piptadenisatrum africanum	813	20	3.000	12						
Bombax	Rhodognaphalon/Bombax brevicuspe					30.000	81	17.500	28	25.000	23
Emire 2/	Terminalia ivorensis	550	14	4.000	16	22.667	61	21.364	35	26.875	25
AVERAGE	**Red Stars**	**743**	**19**	**4.143**	**17**	**22.011**	**60**	**20.579**	**33**	**26.875**	**25**
PINK STARS											
Albizia	Albizia adianthifolia	550	14			15.000	41				
Albizia	Albizia zygia	550	14			15.000	41				
Alstonia	Alstonia boonei					17.500	47				
Yaya	Amphimas pterocarpoides					15.000	41				
Kokote	Anopyxis klaineana					15.000	41				
Berlinia	Berlinia confusa					15.000	41				
Berlinia	Berlinia spp.					15.000	41	12.500	20		
Celtis/Essa	Celtis mildbraedii/zenkeri					15.000	41				
Pamprana	Coryanthe pachyceras					17.500	47				
Denya	Cylicodiscus gabonensis					15.000	41				
Ananta	Cynometra antana					15.000	41				
Duabankye	Dialium aubrevillei					15.000	41	17.500	28		
Potrodum	Erytrophleum sauaveolens					15.000	41				
Nakwa	Holoptelea grandis					15.000	41				
Kroma	Klaindoxa gabonensis					15.000	41				
Kumanini	Lannea welwitschii					15.000	41				
Bompagya	Mammea africana					15.000	41				
Wonton	Morus mesozygia					15.000	41	15.000	24	22.500	21
Danta 2/	Nesogordonia papaverifera	813	20	4.000	16	15.000	41				
Akwana	Ongokea gore					15.000	41				
Afam	Parinari exelsa					15.000	41				
Asoma	Parkia bicolor					15.000	41				
Esia	Petersianthus macrocarpus					17.500	47				
Otie	Pycnanthus angolensis			3.000	12	15.000	41				
Bombax	Rhodognaphalon/Bombax buonopozense					15.000	41				
Wama	Ricinodendron heudelotii					15.000	41				
Sterculia	Sterculia spp.					15.000	41				
Afena	Strombosia glaucescens					17.500	47			35.000	33
Ofram 2a/	Terminalia superba	813	20	4.000	16	15.000	41				
Tanuronini	Trichilia tessmannii					15.000	41				
Okure	Trilepisium madagascariense					22.500	61	25.000	41		
Avodire 1/	Turreanthus africanus	550	14	3.000	12	15.547	42	17.500	28	28.750	27
AVERAGE	**Pink Stars**	**655**	**16**	**3.500**	**14**	**15.564**	**42**	**17.500**	**28**	**28.750**	**27**
Subaha 2/, 4/	Hallea spp./Mitragyna spp.										
Other species											

Notes:
1/ Log export banned in 1979
2/ Log export banned in 1988
2a/ Log export banned in 1988 but lifted in 1990
3/ Log export banned in 1993
4/ Harvest banned in 1993
Log exports was entirely banned in 1995

Sources:
World Bank (1986, Working Paper V p.26) [1986 figures for willingness to pay]
Temporary felling permit to MIM Timber (1991) [1991 figures for willingness to pay]
GTMO (1994) [domestic log prices, 1993]
Treue and Armstrong-Mensah (1995) [domestic log prices, 1994]
Log sales contracts from 14 companies obtained by T.Treue [domestic log prices, 1996]
T. Treue interviews with logging companies. Prices for st. timber is 50% of the domestic log prices (the so-called fifty-fifty arrangement) [1993-1996 figures for willingness to pay]

Appendix 12

GHANA
OBSERVED WILLINGNESS TO PAY FOR STANDING TIMBER COMPARED WITH ACTUAL FOREST FEES (DM/Cu.m.); 1986, 1991, 1993

		1986				1991				1993			
SPECIES		Observed willingness to pay	Royalty rate	Other forest fees A/	Diff.	Observed willingness to pay	Royalty rate	Other forest fees A/	Diff.	Observed willingness to pay	Royalty rate	Other forest fees A/	Diff.
Trade name	**Botanical name**												
SCARLET STARS													
Albizia	Albizia ferruginea	13,84	8,98	0,48	4,37		1,34	0,05		88,06	0,89	0,02	87,15
Asanfina 2/	Aningeria altissima/robusta	20,45	6,62	0,48	13,35	24,53	2,97	0,05	21,50	40,65	1,97	0,02	38,66
Daniella	Daniella ogea/thurifera	13,84	2,52	0,48	10,84		1,59	0,05		88,06	1,06	0,02	86,99
Edinam 1/	Entandophragma angolense	37,73	7,55	0,48	29,70	24,53	3,32	0,05	21,15	121,94	2,20	0,02	119,71
Sapele 1/	Entandophragma cylindricum	37,73	10,06	0,48	27,19	32,70	4,52	0,05	28,13	121,94	3,00	0,02	118,92
Utile 1/	Entandophragma utile	37,73	12,58	0,48	24,67	32,70	5,36	0,05	27,29		3,55	0,02	-3,57
Hyedua 1/	Guibourtia ehie	13,84	16,28	0,48	-2,92	16,35	5,39	0,05	10,91		3,57	0,02	
Mahogany 1/	Khaya anthotheca/grandifolia/ivorensis	37,73	9,22	0,48	28,03	24,53	4,61	0,05	19,87	108,39	3,05	0,02	105,31
Odum 1/	Milicia excelsa/regia	23,90	8,11	0,48	15,30	16,35	3,53	0,05	12,76	40,65	2,34	0,02	38,28
Kusia	Nauclea diderrichii	13,84	8,80	0,48	4,55		2,49	0,05		162,58	1,65	0,02	160,91
Afrormosia 1/	Pericopsis elata		13,32	0,48		32,70	11,54	0,05	21,11	74,52	7,65	0,02	66,84
Kyere	Pterygota macrocarpa	13,84	7,55	0,48	5,81		2,90	0,05		121,94	1,92	0,02	119,99
Makore 1/	Tieghemella heckelii	37,73	11,32	0,48	25,93	24,53	3,62	0,05	20,86	47,42	2,40	0,02	45,00
Wawa 3/	Triplochiton scleroxylon	16,68	3,87	0,48	12,33	12,26	2,01	0,05	10,20	88,59	1,33	0,02	87,23
AVERAGE	**Scarlet Stars**	**24,53**	**9,07**	**0,48**	**14,98**	**24,12**	**3,99**	**0,05**	**20,08**	**91,75**	**2,64**	**0,02**	**89,09**
RED STARS													
Afzelia/Papao	Afzelia africana/bellea	13,84	3,59	0,48	9,76	12,26	2,04	0,05	10,17	47,42	1,35	0,02	46,04
Kyenkyen	Antiaris toxicaria		2,52	0,48		12,26	0,94	0,05	11,27	40,65	0,62	0,02	40,00
Aprokuma	Antrocaryon micraster		3,14	0,48			2,04	0,05		47,42	1,35	0,02	46,04
Canarium	Canarium schweinfurthii		3,77	0,48			1,02	0,05		40,65	0,68	0,02	39,95
Ceiba	Ceiba pentandra		3,77	0,48			0,61	0,05		40,65	0,41	0,02	40,22
Akasa	Chrysophyllum spp.		2,52	0,48			1,02	0,05		40,65	0,68	0,02	39,95
Bonsamdua/Ayan	Distemonanthus benthamianus		5,39	0,48			4,03	0,05		88,06	2,67	0,02	85,37
Candollei 1/	Entandophragma candollei	34,59	10,06	0,48	24,04	24,53	3,41	0,05	21,07	74,52	2,26	0,02	72,24
Guarea	Guarea spp	20,45	9,43	0,48	10,54	16,35	3,88	0,05	12,41	94,84	2,57	0,02	92,24
Niangon 1/	Heritiera utilis	13,84	8,98	0,48	4,37		3,68	0,05		40,65	2,44	0,02	38,19
Kaku/Ekki	Lophira alata		6,99	0,48			2,77	0,05		74,52	1,84	0,02	72,66
Walnut 1/	Lovoa trichilioides		7,55	0,48			5,93	0,05		74,52	3,93	0,02	70,57
Mansonia 1/	Mansonia altissima	13,84	12,58	0,48	0,78	24,53	6,95	0,05	17,52	47,42	4,61	0,02	42,79
Dahoma	Piptadenisatrum africanum	20,45	2,70	0,48	17,27	12,26	2,10	0,05	10,11		1,39	0,02	-1,41
Bombax	Rhodognaphalon/Bombax brevicuspe		3,59	0,48			2,14	0,05			1,42	0,02	
Emire 2/	Terminalia ivorensis	13,84	11,01	0,48	2,35	16,35	3,78	0,05	12,52	61,42	2,51	0,02	58,89
AVERAGE	**Red Stars**	**18,69**	**6,10**	**0,48**	**12,11**	**16,94**	**2,90**	**0,05**	**13,98**	**59,64**	**1,92**	**0,02**	**57,70**
PINK STARS													
Albizia	Albizia adianthifolia	13,84	7,19	0,48	6,17		1,46	0,05		40,65	0,97	0,02	39,66
Albizia	Albizia zygia	13,84	7,19	0,48	6,17		1,46	0,05		40,65	0,97	0,02	39,66
Alstonia	Alstonia boonei		2,52	0,48			1,02	0,05		47,42	0,68	0,02	46,72
Yaya	Amphimas pterocarpoides		2,29	0,48			0,93	0,05		40,65	0,62	0,02	40,01
Kokote	Anopyxis klaineana		6,29	0,48			1,02	0,05		40,65	0,68	0,02	39,95
Berlinia	Berlinia confusa		2,52	0,48			1,02	0,05		40,65	0,68	0,02	39,95
Berlinia	Berlinia spp.		2,52	0,48			1,02	0,05		40,65	0,68	0,02	39,95
Celtis/Essa	Celtis mildbraedii/zenkeri		9,43	0,48			1,74	0,05		40,65	1,15	0,02	39,47
Pamprana	Coryanthe pachyceras		2,52	0,48			1,02	0,05		47,42	0,68	0,02	46,72
Denya	Cylicodiscus gabonensis		7,55	0,48			1,02	0,05		40,65	0,68	0,02	39,95
Ananta	Cynometra antana		2,52	0,48			1,02	0,05		40,65	0,68	0,02	39,95
Duabankye	Dialium aubrevillei		2,52	0,48			1,02	0,05		40,65	0,68	0,02	39,95
Potrodum	Erytrophleum sauaveolens		3,77	0,48			1,02	0,05		40,65	0,68	0,02	39,95
Nakwa	Holoptelea grandis		2,52	0,48			1,02	0,05		40,65	0,68	0,02	39,95
Kroma	Klaindoxa gabonensis		2,52	0,48			1,02	0,05		40,65	0,68	0,02	39,95
Kumanini	Lannea welwitschii		2,52	0,48			1,02	0,05		40,65	0,68	0,02	39,95
Bompagya	Mammea africana		2,52	0,48			1,02	0,05		40,65	0,68	0,02	39,95
Wonton	Morus mesozygia		4,72	0,48			1,28	0,05		40,65	0,85	0,02	39,78
Danta 2/	Nesogordonia papaverifera	20,45	7,55	0,48	12,42	16,35	1,88	0,05	14,42	40,65	1,25	0,02	39,38
Akwana	Ongokea gore		2,52	0,48			1,02	0,05		40,65	0,68	0,02	39,95
Afam	Parinari exelsa		2,52	0,48			1,02	0,05		40,65	0,68	0,02	39,95
Asoma	Parkia bicolor		5,03	0,48			1,02	0,05		40,65	0,68	0,02	39,95
Esia	Petersianthus macrocarpus		2,52	0,48			1,02	0,05		47,42	0,68	0,02	46,72
Otie	Pycnanthus angolensis		3,77	0,48		12,26	0,94	0,05	11,27	40,65	0,62	0,02	40,00
Bombax	Rhodognaphalon/Bombax buonopozense		3,59	0,48			2,14	0,05		40,65	1,42	0,02	39,20
Wama	Ricinodendron heudelotii		2,52	0,48			1,02	0,05		40,65	0,68	0,02	39,95
Sterculia	Sterculia spp.		2,52	0,48			1,02	0,05		40,65	0,68	0,02	39,95
Afena	Strombosia glaucescens		2,52	0,48			1,02	0,05		47,42	0,68	0,02	46,72
Ofram 2a/	Terminalia superba	20,45	8,39	0,48	11,58	16,35	1,91	0,05	14,39	40,65	1,26	0,02	39,36
Tanuronini	Trichilia tessmannii		2,52	0,48			1,02	0,05		40,65	0,68	0,02	39,95
Okure	Trilepisium madagascariense		2,52	0,48			1,02	0,05		60,97	0,68	0,02	60,27
Avodire 1/	Turreanthus africanus	13,84	9,43	0,48	3,92	12,26	3,27	0,05	8,94	42,13	2,17	0,02	39,94
AVERAGE	**Pink Stars**	**16,48**	**4,11**	**0,48**	**11,89**	**14,31**	**1,23**	**0,05**	**13,02**	**42,17**	**0,82**	**0,02**	**41,33**
Subaha 2/, 4/	Hallea spp./Mitragyna spp.		5,03				0,82						

Other species

Notes:
1/ Log export banned in 1979
2/ Log export banned in 1988
2a/ Log export banned in 1988 but lifted in 1990
3/ Log export banned in 1993
4/ Harvest banned in 1993
Log exports was entirely banned in 1995
A/ Other forest fees include
- Concession rent/Contract area rent
- Silvicultural fee (on-reserves)
- Silvicultural fee (off-reserves)
- Property mark
- Conveyance Certificates

Sources:
World Bank (1986, Working Paper V p.26) [1986 figures for willingness to pay]
Temporary felling permit to MIM Timber (1991) [1991 figures for willingness to pay]
GTMO (1994) [domestic log prices, 1993]
Treue and Armstrong-Mensah (1995) [domestic log prices, 1994]
Log sales contracts from 14 companies obtained by T.Treue [domestic log prices, 1996]
T. Treue Interviews with logging companies. Prices for st. timber is 50% of the domestic log prices (the so-called fifty-fifty arrangement) [1993-1996 figures for willingness to pay]
Forest Fees Regulations (1976)
Forest Fees (Amendment) Regulations (1983)
Forest Fees (Amendment) Regulations (1986)
Forest Fees (Amendment) Regulations (1989)
Forest Fees (Amendment) Regulations (1993)

Appendix 12

GHANA
OBSERVED WILLINGNESS TO PAY FOR STANDING TIMBER COMPARED WITH ACTUAL FOREST FEES (DM/Cu.m.); 1994, 1996, 1999

SPECIES		Year 1994				1996				1999			
Trade name	Botanical name	Observed willingness to pay	Royalty rate	Other forest fees A/	Diff.	Observed willingness to pay	Royalty rate	Other forest fees A/	Diff.	Observed willingness to pay 5/	Stumpage fee	Other forest fees A/	Diff.
SCARLET STARS													
Albizia	Albizia ferruginea	52,85	6,63	0,02	46,19	51,64	3,83	5,58	42,23	51,64	15,28	2,95	33,42
Asanfina 2/	Aningeria altissima/robusta		21,14	0,02	-21,15		12,21	5,58	-17,78		59,64	2,95	-62,58
Daniella	Daniella ogea/thurifera		7,49	0,02			4,33	5,58				2,95	-2,95
Edinam 1/	Entandophragma angolense	81,30	9,76	0,02	71,53	93,90	5,63	5,58			18,53	2,95	
Sapele 1/	Entandophragma cylindricum	89,43	16,59	0,02	72,83	93,90	9,58	5,58	78,74	93,90	51,94	2,95	39,01
Utile 1/	Entandophragma utile	48,78	17,76	0,02	31,01		10,25	5,58	-15,83		62,96	2,95	-65,91
Hyedua 1/	Guibourtia ehie	56,91	17,89	0,02	39,01	37,56	10,33	5,58			60,72	2,95	
Mahogany 1/	Khaya anthotheca/grandifolia/ivorensis	65,04	11,71	0,02	53,32	79,81	6,76	5,58	67,47	79,81	45,40	2,95	31,46
Odum 1/	Milicia excelsa/regia	24,39	10,57	0,02	13,80	21,13	6,10	5,58	9,45	21,13	43,28	2,95	-25,10
Kusia	Nauclea diderrichii	97,56	6,63	0,02	90,91		3,83	5,58	-9,41		32,16	2,95	-35,11
Afrormosia 1/	Pericopsis elata	48,78	21,50	0,02	27,27	42,25	12,41	5,58				2,95	-2,95
Kyere	Pterygota macrocarpa	56,91	13,01	0,02	43,89	46,95	7,51	5,58	33,86	46,95		2,95	
Makore 1/	Tieghemella heckelii	28,46	10,70	0,02	17,74	28,17	6,18	5,58	16,41	28,17	51,34	2,95	-26,12
Wawa 3/	Triplochiton scleroxylon	56,91	7,48	0,02	49,41	51,41	4,32	5,58	41,51	51,41	13,59	2,95	34,87
AVERAGE	**Scarlet Stars**	**59,41**	**12,70**	**0,02**	**46,69**	**54,67**	**7,34**	**5,58**	**41,76**	**54,67**	**41,69**	**2,95**	**10,04**
RED STARS													
Afzelia/Papao	Afzelia africana/bellea	24,39	13,01	0,02	11,37	21,13	7,51	5,58			24,01	2,95	
Kyenkyen	Antiaris toxicaria		7,49	0,02	-7,51		4,33	5,58	-9,90		2,52	2,95	-5,47
Aprokuma	Antrocaryon micraster		7,49	0,02			4,33	5,58				2,95	-2,95
Canarium	Canarium schweinfurthii		7,49	0,02		21,13	4,33	5,58			23,52	2,95	
Ceiba	Ceiba pentandra	24,39	7,49	0,02	16,88	21,13	4,33	5,58	11,22	21,13		2,95	
Akasa	Chrysophyllum spp.	24,39	11,38	0,02	12,99	18,78	6,57	5,58	6,63	18,78		2,95	
Bonsamdua/Ayan	Distemonanthus benthamianus	56,91	11,06	0,02	45,84		6,38	5,58	-11,96		19,59	2,95	-22,54
Candollei 1/	Entandophragma candollei	28,46	9,76	0,02	18,68	28,17	5,63	5,58			40,82	2,95	
Guarea	Guarea spp	48,78	10,96	0,02	37,80	46,95	6,33	5,58	35,04	46,95	22,07	2,95	21,93
Niangon 1/	Heritiera utilis		10,41	0,02	-10,42		6,01	5,58	-11,59		26,22	2,95	-29,17
Kaku/Ekki	Lophira alata	48,78	6,63	0,02			3,83	5,58			18,07	2,95	
Walnut 1/	Lovoa trichilioides	28,46	11,51	0,02	16,93	21,13	6,65	5,58			40,71	2,95	
Mansonia 1/	Mansonia altissima		14,54	0,02	-14,55		8,39	5,58	-13,97		26,50	2,95	-29,45
Dahoma	Piptadenisatrum africanum		6,50	0,02			3,76	5,58			7,47	2,95	
Bombax	Rhodognaphalon/Bombax brevicuspe	28,46	6,50	0,02		23,47	3,76	5,58			16,10	2,95	
Emire 2/	Terminalia ivorensis	34,74	10,41	0,02	24,31	25,23	6,01	5,58	13,65	25,23	22,70	2,95	-0,41
AVERAGE	**Red Stars**	**33,46**	**9,54**	**0,02**	**23,90**	**25,23**	**5,51**	**5,58**	**14,15**	**25,23**	**22,33**	**2,95**	**-0,04**
PINK STARS													
Albizia	Albizia adianthifolia		6,50	0,02			3,76	5,58				2,95	
Albizia	Albizia zygia		6,50	0,02			3,76	5,58				2,95	
Alstonia	Alstonia boonei		6,50	0,02			3,76	5,58				2,95	
Yaya	Amphimas pterocarpoides		6,50	0,02			3,76	5,58				2,95	
Kokote	Anopyxis klaineana		6,50	0,02			3,76	5,58				2,95	
Berlinia	Berlinia confusa		6,50	0,02			3,76	5,58				2,95	
Berlinia	Berlinia spp.	20,33	6,50	0,02			3,76	5,58				2,95	
Celtis/Essa	Celtis mildbraedii/zenkeri		7,49	0,02	-7,51		4,33	5,58			9,05	2,95	
Pamprana	Coryanthe pachyceras		6,50	0,02			3,76	5,58				2,95	
Denya	Cylicodiscus gabonensis		6,50	0,02			3,76	5,58			6,55	2,95	
Ananta	Cynometra antana		6,50	0,02			3,76	5,58			5,95	2,95	
Duabankye	Dialium aubrevillei	28,46	6,50	0,02			3,76	5,58				2,95	
Potrodum	Erytrophleum sauaveolens		6,50	0,02	-6,52		3,76	5,58			6,63	2,95	
Nakwa	Holoptelea grandis		6,50	0,02			3,76	5,58				2,95	
Kroma	Klaindoxa gabonensis		6,50	0,02			3,76	5,58				2,95	
Kumanini	Lannea welwitschii		6,50	0,02			3,76	5,58				2,95	
Bompagya	Mammea africana		6,50	0,02			3,76	5,58			5,94	2,95	
Wonton	Morus mesozygia	24,39	11,06	0,02		21,13	6,38	5,58				2,95	
Danta 2/	Nesogordonia papaverifera		9,11	0,02	-9,12		5,26	5,58	-10,84		9,13	2,95	-12,08
Akwana	Ongokea gore		6,50	0,02			3,76	5,58				2,95	
Afam	Parinari exelsa		6,50	0,02			3,76	5,58				2,95	
Asoma	Parkia bicolor		6,50	0,02			3,76	5,58				2,95	
Esia	Petersianthus macrocarpus		6,50	0,02			3,76	5,58			5,95	2,95	
Otie	Pycnanthus angolensis		7,49	0,02			4,33	5,58			7,98	2,95	
Bombax	Rhodognaphalon/Bombax buonopozense		6,50	0,02			3,76	5,58				2,95	
Wama	Ricinodendron heudelotii		6,50	0,02			3,76	5,58				2,95	
Sterculia	Sterculia spp.		7,49	0,02			4,33	5,58			7,93	2,95	
Afena	Strombosia glaucescens		6,50	0,02		32,86	3,76	5,58				2,95	
Ofram 2a/	Terminalia superba		7,15	0,02			4,13	5,58	-9,71		12,76	2,95	-15,71
Tanuronini	Trichilia tessmannii		6,50	0,02			3,76	5,58				2,95	
Okure	Trilepisium madagascariense	40,65	6,50	0,02			3,76	5,58				2,95	
Avodire 1/	Turreanthus africanus	28,46	9,11	0,02	19,33	27,00	5,26	5,58			25,07	2,95	
AVERAGE	**Pink Stars**	**28,46**	**6,92**	**0,02**	**21,52**	**27,00**	**4,00**	**5,58**	**17,42**	**27,00**	**9,36**	**2,95**	**14,69**
Subaha 2/, 4/	Hallea spp./Mitragyna spp.												
Other species													

Notes:
1/ Log export banned in 1979
2/ Log export banned in 1988
2a/ Log export banned in 1988 but lifted in 1990
3/ Log export banned in 1993
4/ Harvest banned in 1993
Log exports was entirely banned in 1995
5/ 1996 observations are used as a proxy
A/ Other forest fees include
- Concession rent/Contract area rent
- Silvicultural fee (on-reserves)
- Silvicultural fee (off-reserves)
- Property mark
- Conveyance Certificates

Sources:
World Bank (1986, Working Paper V p.26) [1986 figures for willingness to pay]
Temporary felling permit to MIM Timber (1991) [1991 figures for willingness to pay]
GTMO (1994) [domestic log prices, 1993]
Treue and Armstrong-Mensah (1995) [domestic log prices, 1994]
Log sales contracts from 14 companies obtained by T.Treue [domestic log prices, 1996]
T. Treue Interviews with logging companies. Prices for st. timber is 50% of the domestic log prices (the so-called fifty-fifty arrangement) [1993-1996 figures for willingness to pay]
Forest Fees Regulations (1976)
Forest Fees (Amendment) Regulations (1983)
Forest Fees (Amendment) Regulations (1986)
Forest Fees (Amendment) Regulations (1989)
Forest Fees (Amendment) Regulations (1993)
Timber Resources Management Regulations (1998) (came into force on 16 December, 1998).

FORESTRY SCIENCES

1. P. Baas (ed.): *New Perspectives in Wood Anatomy.* Published on the Occasion of the 50th Anniversary of the International Association of Wood Anatomists. 1982 ISBN 90-247-2526-7
2. C.F.L. Prins (ed.): *Production, Marketing and Use of Finger-Jointed Sawnwood.* Proceedings of an International Seminar Organized by the Timber Committee of the UNECE (Halmar, Norway, 1980). 1982 ISBN 90-247-2569-0
3. R.A.A. Oldeman (ed.): *Tropical Hardwood Utilization.* Practice and Prospects. 1982 ISBN 90-247-2581-X
4. P. den Ouden (in collaboration with B.K. Boom): *Manual of Cultivated Conifers.* Hardy in the Cold- and Warm-Temperate Zone. 3rd ed., 1982 ISBN Hb 90-247-2148-2; Pb 90-247-2644-1
5. J.M. Bonga and D.J. Durzan (eds.): *Tissue Culture in Forestry.* 1982 ISBN 90-247-2660-3
6. T. Satoo: *Forest Biomass.* Rev. ed. by H.A.I. Madgwick. 1982 ISBN 90-247-2710-3
7. Tran Van Nao (ed.): *Forest Fire Prevention and Control.* Proceedings of an International Seminar Organized by the Timber Committee of the UNECE (Warsaw, Poland, 1981). 1982 ISBN 90-247-3050-3
8. J.J. Douglas: *A Re-Appraisal of Forestry Development in Developing Countries.* 1983 ISBN 90-247-2830-4
9. J.C. Gordon and C.T. Wheeler (eds.): *Biological Nitrogen Fixation in Forest Ecosystems.* Foundations and Applications. 1983 ISBN 90-247-2849-5
10. M. Németh: *Virus, Mycoplasma and Rickettsia Diseases of Fruit Trees.* Rev. (English) ed., 1986 ISBN 90-247-2868-1
11. M.L. Duryea and T.D. Landis (eds.): *Forest Nursery Manual.* Production of Bareroot Seedlings. 1984; 2nd printing 1987 ISBN Hb 90-247-2913-0; Pb 90-247-2914-9
12. F.C. Hummel: *Forest Policy.* A Contribution to Resource Development. 1984 ISBN 90-247-2883-5
13. P.D. Manion (ed.): *Scleroderris Canker of Conifers.* Proceedings of an International Symposium on Scleroderris Canker of Conifers (Syracuse, USA, 1983). 1984 ISBN 90-247-2912-2
14. M.L. Duryea and G.N. Brown (eds.): *Seedling Physiology and Reforestation Success.* Proceedings of the Physiology Working Group, Technical Session, Society of American Foresters National Convention (Portland, Oregon, USA, 1983). 1984 ISBN 90-247-2949-1
15. K.A.G. Staaf and N.A. Wiksten (eds.): *Tree Harvesting Techniques.* 1984 ISBN 90-247-2994-7
16. J.D. Boyd: *Biophysical Control of Microfibril Orientation in Plant Cell Walls.* Aquatic and Terrestrial Plants Including Trees. 1985 ISBN 90-247-3101-1
17. W.P.K. Findlay (ed.): *Preservation of Timber in the Tropics.* 1985 ISBN 90-247-3112-7
18. I. Samset: *Winch and Cable Systems.* 1985 ISBN 90-247-3205-0
19. R.A. Leary: *Interaction Theory in Forest Ecology and Management.* 1985 ISBN 90-247-3220-4
20. S.P. Gessel (ed.): *Forest Site and Productivity.* 1986 ISBN 90-247-3284-0
21. T.C. Hennessey, P.M. Dougherty, S.V. Kossuth and J.D. Johnson (eds.): *Stress Physiology and Forest Productivity.* Proceedings of the Physiology Working Group, Technical Session, Society of American Foresters National Convention (Fort Collins, Colorado, USA, 1985). 1986 ISBN 90-247-3359-6

FORESTRY SCIENCES

22. K.R. Shepherd: *Plantation Silviculture.* 1986 ISBN 90-247-3379-0
23. S. Sohlberg and V.E. Sokolov (eds.): *Practical Application of Remote Sensing in Forestry.* Proceedings of a Seminar on the Practical Application of Remote Sensing in Forestry (Jönköping, Sweden, 1985). 1986 ISBN 90-247-3392-8
24. J.M. Bonga and D.J. Durzan (eds.): *Cell and Tissue Culure in Forestry.* Volume 1: General Principles and Biotechnology. 1987 ISBN 90-247-3430-4
25. J.M. Bonga and D.J. Durzan (eds.): *Cell and Tissue Culure in Forestry.* Volume 2: Specific Principles and Methods: Growth and Development. 1987 ISBN 90-247-3431-2
26. J.M. Bonga and D.J. Durzan (eds.): *Cell and Tissue Culure in Forestry.* Volume 3: Case Histories: Gymnosperms, Angiosperms and Palms. 1987 ISBN 90-247-3432-0
    Set ISBN (Volumes 24–26) 90-247-3433-9
27. E.G. Richards (ed.): *Forestry and the Forest Industries: Past and Future.* Major Developments in the Forest and Forest Industries Sector Since 1947 in Europe, the USSR and North America. In Commemoration of the 40th Anniversary of the Timber Committee of the UNECE. 1987
    ISBN 90-247-3592-0
28. S.V. Kossuth and S.D. Ross (eds.): *Hormonal Control of Tree Growth.* Proceedings of the Physiology Working Group, Technical Session, Society of American Foresters National Convention (Birmingham, Alabama, USA, 1986). 1987 ISBN 90-247-3621-8
29. U. Sundberg and C.R. Silversides: *Operational Efficiency in Forestry.*
    Volume 1: Analysis. 1988 ISBN 90-247-3683-8
30. M.R. Ahuja (ed.): *Somatic Cell Genetics of Woody Plants.* Proceedings of the IUFRO Working Party S2.04-07 Somatic Cell Genetics (Grosshansdorf, Germany, 1987). 1988
    ISBN 90-247-3728-1
31. P.K.R. Nair (ed.): *Agroforestry Systems in the Tropics.* 1989 ISBN 90-247-3790-7
32. C.R. Silversides and U. Sundberg: *Operational Efficiency in Forestry.*
    Volume 2: Practice. 1989 ISBN 0-7923-0063-7
    Set ISBN (Volumes 29 and 32) 90-247-3684-6
33. T.L. White and G.R. Hodge (eds.): *Predicting Breeding Values with Applications in Forest Tree Improvement.* 1989 ISBN 0-7923-0460-8
34. H.J. Welch: *The Conifer Manual.* Volume 1. 1991 ISBN 0-7923-0616-3
35. P.K.R. Nair, H.L. Gholz, M.L. Duryea (eds.): *Agroforestry Education and Training.* Present and Future. 1990 ISBN 0-7923-0864-6
36. M.L. Duryea and P.M. Dougherty (eds.): *Forest Regeneration Manual.* 1991
    ISBN 0-7923-0960-X
37. J.J.A. Janssen: *Mechanical Properties of Bamboo.* 1991 ISBN 0-7923-1260-0
38. J.M. Bonga and P. Von Aderkas: *In Vitro Culture of Trees.* 1992 ISBN 0-7923-1540-5
39. L. Fins, S.T. Friedman and J.V. Brotschol (eds.): *Handbook of Quantitative Forest Genetics.* 1992 ISBN 0-7923-1568-5
40. M.J. Kelty, B.C. Larson and C.D. Oliver (eds.): *The Ecology and Silviculture of Mixed-Species Forests.* A Festschrift for David M. Smith. 1992 ISBN 0-7923-1643-6
41. M.R. Ahuja (ed.): *Micropropagation of Woody Plants.* 1992 ISBN 0-7923-1807-2
42. W.T. Adams, S.H. Strauss, D.L. Copes and A.R. Griffin (eds.): *Population Genetics of Forest Trees.* Proceedings of an International Symposium (Corvallis, Oregon, USA, 1990). 1992
    ISBN 0-7923-1857-9

FORESTRY SCIENCES

43. R.T. Prinsley (ed.): *The Role of Trees in Sustainable Agriculture.* 1993 ISBN 0-7923-2030-1
44. S.M. Jain, P.K. Gupta and R.J. Newton (eds.): *Somatic Embryogenesis in Woody Plants,* Volume 3: Gymnosperms. 1995 ISBN 0-7923-2938-4
45. S.M. Jain, P.K. Gupta and R.J. Newton (eds.): *Somatic Embryogenesis in Woody Plants,* Volume 1: History, Molecular and Biochemical Aspects, and Applications. 1995 ISBN 0-7923-3035-8
46. S.M. Jain, P.K Gupta and R.J. Newton (eds.): *Somatic Embryogenesis in Woody Plants,* Volume 2: Angiosperms. 1995 ISBN 0-7923-3070-6
Set ISBN (Volumes 44–46) 0-7923-2939-2
47. F.L. Sinclair (ed.): *Agroforestry: Science, Policy and Practice.* Selected Papers from the Agroforestry Sessions of the IUFRO 20th World Congress (Tampere, Finland, 6–12 August 1995). 1995 ISBN 0-7923-3696-8
48. J.H. Goldammer and V.V. Furyaev (eds.): *Fire in Ecosystems of Boreal Eurasia.* 1996 ISBN 0-7923-4137-6
49. M.R. Ahuja, W. Boerjan and D.B. Neale (eds.): *Somatic Cell Genetics and Molecular Genetics of Trees.* 1996 ISBN 0-7923-4179-1
50. H.L. Gholz, K. Nakane and H. Shimoda (eds.): *The Use of Remote Sensing in the Modeling of Forest Productivity.* 1996 ISBN 0-7923-4278-X
51. P. Bachmann, M. Köhl and R. Päivinen (eds.): *Assessment of Biodiversity for Improved Forest Planning.* Proceedings of the Conference on Assessment of Biodiversity for Improved Planning (Monte Verità, Switzerland, 1996). 1998 ISBN 0-7923-4872-9
52. G.M.J. Mohren, K. Kramer and S. Sabaté (eds.): *Impacts of Global Change on Tree Physiology and Forest Ecosystems.* Proceedings of the International Conference on Impacts of Global Change on Tree Physiology and Forest Ecosystems (Wageningen, The Netherlands, 26–29 November 1996). 1997 ISBN 0-7923-4921-0
53. P.K.R. Nair and C.R. Latt (eds.): *Directions in Tropical Agroforestry Research.* Selected Papers from the Symposium on Tropical Agroforestry organized in connection with the annual meeting of USA (Indianapolis, U.S.A., 5 November 1996). 1998 ISBN 0-7923-5035-9
54. K. Sassa (ed.): *Environmental Forest Science.* Proceedings of the IUFRO Division 8 Conference Environmental Forest Science. (Kyoto, Japan, 19–23 October 1998). 1998 ISBN 0-7923-5280-7
55. S.M. Jain, P.K Gupta and R.J. Newton (eds.): *Somatic Embryogenesis in Woody Plants,* Volume 4. 1999 ISBN 0-7923-5340-4
56. J.R. Boyle, J.K. Winjum, K. Kavanagh, E.J. Jensen (eds.): *Planted Forests: Contribution to the Quest for Sustainable Societies.* 1999 ISBN 0-7923-5468-0
57. K. von Gadow and G. Hui: *Modelling Forest Development.* 1999 ISBN 0-7923-5488-5
58. J. Abildtrup, F. Helles, P. Holten-Andersen, J.F. Larsen and B.J. Thorsen (eds.): *Modern Time Series Analysis in Forest Products Markets.* 1999 ISBN 0-7923-5524-5
59. S.M. Jain, P.K. Gupta and R.J. Newton (eds.): *Somatic Embryogenesis in Woody Plants,* Volume 5. 1999 ISBN 0-7923-5553-9
60. D. Auclair and C. Dupraz (eds.): *Agroforestry for Sustainable Land-Use.* Fundamental Research and Modelling with Emphasis on Temperate and Mediterranean Applications. 1999 ISBN 0-7923-5799-X

FORESTRY SCIENCES

61. F. Helles, P. Holten-Andersen and L. Wichmann (eds.): *Multiple Use of Forests and Other Natural Resources*. Aspects of Theory and Application. 1999 ISBN 0-7923-5859-7
62. A. Yoshimoto and K. Yukutake (eds.): *Global Concerns for Forest Resource Utilization*. Sustainable Use and Management. 1999 ISBN 0-7923-5968-2
63. C. Mátyás (ed.): *Forest Genetics and Sustainability*. 1999 ISBN 0-7923-6011-7
64. S.M. Jain and S.C. Minocha (eds.): *Molecular Biology of Woody Plants*, Volume 1. 1999 ISBN 0-7923-6012-5
65. E. Mälkonen (eds.): *Forest Condition in a Changing Environment*. The Finnish Case. 2000 ISBN 0-7923-6228-4
66. S.M. Jain and S.C. Minocha (eds.): *Molecular Biology of Woody Plants*, Volume 2. 2000 ISBN 0-7923-6241-1
67. S.M. Jain, P.K. Gupta and R.J. Newton (eds.), *Somatic Embryogenesis in Woody Plants*. Volume 6. 2000 ISBN 0-7923-6419-8
68. T. Treue: *Politics and Economics of Tropical High Forest Management*. A case study of Ghana. 2001 ISBN 0-7923-6931-9

KLUWER ACADEMIC PUBLISHERS – DORDRECHT / BOSTON / LONDON